The Merchant of Venice

Texts and Contexts

OTHER BEDFORD/ST. MARTIN'S TITLES OF INTEREST

William Shakespeare, *The First Part
of King Henry the Fourth: Texts and Contexts*
(The Bedford Shakespeare Series)
EDITED BY BARBARA HODGDON,
DRAKE UNIVERSITY

William Shakespeare, *Hamlet*
(Case Studies in Contemporary Criticism)
EDITED BY SUSAN L. WOFFORD,
UNIVERSITY OF WISCONSIN–MADISON

William Shakespeare,
Macbeth: Texts and Contexts
(The Bedford Shakespeare Series)
EDITED BY WILLIAM C. CARROLL,
BOSTON UNIVERSITY

William Shakespeare,
A Midsummer Night's Dream: Texts and Contexts
(The Bedford Shakespeare Series)
EDITED BY GAIL KERN PASTER,
GEORGE WASHINGTON UNIVERSITY
AND
SKILES HOWARD,
RUTGERS UNIVERSITY AT NEW BRUNSWICK

William Shakespeare, *The Taming
of the Shrew: Texts and Contexts*
(The Bedford Shakespeare Series)
EDITED BY FRANCES E. DOLAN,
MIAMI UNIVERSITY

William Shakespeare, *The Tempest*
(Case Studies in Critical Controversy)
EDITED BY GERALD GRAFF,
UNIVERSITY OF ILLINOIS AT CHICAGO
AND
JAMES PHELAN,
OHIO STATE UNIVERSITY

William Shakespeare, *Twelfth Night,
or What You Will: Texts and Contexts*
(The Bedford Shakespeare Series)
EDITED BY BRUCE R. SMITH,
GEORGETOWN UNIVERSITY

*The Bedford Companion to Shakespeare:
An Introduction with Documents,* Second Edition
BY RUSS MCDONALD,
UNIVERSITY OF NORTH CAROLINA AT GREENSBORO

WILLIAM SHAKESPEARE

The Merchant of Venice

Texts and Contexts

————————— ✄ —————————

Edited by

M. LINDSAY KAPLAN

Georgetown University

palgrave

THE MERCHANT OF VENICE by William Shakespeare, Edited by
M. Lindsay Kaplan

The Library of Congress has catalogued the paperback edition as follows:
2001094884

PALGRAVE, 175 Fifth Avenue, New York, NY 10010

First published by PALGRAVE, 175 Fifth Avenue, New York, NY 10010. Companies and representatives throughout the world. PALGRAVE is the new global
imprint of St. Martin's Press LLC Scholarly and Reference Division and Palgrave
Publishers Ltd. (formerly Macmillan Ltd.).

Manufactured in the United States of America.

7   6   5   4   3   2
f   e   d   c   b   a

ISBN 978-1-349-63494-1   ISBN 978-1-137-07784-4 (eBook)
DOI 10.1007/978-1-137-07784-4

Cover art: Civitatis Orbis Terrarum. By permission of The Folger Shakespeare
Library; Illustration from the codice De Sphaera. Courtesy of the Ministerio per
I Bene e le Attivita Culturali. Biblioteca Estense Universitaria.

Acknowledgments and copyrights can be found at the back of the book on pages 361–62,
which constitute an extension of the copyright page

About the Series

―――――――――――――――――――➤‹―――――――――――――――――――

Shakespeare wrote his plays in a culture unlike, though related to, the culture of the emerging twenty-first century. The Bedford Shakespeare Series resituates Shakespeare within the sometimes alien context of the sixteenth and seventeenth centuries while inviting students to explore ways in which Shakespeare, as text and as cultural icon, continues to be part of contemporary life. Each volume frames a Shakespearean play with a wide range of written and visual material from the early modern period, such as homilies, polemical literature, emblem books, facsimiles of early modern documents, maps, woodcut prints, court records, other plays, medical tracts, ballads, chronicle histories, and travel narratives. Selected to reveal the many ways in which Shakespeare's plays were connected to the events, discourses, and social structures of his time, these documents and illustrations also show the contradictions and the social divisions in Shakespeare's culture and in the plays he wrote. Engaging critical introductions and headnotes to the primary materials help students identify some of the issues they can explore by reading these texts with and against one another, setting up a two-way traffic between the Shakespearean text and the social world these documents help to construct.

Jean E. Howard
Columbia University
Series Editor

About This Volume

————————————————— >< —————————————————

The texts collected in these pages offer a wide range of perspectives from which to consider the issues raised in *The Merchant of Venice*. Some of these texts were written and published before the composition of the play, some did not appear until afterward, and some would never have been available for Shakespeare's perusal. But they all provide a sense of the range of opinions and controversies contemporary to the writing and performance of the play. I have also included texts written by Jews in the period to widen the interpretive frame in which the play could be considered. For all of these texts, my aim has been to provide perspective on the play as well as deepen its complexities and examine its assumptions. The views presented here raise many questions, and I opted in the introductory notes to help articulate those questions rather than try to provide the answers. My hope is that students will bring their own interpretive power both to these texts and to the play and struggle to find satisfctory answers to *The Merchant of Venice*'s challenging questions.

EDITORIAL POLICY

The text of *The Merchant of Venice* reproduced here is that edited by David Bevington in *The Complete Works of Shakespeare,* fourth edition (New York: Addison Wesley Longman, 1997), complete with his notes and glosses, with a few changes in glosses and in lineation. Most of the texts I selected for this volume are the first or only editions, though I occasionally chose editions closer to the date of the play when available. I have modernized the spelling and punctuation of these texts and provided glosses to explain some references as well as archaic terms and phrases. I have relied on the *Oxford English Dictionary* for definitions of obscure terms and phrases. Although I have modernized spelling generally in the texts, I have left varieties of spelling for names that are spelled differently in different texts or in the titles of texts (such as Steven or Stephen Ibara or Ibarra, or *Coryats Crudities*), since spelling was not fixed in the period. I have also included texts in modern editions and provided additional glosses where needed. With regard to dates, I use the more neutral B.C.E. (before the common era) and C.E. (common era) rather than B.C. and A.D. For the manuscript of the Lopez examination, I have modernized the date to 1594, though the original was given in the old style, February 1593/4. *The Dictionary of National Biography (DNB)* provided me with the biographical material in the shorter introductions to the excerpted text, unless otherwise indicated.

ACKNOWLEDGMENTS

I am fortunate to have had the help of a number of students, colleagues, and friends in preparing this volume. Kate Appleton and Claire Leheny provided much needed research and transcription assistance, and Andrew Dinan provided the translation for Andrew Willet's *De Universali*. I owe a large debt of thanks to all of the readers for the project, who offered useful and insightful advice, but especially to Fran Dolan and Jean Howard, who gave invaluable, lengthy responses to the manuscript on more than one occasion. I am also grateful to Kim Hall, Michael Ragussis, Jason Rosenblatt, Bruce Smith, Yehudah Mirsky, Gail Kern Paster, and Leon Wieseltier for suggesting texts and offering practical advice and sympathetic ears. Ruth Luborsky's excellent *Guide to English Illustrated Books, 1536–1603* made it much easier to select illustrations for this volume. The Folger Library staff was especially supportive and helpful to me in my research. I thank Georgetown University for sabbatical leave and the National Endowment for the Humanities and the Memorial Foundation for Jewish Studies for grants that supported this work. Special appreciation goes to Carrie Thompson at Bedford/St. Martin's for her cheerful and effective efforts on my behalf, to Kate Cohen for her wonderful copy editing, to Emily Berleth at Bedford/St. Martin's, and to Leslie Connor, Project Manager at Publisher's Studio/Stratford Publishing. Final thanks go to Karen Henry, who patiently encouraged me throughout the lengthy process of compiling this volume and provided needed advice at crucial moments.

I am especially grateful for the support my mother, Anne Kaplan, and my sister, Rachel Kaplan, provided throughout. Deepest gratitude goes to my husband, Norman Eisen, who gave me the love and encouragement that got me through the most difficult moments of this project. My work on this volume was interrupted by the death of my beloved father, Jerome Eugene Kaplan (z"l). When I returned to working on this play, which represents troubled relations between fathers and daughters, I had numerous occasions to consider how blessed I was and am to have had such a loving, supportive, and generous father, whose unique relationship to Judaism strengthened and fostered my own. My daughter, Tamar Yohanna, was born just as I was completing this manuscript. As a new mother I am grateful to have my father's example of good parenting to follow. This edition is dedicated to his memory.

יהודה יצחק בן אליהו וחנה (ז"ל)

וזאת ליהודה ויאמר שמע יהוה קול יהודה ואל-עמו תביאנו ידיו רב לו ועזר מצריו תהיה.

דברים לג:ז

Yehudah Yitzchak ben Elihu v'Chana (z"l)

"And this [is the blessing] for Yehudah: and [Moses] said, 'Hear, G-d, the voice of Yehudah, and bring him to his people; let his hands be sufficient for him, and be You a help to him from his foes.'"

Deuteronomy 33:7

M. Lindsay Kaplan
Georgetown University

Contents

—>‹—

⇥ 2. *Finance* 187

Usury 188

Jews 189

Merchants 192

Illustrations

><

The Merchant of Venice

Texts and Contexts

Introduction

><<

In a sense, *The Merchant of Venice* provides its own context in which to be viewed or read insofar as the play offers a range of responses for many of the issues it raises. By proposing categories in conflict — tragedy and comedy, law and mercy, Jew and Christian, money and love, "other" and same, female and male — Shakespeare offers his audience opposing perspectives on the action of the play. He also keeps these oppositions from being neat dichotomies by offering exceptions and complexities that multiply the perspectives from which an audience can consider them. However, the play's historical context reveals even more ways to understand the issues Shakespeare represents. The play was probably written in the mid- to late 1590s, near the end of Queen Elizabeth I's reign. The texts selected for this edition express a wide range of ideas available in the early modern period, the era spanning the sixteenth and seventeenth centuries. In offering these contexts, I am less interested in Shakespeare's influences or "intention" than I am in imagining what kind of assumptions and opinions an early modern audience might bring to a performance of the play as well as what kind of views might encourage a modern audience to reexamine its initial response to the play. In this introduction, I identify four central areas of interest — Venice, Finance, Religion, and Love and Gender — considering first how the play presents each topic before turning to a wider view of these ideas provided by the accompanying contextual writings.

Venice

Let's begin by considering the play's setting: how does it represent Venice? The play establishes a primary connection between Venice and trade: act 1 opens with a discussion between Antonio, a merchant of Venice, and his friends Salerio and Salanio about the vicissitudes of the trading profession. The friends imagine that Antonio's vocation brings both wealth and status; they describe his ships as "signors and rich burghers on the flood" who "overpeer the petty traffickers / That curtsy to them, do them reverence" (1.1.10, 12–13). Male, money-centered Venice is contrasted with the almost fairy-tale, feminine world of Belmont presented in scene 2, but the air of wealth and international trade we usually associate with Venice seems transferred to the person of Belmont's Portia. She is a beautiful, rich virgin who is sought, almost as a commodity, by suitors from all over the world. Bassanio, who seeks to woo Portia himself, suggests this perspective in his description of her: "Nor is the wide world ignorant of her worth, / For the four winds blow in from every coast / Renownèd suitors" (1.1.166–68). The diversity of Portia's suitors mirrors the variety of traders drawn to the market of Venice, and it is this diversity that earns Venice its reputation as an international city-state. Portia's descriptions of the men seeking her hand in marriage suggest a certain distaste for their difference: her comments convey humor but also intolerance, as she rejects one suitor after another on the basis of physical or behavioral shortcomings. She considers and dismisses the Prince of Morocco before she even meets him: "If he have the condition of a saint and the complexion of a devil, I had rather he should shrive me than wive me" (1.2.94–96), judging him on what she anticipates to be his appearance rather than on his character.

Not only do people of different national origins, ethnicities, and religions flock to Venice, but its own inhabitants include outsiders, such as the Jewish moneylender Shylock. Venice both needs and distrusts its outsiders; both attitudes are present in the Venetian Christians' treatment of Shylock. Despite their dislike of him, Antonio and Bassanio are forced to tolerate Shylock because they need his money. When Antonio fails to repay the loan to Shylock at the appointed time, the merchant is required to fulfill the terms of the bond he has agreed to — the payment of a pound of his own flesh. Because of the laws protecting aliens in Venice, Shylock can insist on the payment of this bond with apparent impunity, as Antonio himself affirms:

> The Duke cannot deny the course of law;
> For the commodity that strangers have
> With us in Venice, if it be denied,
> Will much impeach the justice of the state,
> Since that the trade and profit of the city
> Consisteth of all nations.   (3.3.26–31)

If the duke refuses to enforce Shylock's agreement with Antonio, all contracts between citizens and strangers could be invalidated, deterring foreign traders from entering into agreements with Venetians and hence threatening the state's prosperity. On the other hand, it is precisely Shylock's alien status that provides Portia with her legal victory over him:

> It is enacted in the laws of Venice,
> If it be proved against an alien
> That by direct or indirect attempts
> He seek the life of any citizen,
> The party 'gainst the which he doth contrive
> Shall seize one half his goods; the other half
> Comes to the privy coffer of the state,
> And the offender's life lies in the mercy
> Of the Duke only, 'gainst all other voice.   (4.1.343–51)

Because the state considers him a stranger, Shylock is vulnerable to its power; while Portia manages to protect Antonio, her exposition of Venetian law reveals its inherent complexities and tensions.

The contextual readings included in this edition both reinforce and call into question the views of Venice presented in the play. In his *History of Italy*, William Thomas reports on the considerable wealth raised by the Venetian state from taxes on trade "after so extreme a sort, that it would make any honest heart sorrowful to hear it" (p. 132 in this volume). He condemns Venetian affluence, the result of excessive taxation and a greedy national character — "the gentleman Venetian is . . . never satisfied with hoarding up of money" (p. 134) — suggesting that Shylock might be more representative of the citizenry than Antonio, who is characterized by self-sacrificing generosity. Robert Wilson's play, *The Three Ladies of London* (p. 154), represents another view of the greedy Italian merchant, but here he is opposed to the generous Gerontus, a kind-hearted Jewish moneylender. Dudley Carleton, writing as the English ambassador to Venice in the early seventeenth century (p. 145), notes the decline of trade there, offering a possible context for understanding Antonio's initial sadness when the play opens and his somewhat marginalized status at the play's end. Thomas Coryate affirms Venice as a center of international trade, marvelling at the variety of merchants who congregate there: "many Polonians, Slavonians, Persians, Grecians, Turks, Jews, Christians of all the famousest regions of Christendom" (p. 139). Like Shakespeare, he suggests a parallel in the attraction of visitors to both the wealth and the women of Venice, although the women he refers to are prostitutes: "For so infinite are the allurements of the famous Calypsos, that the fame of them hath drawn many to Venice from some of the remotest parts of Christendom, to contemplate their beauties, and enjoy their pleasing dalliances" (p. 145). This popular view of Venetian women plays under the surface of the play's more positive representation of Portia, perhaps raising questions about her virtue.

The problems of understanding and negotiating national, religious, and racial difference, which the play stages, is one England also faced in the period as it sought to define its national character in relation to both neighboring countries and new populations met in the course of exploration and trade. Portia's opinions of her suitors, when compared to early modern travel writing such as Fynes Moryson's, turn out to be commonly held stereotypes about various nationalities, including the English. A London audience might have recognized and been amused by her characterization of the grumpy, puritanical County Palatine, whom Moryson describes as "brought up in the reformed religion . . . [and who] carried himself like a grave and noble Prince" (see p. 166), but would they have been able to laugh at their own expense when she rehearses clichés about the English themselves?

The status of the other, or alien, had important legal ramifications that can be considered from a number of perspectives. Venetian law was famous for its protection of the rights of foreigners; William Thomas's discussion of the freedom of strangers there, as well as the potential for the corruption of the legal system, suggests a different perspective for viewing the play's trial scene — does Shylock receive a fair trial? Should he? Early modern English laws protecting resident aliens also support a notion of an equal standard of justice for strangers and native citizens. However, the prominent jurist and legal historian Sir Edward Coke, in his discussion of this subject, categorizes Jews as perpetual enemies who are not allowed the redress enjoyed by friendly aliens, suggesting that Shylock might be perceived by an early modern audience as receiving a more merciful sentence than he deserves. On the other hand, Coryate's provocation of the Jews of Venice in "reprehend[ing] their religion [which results in] . . . fearing lest they would have offered me some violence" (p. 144) raises the question of whether Shylock is similarly provoked by Antonio, and therefore less culpable in the threat he makes against Antonio's life.

Finance

A consideration of the financial workings of various characters' professions offers another context from which to evaluate the play's action. While the world of a Venetian merchant might be imagined as one of wealth, status, and exotic travel, Salerio and Solanio counterbalance this fantasy of prosperity and power by another, opposing fantasy of the merchant's life as one filled with anxiety and preoccupation: "every object that might make me fear / Misfortune to my ventures, out of doubt / would make me sad" (1.1.20–22). Antonio dismisses this view, explaining that he reduces his risks by diversifying his business: "My ventures are not in one bottom[1] trusted, / Nor to one place; nor is my whole estate / Upon the fortune of this present year" (1.1.42–44). However, when his investment in Bassanio's trip to woo and wed Portia does not pay off before the loan to Shylock comes due, Antonio's other losses at sea put not only his fortune but his life in jeopardy. Upon reading the letter Antonio writes regarding his misfortunes, Bassanio describes its very words as wounds "Issuing lifeblood."

> But is it true, Salerio?
> Hath all his ventures failed? What, not one hit?
> From Tripolis, from Mexico, and England,
> From Lisbon, Barbary, and India,
> And not one vessel scape the dreadful touch
> Of merchant-marring rocks? (3.2.264–69)

In spite of Antonio's prudence and care, he cannot escape the risks that attend the merchant's voyages.

Moneylending appears to provide a surer return on one's investment. Shylock considers both Antonio's worth and the hazards his various ventures face before deciding to lend him money:

> ships are but boards, sailors but men. There be land rats and water rats, water thieves and land thieves — I mean pirates — and then there is the peril of waters, winds, and rocks. The man is, notwithstanding, sufficient. Three thousand ducats. I think I may take his bond. (1.3.17–20)

[1] **bottom**: ship's hold.

In lending Antonio money, Shylock considers the risk, but the risk appears shouldered primarily by the merchant; the loan agreement includes no provision for forgiveness of the debt if Antonio's ships all fail. The play contrasts Antonio and Shylock with regard to the way each character lends money. Although a merchant by profession, Antonio also acts as a lender in Venice; his practice of providing interest-free loans ignites Shylock's anger: "He lends out money gratis and brings down / The rate of usance here with us in Venice" (1.3.34–35). By not charging interest on the loans he offers, Antonio reduces what Shylock can charge on interest-bearing loans. The play attaches moral valences to these financial practices, associating free loans with generosity, love, and self-sacrifice and usury with greed, hatred, and selfishness. Antonio assures Bassanio that "My purse, my person, my extremest means / Lie all unlocked to your occasions" (1.1.137–38). After he has apparently lost his fortune and forfeited the loan, the merchant regrets the loss neither of money nor of life; all that matters to him is that Bassanio loves him: "Repent but you that you shall lose your friend. / And he repents not that he pays your debt" (4.1.273–74). In contrast, Shylock obsesses about his money, resenting the costs of keeping a servant, dreaming of his money bags, and encouraging his clients to waste the money loaned them (act 2, scene 5); after Jessica elopes with Lorenzo and steals a portion of her father's wealth, Shylock seems to mourn his financial loss more than his filial one:

A diamond gone, cost me two thousand ducats in Frankfort . . . and other precious, precious jewels. I would my daughter were dead at my foot, and the jewels in her ear! Would she were hearsed at my foot, and the ducats in her coffin! (3.1.63–68)

Shylock's greed and profession are clearly linked in the play to his Jewishness, a fact that he himself recognizes. He explains his hatred for Antonio in precisely these terms:

He hates our sacred nation, and he rails,
Even there where merchants most do congregate,
On me, my bargains, and my well-won thrift,
Which he calls interest. Cursèd be my tribe
If I forgive him! (1.3.38–42)

The Christians in the play equate avarice with Judaism; in those moments when Jews seem to offer money freely, their actions are described as Christian, suggesting that the concept of a generous Jew is inconceivable (see 1.3.169–70, and 2.6.52 where "gentle" can carry the sense of "gentile" or "Christian").

Early modern writings on merchants and moneylenders indicate that attitudes toward these professions were in flux at the end of the sixteenth century. Earlier views of merchants, who also served as moneylenders, characterized them much as Shylock is represented in the play: greedy and self-serving. In Robert Wilson's *Three Ladies*, the Italian merchant Mercadorus refuses to repay money he borrowed; fleeing to Turkey, he threatens to convert to Islam, which by Turkish law frees him from his debts.

> Arrest me dou skal knave,[2] marry do and if thou dare,
> Me will not pay de one penny, arrest me, do, me do not care.
> Me will be a Turk, me came hedar[3] for dat cause,
> Darefore me care not for de so much as two straws. (p. 156)

Rather than repay the money he owes, this merchant would callously abandon his faith. However, as more Englishmen took up the trading profession, which became more important to the national economy, views of merchants improved. John Wheeler, secretary of the Merchant Adventurer's Company, insists on the dignity and value of his profession:

> Whereas indeed merchandise, which is used by way of proper vocation, being rightly considered of, is not to be despised, or accounted base by men of judgement, but to the contrary, by many reasons and examples it is to be proved that the estate is honorable, and may be exercised not only of those of the third estate[4] (as we term them), but also by the nobles and chiefest men of this realm with commendable profit, and without any derogation to their nobilities, high degrees, and conditions, with what great good to their states, honor, and enriching of themselves and their countries. . . . (p. 232)

The status of gentlemen merchants is enhanced by their profession, which enriches not only them personally but also the state. Daniel Price, chaplain to King James I, preaches a sermon in which he characterizes the risk-filled life of the merchant as "the most diligent, careful, assiduous, industrious, laborious, and indefatigable of all other kinds of life" and compares it to "the difficulty of obtaining the kingdom of heaven," suggesting that he represents the model Christian (p. 236). This would seem to affirm the play's representation of Antonio as merchant and exemplary Christian.

Contemporary views of usurers echo the play's condemnation of Shylock's vocation and his greed; Thomas Wilson contrasts the admirable risks the merchant takes with the deplorable parasitic laziness of usurers (p. 197).

---

[2] **dou skal knave:** you contemptible rascal.
[3] **hedar:** hither, here.
[4] **the third estate:** commoners; aristocrats were thought to be above work in this period.

Like many others who write on the subject, he associates Judaism with charging interest:

> What is the matter that Jews are so universally hated wheresoever they come? Forsooth, usury is one of the chief causes, for they rob all men that deal with them, and undo them in the end. And ... all these Englishmen ... that lend their money or their goods whatsoever for gain, for I take them to be no better than Jews. Nay, shall I say: they are worse than Jews. (pp. 197–98)

The traditional identification of Jews as usurers developed in part because it was one of the few careers permitted to Jews, as Fynes Moryson points out: "And where [the Jews] are allowed to dwell, they live upon usury ... but are not allowed to buy any lands, houses, or stable inheritances, neither have they any coin of their own, but use the coins of Princes where they live" (pp. 171–72). Another reason for linking Jews with usury lay in the interpretation of biblical law, which permits charging interest from a stranger, but prohibits usury between "brothers" (Deuteronomy 23:19–20). Christian doctrine prohibited lending on interest between Christians but was more permissive in allowing Jews to charge interest of Christians; as a result, usury tended to be seen as a Jewish practice. Surprisingly, the early modern Jewish perspective on usury was similarly negative. Jewish law forbids usury between Jews, and even discouraged lending on interest to non-Jews, even though the Bible permitted it, fearing that business dealings with non-Jews would have a corrupting influence, a point discussed by Yehiel Nissim da Pisa in *The Eternal Life*, his treatise on the laws of charging interest. Aware of Christian condemnations of "Jewish usury," da Pisa distinguishes it from fraud; he disapproves of charging interest, but acknowledges that it may be a necessary evil.

> Though we say that one may not defraud a gentile, one may lend him on interest since he has voluntarily and of his own accord chosen to pay the sum. ... My opinion is that this is merely a permissive act and even though the Torah permitted it, our sages prohibited us from lending to gentiles on interest as a precautionary measure. They later eased the law so that we may lend to gentiles in order to make a living. (pp. 216–17)

In both Christian and Jewish opinion and practice up to the sixteenth century, usury within and between the two groups tends to be both condemned and tolerated. However, by the end of the sixteenth century, Jewish moneylenders became increasingly rare, and the stigma associated with usury began to lessen. As charging interest became more accepted between Christians — in England, a law passed in 1571 permitted charging interest of up to 10 percent — the need for Jewish moneylenders decreased. During the same period, Jews increasingly turned to trade as a vocation, and the Jewish merchant became a more common figure, as the French explorer Nicolas de Nicolay points out in his discussion of the Jewish community of Constantinople. While the play seems to present Shylock and Antonio as opposites, changes in early modern attitudes toward merchants and moneylending, as well as in the practices of Christians and Jews in these professions, suggest significant areas of overlap between the two. Indeed, these shifts may be registered in the play in Portia's inability to tell Antonio and Shylock apart: "Which is the merchant here, and which the Jew?" (4.1.169).

Religion

The play's consideration of financial matters in the context of conflicts between Christians and Jews necessarily raises questions about its representation of religion. While never explicitly engaging the biblical texts regulating usury, the economic disagreement between Antonio and Shylock is depicted in terms of competing interpretations of the Bible. Shylock attempts to justify charging interest with reference to an episode in Genesis, in which Jacob, who is to be paid for his services in multicolored sheep, sets peeled twigs in front of the copulating flocks, which supposedly results in the birth of speckled sheep, hence increasing his wages (1.3.62–81). Antonio not only rejects the relevance of the episode to the issue at hand, but dismisses Shylock's legitimacy as an interpreter of the Bible.

> The devil can cite Scripture for his purpose.
> An evil soul producing holy witness
> Is like a villain with a smiling cheek . . .
> O, what a goodly outside falsehood hath! (1.3.89–93)

Antonio sees his disagreement with Shylock in terms of absolute good and evil, suggesting that no common ground exists between Christians and Jews. When Shylock offers Antonio a loan without interest, the merchant responds: "Hie thee, gentle Jew / The Hebrew will turn Christian; he grows kind" (1.3.168–69). The "Hebrew" can be viewed positively only to the extent that he appears to behave like or become a Christian. The trial in act 4, which pits Portia against Shylock, sets up another apparent binary between Christians and Jews in the opposition of mercy and justice. Though Shylock craves "the law, / The penalty and forfeit of my bond" (4.1.201–02) and rejects Portia's request for mercy in his proceeding against Antonio, when Shylock loses his suit, she asks the merchant "What mercy can you render

FIGURE 1 *Jacob Breeding Spotted Sheep. Illustrating the ninth commandment (prohibiting covetousness), from Thomas Cranmer,* Catechismus *(1548). While Antonio denies any connection between the story in Genesis and the acquisition of wealth, Cranmer clearly uses this episode to illustrate greed. Referring to the story therefore works to convict Shylock of avarice rather than, as he would have it, excuse it.*

him, Antonio?" (4.1.373). However, this association of justice with the Jew Shylock and mercy with the Christian Portia is less than firm. Portia wins the case not by employing mercy but by applying the law more stringently and deftly than her opponent, as she informs him: "be assured / Thou shalt have justice, more than thou desir'st" (4.1.310–11). She gets the better of Shylock by reading the bond in a strictly legal, literal manner, which prohibits its being fulfilled, and while she assures him at the outset of the trial that "the Venetian law / Cannot impugn you as you do proceed" (4.1.173–74) she introduces two laws at the end of the trial which he has transgressed in pursuing his bond (4.1.300–09, 342–51). Shylock's punishment and pardon at the end of the trial similarly complicate the categories of mercy and justice; although he does not suffer the fullest force of the law, the conditions of his pardon — converting to Christianity and eventually deeding all of his property to his estranged daughter and son-in-law — hardly constitute mercy, though he does escape with his life and half of his property.

The issue of conversion similarly blurs distinctions between Christians and Jews. If Shylock's wickedness is a function of his Jewishness, as Antonio and others in the play suggest, then his conversion to Christianity would solve the problem. In the case of Shylock's daughter Jessica, her goodness seems directly correlated with the possibility that she already is or plans to become a Christian (2.3.10–20); she is referred to as "gentle," that is, "gentile," Jessica on several occasions (2.4.19, 34; 2.6.52). Her conversion appears to be unproblematic, as she undertakes it willingly and already embodies what other characters in the play perceive as "Christian" characteristics. She is even represented as racially distinct from her father, possessing whiter skin and even lighter blood than Shylock. Her beauty or "fairness" is associated with whiteness — "whiter than the paper it writ on / Is the fair hand that writ" (2.4.12–13), while her difference from her father is described in terms of contrasting light and dark flesh and blood: "There is more difference between [Shylock's] flesh and hers than between jet and ivory, more between [their] bloods than there is between red wine and Rhenish⁵" (3.1.29–31). However, both Jessica and Shylock insist that they share the same blood (2.3.17–18; 3.1.28), and Jessica is still perceived as Jewish because she is Shylock's daughter (2.4.34–37; act 3, scene 5 passim). These contradictory views raise questions about whether Jewish difference is a matter of race or belief, whether conversion is effective, and whether the threat of Shylock's Judaism can be nullified.

Finally, it is worth considering how the play represents the complexity of Shylock's Jewish difference. While his financial losses seem to spur his hatred of Christians (1.1.39–50; 3.1.60 ff), these two defining motives, desire for money and revenge, seem to work at cross purposes by act 4 when he refuses repayment of the loan several times over in order to pursue his murderous bond with Antonio. Further adding to the inconsistency of his character are the moments when Shylock evokes the audience's sympathy by describing his persecution by Antonio and insisting on the common humanity of Jews and Christians (3.1.41 ff). Shylock's speech argues that Jews resemble Christians in basic human traits; further consideration of the behavior of the Christians in the play suggests that they may also share some of Shylock's negative behaviors in their preoccupation with wealth and their reflexive hatred of religious others. Shylock's character suggests an amalgam of beliefs about Jews, rather than a psychologically consistent persona, and his interactions with Venetian Christians evoke a variety of perspectives on Jewish-Christian relations.

⁵ Rhenish: white wine.

Conflict between Christians and Jews is as old as Christianity itself and manifests itself early in debates about scriptural interpretation, an issue on which Shylock and Antonio also disagree in the play. One crux in this debate lay in the interpretation of the laws set out in the Hebrew bible. While Jews interpreted them as still valid and binding, Christians understood the civil and ritual laws as fulfilled by the coming of Jesus and therefore no longer in force. According to this view, articulated in the Christian scripture by St. Paul, sinners relied on the mercy of God's saving grace, rather than adherence to the law, for salvation. Hence, Christians understood observance of the biblical law as evidence that Jews misunderstood scripture and rejected God's grace or mercy as embodied in Jesus. In his disputation with a rabbi of Venice, Thomas Coryate's argument hinges precisely on what he perceives to be Judaism's failure to understand its own scripture, which he believes identifies Jesus as the Messiah:

> Withal I added that the predictions and sacred oracles both of Moses, and all the holy Prophets of God, aimed altogether at Christ as their only mark, in regard he was the full consummation of the law and the Prophets, and I urged a place of Isaiah (17:14) unto him concerning the name *Emmanuel,* and a virgin's conceiving and bearing of a son; and at last descended to the persuasion of him to abandon and renounce his Jewish religion and to undertake the Christian faith, without the which he should be eternally damned. He again replied that we Christians do misinterpret the Prophets, and very perversely wrest them to our own sense, and for his own part he had confidently resolved to live and die in his Jewish faith, hoping to be saved by the observations of Moses's Law. (p. 143)

For rejecting Jesus, the Jews were perceived as damned, which accounts in part for their being likened to the devil and associated with wickedness, as Shylock is at numerous points in the play.

In considering the religious contexts for the play, however, we need to look not only at early modern attitudes about the relationship of Christianity to Judaism, but also at the religious controversies of the Reformation that helped shape that relationship. Because England had been without a Jewish community for three centuries — the Jews having been expelled from the country near the end of the thirteenth century — English acquaintance with Jews was limited to representations of Jews, and contact with a few visiting Jews or converted Jewish immigrants and with Jewish communities abroad. The dominant religious conflict in early modern England took place not between Jews and Christians, but between Protestants and Catholics, in the discord ensuing from Henry VIII's decision to leave the Catholic Church and establish the Anglican Church.[6] As partisans struggled to determine the legitimacy and primacy of one over the other, *Jew* was often used as a term of disparagement in characterizing the shortcomings of the opposing side, as Catholic priest Richard Bristow does in his treatise *Demands to Be Proponed of Catholics to the Heretics:*

> [Protestants] must grant . . . [the Catholic Church] now to be the true Church, as being alone with that.[7] If then they will say, that this is not a good argument, let they be further demanded, whether they dare take part also with the very Jews and Paynims[8] against the Christians, yea and against the Godhead[9] of Christ himself . . . (p. 275)

In rejecting the Catholic Church, Protestants are likened to Jews and other nonbelievers who reject the divinity of Jesus himself.

While attitudes toward contemporary Jews could be used to malign a Christian opponent, views of the biblical Jews as the chosen people were often employed in a more positive context, and frequently served as a model for Protestant England, which saw itself as a New Israel (Greenfield 52). In a sermon on repentance, the Protestant preacher William Perkins suggests that the members of his audience not only identify with Israel, but see themselves as even worse sinners:

> But now England, how hast thou requited this kindness of the Lord? Certainly even with a great measure of unkindness: that is, with more and greater sins than ever Israel did. So that if Moses spake true of them, then may our Moses much more truly cry out against England: "Do thou thus requite the Lord thou foolish people?" And if this Prophet said thus of Israel for three sins, then may it be said of England, for 300 sins (O England) a nation not worthy to be beloved: for thou hast multiplied thy transgressions, above theirs of Israel . . . (p. 279)

Hence, early modern views of Jews ranged from extremely negative to quite positive, enabling the English both to contrast themselves with Jews and to identify with them. This mixed conception of Jews is reflected in the play's depictions of Shylock and Jessica as sometimes sympathetic, sometimes antipathetic characters. Further complicating Christian attitudes toward Jews was the increasingly popular idea that conversion of all Jews was a

---

[6] The Anglican Church is the official Protestant church of England.
[7] **alone with that:** the same as the early Christian Church.
[8] **Paynims:** pagans.
[9] **Godhead:** divinity.

necessary precursor to the second coming of Jesus and the establishment of his kingdom on earth. During the period of religious turmoil in the wake of the Reformation, Catholic and Protestant authors alike considered both the practical and theological aspects of this issue. In *Roma Sancta*, the English Catholic priest Gregory Martin discusses attempts to convert the Jewish community of Rome through preaching:

> What need I stand here in the commendation of this exercise? It concerneth the conversion of [the Jews] that were always the greatest enemies to Christ and Christian Religion. . . . And during the time of infidelity they were permitted to live . . . principally for that God would have many of them also saved from time to time by few and few, which in scripture are called *reliquiae* (the remnant) [Romans 11]. . . . In these respects the Holy Church by God's providence spared their lives, as also for that they were never children of the Church, nor made profession in baptism, and therefore may not be compelled to the faith. . . .
>
> This people therefore thus hitherto preserved in the world as they are not forced, so by all charitable means they are invited and persuaded to forsake obstinate Judaism and to become Christians, as every year they do. (pp. 281–82)

While Christians understood the conversion of the Jews as an admirable and inevitable undertaking, part of God's divine plan, most stopped short, as Martin does, of advocating coercion as a means of effecting this goal.

Contemporary Jews who wrote on the subject of forced conversion, such as Samuel Usque in his history, *Consolation for the Tribulations of Israel*, also condemned the practice:

> But after the Jews were forcibly converted to Christianity, . . . calamities multiplied to such an extent that the land was almost desolate. . . . The king again took counsel, and there were many who voiced the opinion that the calamities had doubled because of the force which had been used to bring the Jews to the Christian faith; spiritual matters, they asserted, should be freely decided, since our Lord had granted freedom to the human will. (p. 290)

Usque suggests that God punishes the perpetrators of forced conversions, and imagines that the Christians in his fictionalized narrative acknowledge this view. Early modern Christian writers also wrote passionately against religious persecution, which attempted to force the beliefs of those who dissented on the basis of conscience. English Catholics and Protestants both deplored the use of force in shaping religious belief; while each side did actively seek to convert the other, only voluntary conversions were considered valid. In his *Acts and Monuments*, a history of Protestant martyrdom,

John Foxe includes an account from the reign of Catholic Queen Mary Tudor of the forced conversion to Catholicism and subsequent recantation of Archbishop Thomas Cranmer:

> With these and like provocations, these fair flatterers ceased not to solicit and urge him, using all means they could to draw him to their side, whose force his manly constancy did a great while resist. But at last, when they made no end of calling and crying upon him, the archbishop, being over-come, whether through their importunity,[10] or by his own imbecility,[11] or of what mind I cannot tell, at length gave his hand.
>
> It might be supposed that it was done for the hope of life, and better days to come. . . . But howsoever it was, plain it was, to be against his conscience. (p. 293)

Although Cranmer eventually agrees to convert, the fact that he does so against his conscience invalidates the decision for Foxe; the account continues with an exalted description of the archbishop's public rejection of this forced conversion shortly before his execution. Such examples provide a troubling context for Shylock's conversion at the close of the trial. Shylock states that he is "content" with the conditions of his sentence, a word that could indicate satisfaction, containment, or contention; if he refuses Antonio's condition, Shylock faces death as the Duke promises to "recant / The pardon that I late pronouncèd here" (4.1.386–87). While some viewers might have been happy to see the villainous Jew punished and transformed into a Christian, others might have been troubled by the punitive force driving this conversion. In what ways do Shylock's illness and subsequent disappearance from the play shape our understanding of this conversion?

Love and Gender

The final focus of the play considered in this edition is the nexus of love and gender: how are gender roles represented in the play and what is the nature of the love relationships between and among genders? To return once more to the opening of the play, we should consider another possible reason Solanio offers for Antonio's sadness: "Why then you are in love" (1.1.46). Antonio's response — "Fie, fie!" — carries a stronger meaning than a simple denial,[12] suggesting that Antonio may be in love but wishes to hide it. The

---

[10] importunity: ceaseless requests.

[11] imbecility: weakness.

[12] The *Oxford English Dictionary* (*OED*) defines *fie* as "[a]n exclamation expressing, in early use, disgust or indignant reproach."

merchant appears to have a particular attachment to Bassanio (1.1.57–59), who seems aware of this in responding indirectly to Antonio's question about Portia. Bassanio emphasizes the primacy of the relationship with Antonio by focusing on the financial purpose of the "pilgrimage" (1.1.119), which would enable the young man to pay back the large debt he owes his friend. Antonio goes out of his way to help Bassanio, offering "[his] purse, [his] person" (1.1.137), and breaking his practice of never borrowing on interest in seeking a loan from Shylock (1.3.51–54). In a discussion of Bassanio's departure from Antonio, Salerio repeats the emotional (and perhaps manipulative) farewell speech given by the merchant, while Solanio comments: "I think he only loves the world for [Bassanio]" (2.8.36–50). While Antonio is willing to suffer the penalty of the bond — die for his friend — his sacrifice does not appear entirely selfless:

> Commend me to your honorable wife.
> Tell her the process of Antonio's end,
> Say how I loved you, speak me fair in death;
> And, when the tale is told, bid her be judge
> Whether Bassanio had not once a love.
> Repent but you that you shall lose your friend.
> And he repents not that he pays your debt.   (4.1.268–74)

Antonio seems to set up a competition here between himself and Portia, one he apparently wins, given Bassanio's response that he esteems his friend above his wife and his subsequent willingness to give up Portia's ring to thank the "doctor of law" for saving his friend. When Portia confronts Bassanio with the loss of the ring at the play's conclusion, however, she appears to gain the upper hand, shaming Antonio into insuring that her husband will remain true to her: "I dare be bound again, / My soul upon the forfeit, that your lord / Will never more break faith advisedly" (5.1.249–51).

If the conclusion of the play gives greater weight to the heterosexual bond, what is the nature of this relationship? Bassanio's pursuit of Portia is described alternately as "a secret pilgrimage" (1.1.119), a financial venture undertaken to repay a debt, and a heroic quest, suggesting devotion, wealth, and fame as possible motives (1.1.118 ff). In fact, Portia is sought, as any valuable commodity might be, by many admirers; her father instituted the casket test precisely to insure that the correct box "will no doubt never be chosen by any rightly but one who you shall rightly love" (1.2.24–25). Portia makes clear early in the play that she has no affection for her current roster of suitors, though she intimates that she may already have feelings for Bassanio (1.2.82 ff). Miraculously, or inevitably, he chooses the right casket, though the significance of this remains elusive. Did he select correctly because he is the best candidate for husband, because he truly loves her and not her wealth? Or does she influence his choice, suggesting that she selects *him*, and in the process may be undermining the purpose of the lottery? Their marriage begins with Portia's submission to her husband, citing her unworthiness and committing "[Herself] and what is [hers] to [Bassanio]" (3.2.166). However she does so conditionally, with the gift of the ring, suggesting that when he gives it away (that is, back to her when she is disguised as the lawyer), he loses his claim to her and her property. Her teasing claim to having been unfaithful to Bassanio by having sex with the lawyer, though untrue, further suggests his loss of control over her body and fidelity.

The play explores issues of gender and hierarchy not just in marriage but also in the relationship between parents and children, especially fathers and daughters. Portia's father, though dead, still exercises control over her choice of husband through the institution of the casket lottery. Portia complains about this imposed limitation: "I may neither choose who I would nor refuse who I dislike; so is the will of a living daughter curbed by the will of a dead father" (1.2.17–19). Furthermore, in her discussion with Bassanio before he makes his choice, she raises the possibility of teaching him "How to choose right" (3.2.10–11) only to dismiss it. Yet elsewhere, she suggests that she would manipulate a suitor's choice in order to evade an unwanted marriage: "I will do anything, Nerissa, ere I will be married to a sponge" (1.2.72–73). The circumstances leading up to Bassanio's selection also differ significantly from those for the two previous suitors who entered into the lottery; does Portia provide hints to her lover through the meaning and the rhyme of the song she orders played before he chooses? In other words, how obedient is Portia to her father's will and how obedient should she be? The same question should be asked about Jessica, who violates her father's "will," or authority, and his casket in forming her marriage. On the one hand, her theft and elopement could be seen as a terrible rending of the bond between daughter and father. On the other hand, others in the play praise her for her actions; Gratiano and Lorenzo admire Jessica, even as she robs her father:

> GRATIANO: Now, by my hood, a gentle and no Jew.
> LORENZO: Beshrew me but I love her heartily,
>   For she is wise, if I can judge of her,
>   And fair she is, if that mine eyes be true,
>   And true she is, as she hath proved herself;   (2.6.52–56)

Do Jessica and Portia both disobey their fathers? If so, why might Jessica's disobedience be approved while Portia's might not? The play also raises interesting questions about the happiness of these marriages founded with or without parental approval. Shylock, for all his shortcomings, provides a model of marital devotion, valuing his wife's ring above any material good (3.1.90–92) and criticizing Bassanio and Gratiano — who later give up their wives' rings — for offering to sacrifice their wives for Antonio: "These be the Christian husbands. I have a daughter; / Would any of the stock of Barabbas / Had been her husband rather than a Christian" (4.1.290–92). Lorenzo's late arrival for his own elopement calls into question his devotion to his new bride (2.6.1–20); the conversations between these newlyweds appear to grow more contentious as the play proceeds. We might consider this a loving flirtatious banter, though the lovers to whom they compare themselves in their romantic moonlight exchange at the beginning of act 5 all ended betrayed and/or dead. Similarly, we wonder about the happiness of the two couples at the conclusion of the play. On the one hand, all seems to end well with both husbands and wives assured of their spouse's fidelity and love; on the other hand, how do we feel about marriages that begin with disguise, deception, and breaking of promises? Do Bassanio and Gratiano's allegiances lie with their wives or with other males at the play's end?

Early modern discourse on same-sex friendships often emphasizes the virtue of such relationships, suggesting that they involve a greater selflessness and purer form of affection than the love between a man and a woman. In his discussion of friendship, Sir Thomas Elyot draws on the writings of the Roman orator Cicero to define true friends

> to be those persons, which so do bear themselves and in such wise do live, that their faith, surety, equality, and liberality be sufficiently proved. Ne[13] that there is in them any covetise, willfulness, or foolhardiness, and that in them is great stability or constancy. (p. 343)

Antonio and Bassanio appear to meet these criteria in their mutual devotion; each makes sacrifices for the other and generously offers significant financial resources to the other in times of trouble (see act 1, scene 1, and act 4, scene 1). Financial matters govern their relationship, however: Bassanio seeks Antonio out in order to borrow money so that he can eventually pay back other debts he owes his friend. Is he a true friend to Antonio, or does he take advantage of the merchant's affection for him in asking for loans? Why is Bassanio in debt to begin with? Does his behavior in the play suggest that he has altered his habits to avoid future debt? Antonio, on the other hand, may not care about the loans being repaid, but does he offer the money with no strings attached, or does he expect to receive affection from Bassanio in return for his investment? Both Elyot and Philemon Holland,

---

[13] Ne: nor.

another author of the period, consider friendship in relation to marriage, suggesting that both types of attachments can involve deep emotional and erotic ties. These two types of relationships can be seen in competition with each other, and both authors offer opinions on the superiority of homoerotic attachments over heterosexual ones. The play does not specify the exact nature of Antonio and Bassanio's relationship: is it simply a close friendship, or is it a loving, erotic bond? Does it ultimately facilitate and reinforce the marriage between Bassanio and Portia or challenge it? If the latter is the case, how would an early modern audience decide which is the prevailing alliance at the play's end?

The early modern writings on marriage also offer a variety of opinions on the proper motives for its formation as well as on its proper functioning. The legal definition of a valid union between husband and wife required it to be entered into freely by both parties, emphasizing the importance of mutual affection. However, marriages formed primarily on the basis of love or lust were discouraged; authors writing on the subject emphasized pragmatic concerns such as economics, similarity of social status, and parental approval. Nevertheless, while a person considering marriage might take into account the wealth of a potential spouse, writers consistently condemned the notion of deciding on the sole basis of property. Alexander Niccholes uses sarcasm to emphasize this point in his *Discourse of Marriage and Wiving*:

> It is a fashion much in use in these times to choose wives as chapmen[14] sell their wares, with *Quantum dabitis?* What is the most you will give? And if their parents, or guardians shall reply their virtues are their portions, and others have they none, . . . these . . . shall be nothing valued, or make up where wealth is wanting; these may be adjuncts or good additions, but money must be the principal, of all that marry, and . . . there are but few that undergo it for the right end and use, whereby it comes to pass that many attain not to the blessedness therein. (p. 337)

If financial considerations alone are rejected as a sufficient reason for marrying, so too is physical attraction. Juan Luis Vives emphasizes the dangers of affection that is based on appearances:

> Love of the beauty is a forgetting of reason and the next thing unto frenzy, a foul vice, and an unmannerly for an whole mind. It troubleth all the wits, it breaketh and abateth high and noble stomachs, and draweth them down from the study and thinking of high and excellent things unto low and vile, and causeth them to be full of groaning and complaining, to be angry, hasty,

---

[14] **chapmen:** merchants, dealers.

> foolhardy, strait in ruling, full of vile and servile flattering, unmeet for every-
> thing, and at the last unmeet for love itself. (p. 322)

Such allurement ultimately undermines true love, in Vives's opinion, rather than establishes it. The range of motives for marrying in early modern writings mirrors those suggested in the play, although Shakespeare does not seem to give priority to any particular one. While Portia and Bassanio clearly enter into their marriage willingly, the grounds for their union are several. Both profess deep affection for each other and, on the basis of their speeches as well as the significance of the lottery, choose the other for their virtuous qualities, rather than their attractiveness, wealth, or status. However, Portia clearly rejects other suitors on the basis of appearance, and Bassanio first presents his marriage as a means to gain capital to repay Antonio. How should we weigh the various reasons motivating their alliance? In the quotation above, Niccholes focuses on the wealth of potential wives rather than husbands in part because the wife's property transferred to the husband upon marriage, according to the common law doctrine of unity of person. The husband permanently acquired the movable goods of his wife, while, for the duration of the marriage, he gained control over any real estate she owned; most of her legal rights were subsumed into those of her husband, who was empowered to represent her interests in court as he saw fit. This gave considerable power to the husband and subordinated his wife. What role does Portia's property play in the plot of the drama? How are the power relations between husbands and wives finally negotiated and resolved? Who has control over Portia's wealth at the end of the play and how is this accomplished?

Another factor influencing the selection of an appropriate spouse, both in early modern culture and within the plot of the play, is the role of parental guidance. Although parents were discouraged from coercing their children into marriage, children were expected to solicit and follow their parents' advice, as Thomas Becon advises in his *Catechism*:

> For this is part of the honor that the children owe to their parents and tutors
> by the commandment of God, even to be bestowed in marriage as it pleaseth
> the godly, prudent, and honest parents or tutors to appoint; with this persua-
> sion, that they,[15] for their age, wisdom, and experience, yea, and also for the
> tender love, singular benevolence, and hearty goodwill that they bear toward
> them, both know and will better provide for them than they be able to pro-
> vide for themselves. (pp. 330–31)

---

[15] **they:** parents.

He argues that the superior wisdom and experience of parents makes them better able than their children to choose their spouses. In this context, how would the actions of Shylock and Portia's father with regard to their daughters be understood by an early modern audience? How would the audience evaluate the role each daughter played in forming her own marriage?

The texts offered in this volume should help deepen and perhaps even complicate a reader's understanding of the play rather than easily resolve questions. By providing contexts within four categories, I hope to demonstrate the range of opinions within each rubric, which offer a number of ways to view related issues in the play. However, the ideas in each section not only influence how we understand that topic, but also shape our interpretation of the concepts framed in other sections. The suggestions offered in this general introduction and in the introductory material throughout the book aim to provide a starting point for your own interpretation of the play in its various cultural contexts.

PART ONE

WILLIAM SHAKESPEARE

The Merchant of Venice

Edited by David Bevington

The Merchant
of
Venice

SHYLOCK, *a rich Jew*
JESSICA, *his daughter*
TUBAL, *a Jew, Shylock's friend*
LAUNCELOT GOBBO, *a clown, servant to Shylock and then to Bassanio*
OLD GOBBO, *Launcelot's father*

Magnificoes of Venice, Officers of the Court of Justice, Jailor, Servants to Portia, and other Attendants

SCENE: *Partly at Venice and partly at Belmont, the seat of Portia*]

ACT I, SCENE I°

Enter Antonio, Salerio, and Solanio.

ANTONIO:
    In sooth, I know not why I am so sad.°
    It wearies me, you say it wearies you;
    But how I caught it, found it, or came by it,
    What stuff 'tis made of, whereof it is born,
    I am to learn;°   5
    And such a want-wit sadness makes of me°
    That I have much ado to know myself.

SALERIO:
    Your mind is tossing on the ocean,
    There where your argosies° with portly° sail,
    Like signors° and rich burghers on the flood,   10
    Or as it were the pageants° of the sea,
    Do overpeer° the petty traffickers
    That curtsy° to them, do them reverence,
    As they fly by them with their woven wings.

SOLANIO:
    Believe me, sir, had I such venture forth,°   15
    The better part of my affections would
    Be with my hopes abroad. I should be still°

ACT I, SCENE I. **Location:** A street in Venice. **1. sad:** morose, dismal-looking. **5. am to learn:** have yet to learn. **6. such . . . of me:** sadness makes me so distracted, lacking in good sense. **9. argosies:** large merchant ships. (So named from *Ragusa,* the modern city of Dubrovnik.) **portly:** majestic. **10. signors:** gentlemen. **11. pageants:** mobile stages used in plays or processions. **12. overpeer:** look down upon. **13. curtsy:** i.e., bob up and down, or lower topsails in token of respect (*reverence*). **15. venture forth:** investment risked. **17. still:** continually.

Plucking the grass to know where sits the wind,
Peering in maps for ports and piers and roads;°
And every object that might make me fear   20
Misfortune to my ventures, out of doubt
Would make me sad.
SALERIO: My wind cooling my broth
Would blow me to an ague° when I thought
What harm a wind too great might do at sea.
I should not see the sandy hourglass run   25
But° I should think of shallows and of flats,°
And see my wealthy *Andrew*° docked in sand,
Vailing° her high-top° lower than her ribs
To kiss her burial.° Should I° go to church
And see the holy edifice of stone   30
And not bethink me straight° of dangerous rocks
Which, touching but my gentle vessel's side,
Would scatter all her spices on the stream,
Enrobe the roaring waters with my silks,
And, in a word, but even now° worth this,°   35
And now worth nothing? Shall I have the thought
To think on this, and shall I lack the thought
That such a thing bechanced° would make me sad?
But tell not me. I know Antonio
Is sad to think upon his merchandise.   40
ANTONIO:
Believe me, no. I thank my fortune for it,
My ventures are not in one bottom° trusted,
Nor to one place; nor is my whole estate
Upon the fortune of this present year.°
Therefore my merchandise makes me not sad.   45
SOLANIO:
Why then, you are in love.
ANTONIO: Fie, fie!

---

19. **roads:** anchorages, open harbors.   23. **blow . . . ague:** i.e., start me shivering.   26. **But:** without it happening that.   **flats:** shoals.   27. *Andrew:* name of a ship (perhaps after the *St. Andrew,* a Spanish galleon captured at Cadiz in 1596).   28. **Vailing:** lowering. (Usually as a sign of submission.)   **high-top:** topmast.   29. **burial:** burial place.   **Should I:** if I should.   31. **bethink me straight:** be put in mind immediately.   35. **even now:** a short while ago. **this:** i.e., the cargo of spices and silks.   38. **bechanced:** having happened.   42. **bottom:** ship's hold.   44. **Upon . . . year:** i.e., risked upon the chance of the present year.

SOLANIO:

Not in love neither? Then let us say you are sad
Because you are not merry; and 'twere as easy
For you to laugh and leap, and say you are merry
Because you are not sad. Now, by two-headed Janus,°   50
Nature hath framed strange fellows in her time:
Some that will evermore peep through their eyes°
And laugh like parrots at a bagpiper,°
And other° of such vinegar aspect°
That they'll not show their teeth in way of smile   55
Though Nestor° swear the jest be laughable.

Enter Bassanio, Lorenzo, and Gratiano.

Here comes Bassanio, your most noble kinsman,
Gratiano, and Lorenzo. Fare ye well.
We leave you now with better company.

SALERIO:

I would have stayed till I had made you merry,   60
If worthier friends had not prevented° me.

ANTONIO:

Your worth is very dear in my regard.
I take it your own business calls on you,
And you embrace th' occasion° to depart.

SALERIO: Good morrow, my good lords.   65

BASSANIO:

Good signors both, when shall we laugh?° Say, when?
You grow exceeding strange.° Must it be so?°

SALERIO:

We'll make our leisures to attend on° yours.

Exeunt Salerio and Solanio.

LORENZO:

My lord Bassanio, since you have found Antonio,
We two will leave you, but at dinnertime,   70
I pray you, have in mind where we must meet.

---

50. **two-headed Janus:** a Roman god of all beginnings, represented by a figure with two faces.
52. **peep . . . eyes:** i.e., look with eyes narrowed by laughter.   53. **at a bagpiper:** i.e., even at a
bagpiper, whose music was regarded as melancholic.   54. **other:** others.   **vinegar aspect:**
sour, sullen looks.   56. **Nestor:** venerable senior officer in the *Iliad*, noted for gravity.
61. **prevented:** forestalled.   64. **occasion:** opportunity.   66. **laugh:** i.e., be merry together.
67. **strange:** distant.   **Must it be so:** must you go? or, must you show reserve?   68. **attend on:**
wait upon, i.e., suit.

BASSANIO: I will not fail you.

GRATIANO:

You look not well, Signor Antonio.
You have too much respect upon the world.°
They lose it that do buy it with much care.   75
Believe me, you are marvelously changed.

ANTONIO:

I hold the world but as the world, Gratiano —
A stage where every man must play a part,
And mine a sad one.

GRATIANO: Let me play the fool.
With mirth and laughter let old wrinkles come,   80
And let my liver rather heat with wine°
Than my heart cool with mortifying° groans.
Why should a man whose blood is warm within
Sit like his grandsire cut in alabaster?°
Sleep when he wakes, and creep into the jaundice°   85
By being peevish? I tell thee what, Antonio —
I love thee, and 'tis my love that speaks —
There are a sort of men whose visages
Do cream and mantle° like a standing° pond,
And do a willful stillness entertain°   90
With purpose to be dressed in an opinion°
Of° wisdom, gravity, profound conceit,°
As who should say,° "I am Sir Oracle,
And when I ope my lips let no dog bark!"°
O my Antonio, I do know of these   95
That therefore only are reputed wise
For saying nothing, when, I am very sure,
If they should speak, would almost damn those ears

---

74. **respect . . . world:** concern for worldly affairs of business.   81. **heat with wine:** (The liver was regarded as the seat of the passions and wine as an agency for inflaming them.)   82. **mortifying:** penitential and deadly. (Sighs were thought to cost the heart a drop of blood.)   84. **in alabaster:** i.e., in a stone effigy upon a tomb.   85. **jaundice:** (Regarded as arising from the effects of too much choler or yellow bile, one of the four humors, in the blood.)   89. **cream and mantle:** become covered with scum, i.e., acquire a lifeless, stiff expression.   **standing:** stagnant.   90. **And . . . entertain:** and who maintain or assume a self-imposed, obstinate silence.   91. **opinion:** reputation.   91–92. **be . . . Of:** i.e., gain a reputation for.   92. **profound conceit:** deep thought.   93. **As . . . say:** as if to say.   94. **And . . . bark:** i.e., and I am worthy of great respect.

Which, hearing them, would call their brothers fools.°
I'll tell thee more of this another time.   100
But fish not with this melancholy bait
For this fool gudgeon, this opinion.° —
Come, good Lorenzo. — Fare ye well awhile.
I'll end my exhortation after dinner.

LORENZO: [to Antonio and Bassanio]
Well, we will leave you then till dinnertime.   105
I must be one of these same dumb° wise men,
For Gratiano never lets me speak.

GRATIANO:
Well, keep° me company but two years more,
Thou shalt not know the sound of thine own tongue.

ANTONIO:
Fare you well. I'll grow a talker for this gear.°   110

GRATIANO:
Thanks, i' faith, for silence is only commendable
In a neat's° tongue dried and a maid not vendible.°

Exeunt [Gratiano and Lorenzo].

ANTONIO: Is that anything now?°

BASSANIO: Gratiano speaks an infinite deal of nothing, more than any
man in all Venice. His reasons° are as two grains of wheat hid in two   115
bushels of chaff; you shall seek all day ere you find them, and when you
have them they are not worth the search.

ANTONIO:
Well, tell me now what lady is the same°
To whom you swore a secret pilgrimage,
That you today promised to tell me of.   120

BASSANIO:
'Tis not unknown to you, Antonio,
How much I have disabled mine estate
By something showing a more swelling port°

98–99. would . . . fools: i.e., would virtually condemn their hearers into calling them fools.
(Compare Matthew 5:22, in which anyone calling another a fool is threatened with damnation.)
101–102. fish . . . opinion: i.e., don't go fishing for a reputation of being wise, using your
melancholy silence as the bait to fool people. (*Gudgeon*, a small fish, was thought of as a type of
gullibility.)   106. dumb: mute, speechless.   108. keep: if you keep.   110. for this gear: as a
result of this business, i.e., your talk.   112. neat's: ox's.   vendible: salable, i.e., in the marriage
market.   113. Is . . . now: i.e., was all that talk about anything?   115. reasons: reasonable
ideas.   118. the same: i.e., the one.   123. By . . . port: by showing a somewhat more lavish
style of living.

Than my faint means would grant continuance.°
Nor do I now make moan to be abridged   125
From such a noble rate;° but my chief care
Is to come fairly off° from the great debts
Wherein my time,° something too prodigal,
Hath left me gaged.° To you, Antonio,
I owe the most, in money and in love,   130
And from your love I have a warranty°
To unburden° all my plots and purposes
How to get clear of all the debts I owe.

ANTONIO:
I pray you, good Bassanio, let me know it;
And if it stand, as you yourself still do,   135
Within the eye of honor,° be assured
My purse, my person, my extremest means
Lie all unlocked to your occasions.

BASSANIO:
In my schooldays, when I had lost one shaft,°
I shot his° fellow of the selfsame flight°   140
The selfsame way with more advisèd° watch
To find the other forth,° and by adventuring° both
I oft found both. I urge this childhood proof
Because what follows is pure innocence.°
I owe you much, and, like a willful youth,   145
That which I owe is lost; but if you please
To shoot another arrow that self° way
Which you did shoot the first, I do not doubt,
As I will watch the aim, or° to find both
Or bring your latter hazard° back again   150
And thankfully rest° debtor for the first.

---

124. **grant continuance:** allow to continue.   125–26. **make . . . rate:** complain at being cut
back from such a high style of living.   127. **to . . . off:** honorably to extricate myself.
128. **time:** youthful lifetime.   129. **gaged:** pledged, in pawn.   131. **warranty:** authorization.
132. **unburden:** disclose.   135–36. **if . . . honor:** if it looks honorable, as your conduct has
always done.   139. **shaft:** arrow.   140. **his:** its.   **selfsame flight:** same kind and range.
141. **advisèd:** careful.   142. **forth:** out.   **adventuring:** risking.   144. **innocence:** ingenuous-
ness, sincerity.   147. **self:** same.   149. **or:** either.   150. **hazard:** that which was risked.
151. **rest:** remain.

ANTONIO:
You know me well, and herein spend but time°
To wind about my love with circumstance;°
And out of° doubt you do me now more wrong
In making question of my uttermost°   155
Than if you had made waste of all I have.
Then do but say to me what I should do
That in your knowledge may by me be done,
And I am prest° unto it. Therefore speak.

BASSANIO:
In Belmont is a lady richly left;°   160
And she is fair and, fairer than that word,
Of wondrous virtues. Sometime° from her eyes
I did receive fair speechless messages.
Her name is Portia, nothing undervalued
To° Cato's daughter, Brutus' Portia.°   165
Nor is the wide world ignorant of her worth,
For the four winds blow in from every coast
Renownèd suitors, and her sunny locks
Hang on her temples like a golden fleece,
Which makes her seat of Belmont Colchis'° strand,°   170
And many Jasons come in quest of her.
O my Antonio, had I but the means
To hold a rival place with one of them,
I have a mind presages° me such thrift°
That I should questionless be fortunate.   175

ANTONIO:
Thou know'st that all my fortunes are at sea;
Neither have I money nor commodity°
To raise a present sum.° Therefore go forth.
Try what my credit can in Venice do;
That shall be racked° even to the uttermost   180
To furnish thee to Belmont, to fair Portia.

---

152. **spend but time:** only waste time.   153. **To . . . circumstance:** i.e., in not asking plainly what you want. (*Circumstance* here means "circumlocution.")   154. **out of:** beyond.   155. **In . . . uttermost:** in showing any doubt of my intention to do all I can.   159. **prest:** ready.   160. **richly left:** left a large fortune (by her father's will).   162. **Sometime:** once.   164–65. **nothing undervalued To:** of no less worth than.   165. **Portia:** (The same Portia as in Shakespeare's *Julius Caesar.*)   170. **Colchis:** (Jason adventured for the golden fleece in the land of Colchis, on the Black Sea.)   **strand:** shore.   174. **presages:** i.e., which presages.   **thrift:** profit and good fortune.   177. **commodity:** merchandise.   178. **a present sum:** ready money.   180. **racked:** stretched.

Go presently° inquire, and so will I,
Where money is, and I no question make°
To have it of my trust° or for my sake.°   *Exeunt.*

ACT I, SCENE 2°

Enter Portia with her waiting woman, Nerissa.

PORTIA: By my troth,° Nerissa, my little body is aweary of this great
world.

NERISSA: You would be,° sweet madam, if your miseries were in the same
abundance as your good fortunes are; and yet, for aught I see, they are as
sick that surfeit° with too much as they that starve with nothing. It is no   5
mean° happiness, therefore, to be seated in the mean.° Superfluity comes
sooner by° white hairs, but competency° lives longer.

PORTIA: Good sentences,° and well pronounced.°

NERISSA: They would be better if well followed.

PORTIA: If to do were as easy as to know what were good to do, chapels   10
had been churches and poor men's cottages princes' palaces. It is a good
divine° that follows his own instructions. I can easier teach twenty what
were good to be done than to be one of the twenty to follow mine own
teaching. The brain may devise laws for the blood,° but a hot temper
leaps o'er a cold decree — such a hare is madness, the youth, to skip o'er   15
the meshes° of good counsel, the cripple.° But this reasoning is not in the
fashion to choose me a husband.° O me, the word "choose"! I may neither choose who I would nor refuse who I dislike; so is the will of a living
daughter curbed by the will° of a dead father. Is it not hard, Nerissa, that
I cannot choose one nor refuse none?   20

NERISSA: Your father was ever virtuous, and holy men at their death have
good inspirations; therefore the lottery that he hath devised in these
three chests of gold, silver, and lead, whereof who° chooses his° meaning

182. **presently:** immediately.   183. **no question make:** have no doubt.   184. **of my trust:** on
the basis of my credit as a merchant.   **sake:** i.e., personal stake.   ACT I, SCENE 2. **Location:**
Belmont. Portia's house.   1. **troth:** faith.   3. **would be:** would have reason to be (weary).
5. **surfeit:** overindulge.   6. **mean:** small (with a pun; see next note).   **in the mean:** having
neither too much nor too little.   6–7. **comes sooner by:** acquires sooner.   7. **competency:**
modest means.   8. **sentences:** maxims.   **pronounced:** delivered.   12. **divine:** clergyman.
14. **blood:** (Thought of as a chief agent of the passions, which in turn were regarded as the enemies of reason.)   16. **meshes:** nets (used here for hunting hares).   **good counsel, the cripple:**
(Wisdom is portrayed as old and no longer agile.)   16–17. **But . . . husband:** but this talk
is not the way to help me choose a husband.   18–19. **will . . . will:** volition . . . testament.
23. **who:** whoever.   **his:** i.e., the father's.

chooses you, will no doubt never be chosen by any rightly but one who you shall rightly° love. But what warmth is there in your affection 25 towards any of these princely suitors that are already come?

PORTIA: I pray thee, overname them,° and as thou namest them I will describe them; and according to my description level° at my affection.

NERISSA: First, there is the Neapolitan prince.

PORTIA: Ay, that's a colt° indeed, for he doth nothing but talk of his horse, 30 and he makes it a great appropriation° to his own good parts° that he can shoe him himself. I am much afeard my lady his mother played false with a smith.

NERISSA: Then is there the County° Palatine.°

PORTIA: He doth nothing but frown, as who should say,° "An° you will not 35 have me, choose."° He hears merry tales and smiles not. I fear he will prove the weeping philosopher° when he grows old, being so full of unmannerly sadness° in his youth. I had rather be married to a death's-head with a bone in his mouth than to either of these. God defend me from these two! 40

NERISSA: How say you by° the French lord, Monsieur Le Bon?

PORTIA: God made him, and therefore let him pass for a man. In truth, I know it is a sin to be a mocker, but he! Why, he hath a horse better than the Neapolitan's, a better bad habit of frowning than the Count Palatine; he is every man in no man.° If a throstle° sing, he falls straight° a-capering. 45 He will fence with his own shadow. If I should marry him, I should marry twenty husbands. If he would despise me, I would forgive him, for if° he love me to madness, I shall never requite him.

NERISSA: What say you, then, to° Falconbridge, the young baron of England? 50

PORTIA: You know I say nothing to him, for he understands not me, nor I him. He hath neither Latin, French, nor Italian, and you will come into the court and swear that I have a poor pennyworth in the English.° He is

---

24-25. **rightly ... rightly:** correctly ... truly. 27. **overname them:** name them over. 28. **level:** aim, guess. 30. **colt:** i.e., wanton and foolish young man (with a punning appropriateness to his interest in horses). 31. **appropriation:** addition. **good parts:** accomplishments. 34. **County:** count. **Palatine:** one possessing royal privileges. 35. **who should say:** one might say. **An:** if. 36. **choose:** i.e., do as you please. 37. **the weeping philosopher:** i.e., Heraclitus of Ephesus, a melancholic and retiring philosopher of about 500 B.C.E., often contrasted with Democritus, the "laughing philosopher." 38. **sadness:** melancholy. 41. **by:** about. 45. **he is ... no man:** i.e., he borrows aspects from everyone but has no character of his own. **throstle:** thrush. **straight:** at once. 48. **if:** even if. 49. **say you ... to:** do you say about. (But Portia wittily puns, in her reply, on the literal sense of "speaking to.") 52-53. **come ... English:** i.e., bear witness that I can speak very little English.

a proper man's picture,° but alas, who can converse with a dumb show?°
How oddly he is suited!° I think he bought his doublet° in Italy, his round   55
hose° in France, his bonnet° in Germany, and his behavior everywhere.

NERISSA: What think you of the Scottish lord, his neighbor?

PORTIA: That he hath a neighborly charity in him, for he borrowed° a box
of the ear of the Englishman and swore he would pay him again when he
was able. I think the Frenchman became his surety and sealed under for   60
another.°

NERISSA: How like you the young German, the Duke of Saxony's nephew?

PORTIA: Very vilely in the morning, when he is sober, and most vilely in
the afternoon, when he is drunk. When he is best he is a little worse than
a man, and when he is worst he is little better than a beast. An° the worst   65
fall° that ever fell, I hope I shall make shift° to go without him.

NERISSA: If he should offer° to choose, and choose the right casket, you
should refuse to perform your father's will if you should refuse to accept
him.

PORTIA: Therefore, for fear of the worst, I pray thee, set a deep glass of   70
Rhenish wine° on the contrary° casket, for if° the devil be within and
that temptation without, I know he will choose it. I will do anything,
Nerissa, ere I will be married to a sponge.

NERISSA: You need not fear, lady, the having any of these lords. They have
acquainted me with their determinations, which is indeed to return to their   75
home and to trouble you with no more suit, unless you may be won by
some other sort° than your father's imposition° depending on the caskets.

PORTIA: If I live to be as old as Sibylla,° I will die as chaste as Diana,°
unless I be obtained by the manner of my father's will. I am glad this
parcel° of wooers are so reasonable, for there is not one among them but I   80
dote on his very absence, and I pray God grant them a fair departure.

NERISSA: Do you not remember, lady, in your father's time, a Venetian, a
scholar and a soldier, that came hither in company of the Marquess of
Montferrat?

53–54. **He . . . picture:** i.e., he looks handsome.   54. **dumb show:** pantomime.   55. **suited:**
dressed.   **doublet:** upper garment corresponding to a jacket.   55–56. **round hose:** short,
puffed-out breeches.   56. **bonnet:** hat.   58. **borrowed:** received. (But with a play on the idea
of something that must be repaid.)   60–61. **became . . . another:** guaranteed the Scot's payment (of a box on the ear) and promised (literally, affixed his seal to a written promise) to give
the Englishman yet another on his own behalf. (An allusion to the age-old alliance of the
French and the Scots against the English.)   65. **An:** if.   66. **fall:** befall.   **make shift:** manage.   67. **offer:** undertake.   71. **Rhenish wine:** a German white wine from the Rhine Valley.
contrary: i.e., wrong.   **if:** even if.   77. **sort:** means (with perhaps a suggestion too of "casting"
or "drawing of lots").   **imposition:** command, charge.   78. **Sibylla:** the Cumaean Sibyl, to
whom Apollo gave as many years as there were grains in her handful of sand.   **Diana:** goddess
of chastity and of the hunt.   80. **parcel:** assembly, group.

PORTIA: Yes, yes, it was Bassanio — as I think, so was he called.   85
NERISSA: True, madam. He, of all the men that ever my foolish eyes
looked upon was the best deserving a fair lady.
PORTIA: I remember him well, and I remember him worthy of thy praise.

Enter a Servingman.

How now, what news?
SERVINGMAN: The four° strangers seek for you, madam, to take their   90
leave; and there is a forerunner° come from a fifth, the Prince of
Morocco, who brings word the Prince his master will be here tonight.
PORTIA: If I could bid the fifth welcome with so good heart as I can bid
the other four farewell, I should be glad of his approach. If he have the
condition° of a saint and the complexion of a devil,° I had rather he   95
should shrive me° than wive me.
Come, Nerissa. [*To Servingman.*] Sirrah,° go before. Whiles we shut the
gate upon one wooer, another knocks at the door.   *Exeunt.*

ACT 1, SCENE 3°

Enter Bassanio with Shylock the Jew.

SHYLOCK: Three thousand ducats,° well.
BASSANIO: Ay, sir, for three months.
SHYLOCK: For three months, well.
BASSANIO: For the which, as I told you, Antonio shall be bound.
SHYLOCK: Antonio shall become bound, well.   5
BASSANIO: May you stead° me? Will you pleasure° me? Shall I know your
answer?
SHYLOCK: Three thousand ducats for three months and Antonio bound.
BASSANIO: Your answer to that.
SHYLOCK: Antonio is a good° man.   10
BASSANIO: Have you heard any imputation to the contrary?
SHYLOCK: Ho, no, no, no, no! My meaning in saying he is a good man
is to have you understand me that he is sufficient.° Yet his means are in

---

90. **four:** (Nerissa actually names six suitors; possibly a sign of revision or the author's early draft.) 91. **forerunner:** herald. 95. **condition:** disposition, character. **complexion of a devil:** (Devils were thought to be black; but *complexion* can also mean "temperament," "disposition.") 96. **shrive me:** act as my confessor and give absolution. 97. **Sirrah:** (Form of address to social inferior.) ACT 1, SCENE 3. Location: Venice. A public place. 1. **ducats:** gold coins. 6. **stead:** supply, assist. **pleasure:** oblige. 10. **good:** (Shylock means "solvent," a good credit risk; Bassanio interprets it in the moral sense.) 13. **sufficient:** i.e., a good security.

supposition.° He hath an argosy bound to Tripolis, another to the Indies. I understand, moreover, upon the Rialto,° he hath a third at Mexico, a fourth for England, and other ventures he hath squandered° abroad. But ships are but boards, sailors but men. There be land rats and water rats, water thieves and land thieves — I mean pirates — and then there is the peril of waters, winds, and rocks. The man is, notwithstanding, sufficient. Three thousand ducats. I think I may take his bond. 20

BASSANIO: Be assured° you may.

SHYLOCK: I will be assured° I may; and that I may be assured, I will bethink me. May I speak with Antonio?

BASSANIO: If it please you to dine with us.

SHYLOCK: Yes, to smell pork, to eat of the habitation which your prophet 25 the Nazarite° conjured the devil into. I will buy with you, sell with you, talk with you, walk with you, and so following, but I will not eat with you, drink with you, nor pray with you. What news on the Rialto? Who is he comes here?

Enter Antonio.

BASSANIO: This is Signor Antonio. 30

SHYLOCK: [*aside*]
How like a fawning publican° he looks!
I hate him for° he is a Christian,
But more for that in low simplicity°
He lends out money gratis° and brings down
The rate of usance° here with us in Venice. 35
If I can catch him once upon the hip,°
I will feed fat° the ancient grudge I bear him.
He hates our sacred nation,° and he rails,
Even there where merchants most do congregate,
On me, my bargains, and my well-won thrift,° 40
Which he calls interest. Cursèd be my tribe

13–14. **in supposition:** doubtful, uncertain.   15. **the Rialto:** the merchants' exchange in Venice and the center of commercial activity.   16. **squandered:** scattered.   21, 22. **assured:** (Bassanio means that Shylock may trust Antonio, whereas Shylock means that he will provide legal assurances.)   26. **Nazarite:** Nazarene. (For the reference to Christ's casting evil spirits into a herd of swine, see Matthew 8:30–32, Mark 5:1–13, and Luke 8:32–33.)   31. **publican:** Roman tax gatherer (a term of opprobrium; see Luke 18:9–14); or, innkeeper.   32. **for:** because.   33. **low simplicity:** humble foolishness.   34. **gratis:** without charging interest.   35. **usance:** usury, interest.   36. **upon the hip:** i.e., at my mercy. (A figure of speech from wrestling; see Genesis 32:24–29.)   37. **fat:** until fatted for the kill.   38. **our sacred nation:** i.e., the Hebrew people.   40. **thrift:** thriving.

If I forgive him!

BASSANIO: Shylock, do you hear?

SHYLOCK:

    I am debating of my present store,°

    And, by the near guess of my memory,

    I cannot instantly raise up the gross°   45

    Of full three thousand ducats. What of that?

    Tubal, a wealthy Hebrew of my tribe,

    Will furnish me. But soft,° how many months

    Do you desire? [*To Antonio.*] Rest you fair, good signor!

    Your worship was the last man in our mouths.°   50

ANTONIO:

    Shylock, albeit I neither lend nor borrow

    By taking nor by giving of excess,°

    Yet, to supply the ripe wants° of my friend,

    I'll break a custom. [*To Bassanio.*] Is he yet possessed°

    How much ye would?°   55

SHYLOCK: Ay, ay, three thousand ducats.

ANTONIO: And for three months.

SHYLOCK:

    I had forgot — three months, you told me so.

    Well then, your bond. And let me see — but hear you,

    Methought you said you neither lend nor borrow   60

    Upon advantage.°

ANTONIO: I do never use it.

SHYLOCK:

    When Jacob grazed his uncle Laban's sheep —

    This Jacob° from our holy Abram° was,

    As his wise mother wrought in his behalf,

    The third° possessor;° ay, he was the third —   65

ANTONIO:

    And what of him? Did he take interest?

SHYLOCK:

    No, not take interest, not as you would say

    Directly interest. Mark what Jacob did.

---

43. I am . . . store: I am considering my current supply of money (and that is why I didn't answer you right away). **45. gross:** total. **48. soft:** i.e., wait a minute. **50. Your . . . mouths:** i.e., we were just speaking of you. (But with ominous connotation of devouring; compare line 37.) **52. excess:** interest. **53. ripe wants:** pressing needs. **54. possessed:** informed. **55. ye would:** you want. **61. advantage:** interest. **63. Jacob:** (See Genesis 27, 30:25-43.) **Abram:** Abraham. **65. third:** i.e., after Abraham and Isaac. **possessor:** i.e., of the birthright of which, with the help of Rebecca, he was able to cheat Esau, his elder brother.

When Laban and himself were compromised°
That all the eanlings° which were streaked and pied°   70
Should fall as Jacob's hire,° the ewes, being rank,°
In end of autumn turnèd to the rams,
And when the work of generation° was
Between these woolly breeders in the act,
The skillful shepherd peeled me certain wands,°   75
And in the doing of the deed of kind°
He stuck them up before the fulsome° ewes,
Who then conceiving did in eaning° time
Fall° parti-colored° lambs, and those were Jacob's.
This was a way to thrive, and he was blest;   80
And thrift° is blessing, if men steal it not.

ANTONIO:
This was a venture, sir, that Jacob served for,°
A thing not in his power to bring to pass,
But swayed and fashioned by the hand of heaven.
Was this inserted to make interest good?°   85
Or is your gold and silver ewes and rams?

SHYLOCK:
I cannot tell.° I make it breed as fast.
But note me, signor —

ANTONIO: Mark you this, Bassanio,
The devil can cite Scripture° for his purpose.
An evil soul producing holy witness   90
Is like a villain with a smiling cheek,
A goodly apple rotten at the heart.
O, what a goodly outside falsehood hath!

SHYLOCK:
Three thousand ducats. 'Tis a good round sum.
Three months from twelve, then let me see, the rate —   95

ANTONIO:
Well, Shylock, shall we be beholding° to you?

---

69. **compromised:** agreed.   70. **eanlings:** young lambs or kids.   **pied:** spotted.   71. **hire:** wages, share.   **rank:** in heat.   73. **work of generation:** mating.   75. **peeled . . . wands:** i.e., partly stripped the bark of some sticks. (*Me* is used colloquially.)   76. **deed of kind:** i.e., copulation.   77. **fulsome:** lustful, well-fed.   78. **eaning:** lambing.   79. **Fall:** give birth to. **parti-colored:** multicolored.   81. **thrift:** thriving, profit.   82. **venture . . . for:** uncertain commercial venture on which Jacob risked his wages.   85. **inserted . . . good:** brought in to justify the practice of usury.   87. **I cannot tell:** i.e., I don't know about that.   89. **devil . . . Scripture:** (See Matthew 4:6.)   96. **beholding:** beholden, indebted.

SHYLOCK:

    Signor Antonio, many a time and oft
    In the Rialto you have rated° me
    About my moneys and my usances.
    Still have I borne it with a patient shrug,   100
    For sufferance° is the badge of all our tribe.
    You call me misbeliever, cutthroat dog,
    And spit upon my Jewish gaberdine,°
    And all for use of that which is mine own.
    Well then, it now appears you need my help.   105
    Go to,° then. You come to me and you say,
    "Shylock, we would have moneys"— you say so,
    You, that did void your rheum° upon my beard
    And foot me as you spurn° a stranger cur
    Over your threshold. Moneys is your suit.°   110
    What should I say to you? Should I not say,
    "Hath a dog money? Is it possible
    A cur can lend three thousand ducats?" Or
    Shall I bend low, and in a bondman's key,°
    With bated° breath and whispering humbleness,   115
    Say this:
    "Fair sir, you spit on me on Wednesday last,
    You spurned me such a day, another time
    You called me dog, and for these courtesies
    I'll lend you thus much moneys"?   120

ANTONIO:

    I am as like° to call thee so again,
    To spit on thee again, to spurn thee too.
    If thou wilt lend this money, lend it not
    As to thy friends, for when did friendship take
    A breed for barren metal° of° his friend?   125
    But lend it rather to° thine enemy,
    Who,° if he break,° thou mayst with better face

---

98. **rated:** berated, rebuked.   101. **sufferance:** endurance.   103. **gaberdine:** loose outer garment like a cape or mantle.   106. **Go to:** (An exclamation of impatience or annoyance.) 108. **rheum:** spittle.   109. **spurn:** kick.   110. **suit:** request.   114. **bondman's key:** serf's tone of voice.   115. **bated:** subdued.   121. **like:** likely.   125. **A breed . . . metal:** offspring from money, which cannot naturally breed. (One of the oldest arguments against usury was that it was thereby "unnatural.")   **of:** from.   126. **to:** as if to.   127. **Who:** from whom.   **break:** fail to pay on time.

Exact the penalty.

SHYLOCK: Why, look you how you storm!
I would be friends with you and have your love,
Forget the shames that you have stained me with,   130
Supply your present wants, and take no doit°
Of usance for my moneys, and you'll not hear me.
This is kind° I offer.

BASSANIO: This were° kindness.

SHYLOCK: This kindness will I show.   135
Go with me to a notary, seal me there
Your single bond;° and, in a merry sport,
If you repay me not on such a day,
In such a place, such sum or sums as are
Expressed in the condition, let the forfeit   140
Be nominated for° an equal° pound
Of your fair flesh, to be cut off and taken
In what part of your body pleaseth me.

ANTONIO:
Content, in faith. I'll seal to such a bond
And say there is much kindness in the Jew.   145

BASSANIO:
You shall not seal to such a bond for me!
I'll rather dwell° in my necessity.

ANTONIO:
Why, fear not, man, I will not forfeit it.
Within thee two months — that's a month before
This bond expires — I do expect return   150
Of thrice three times the value of this bond.

SHYLOCK:
O father Abram, what these Christians are,
Whose own hard dealings teaches them suspect
The thoughts of others! Pray you, tell me this:
If he should break his day, what should I gain   155
By the exaction of the forfeiture?
A pound of man's flesh taken from a man

---

131. **doit:** a Dutch coin of very small value.   133. **kind:** kindly.   134. **were:** would be (if seriously offered).   137. **single bond:** bond signed alone without other security; unconditional. (Shylock pretends the *condition*, line 139, is only a joke.)   141. **nominated for:** named, specified as.   **equal:** exact.   147. **dwell:** remain.

Is not so estimable,° profitable neither,
As flesh of muttons, beefs, or goats. I say
To buy his favor I extend this friendship.   160
If he will take it, so;° if not, adieu.
And for my love, I pray you, wrong me not.°

ANTONIO:
Yes, Shylock, I will seal unto this bond.

SHYLOCK:
Then meet me forthwith at the notary's.
Give him direction for this merry bond,   165
And I will go and purse the ducats straight,
See to my house, left in the fearful° guard
Of an unthrifty knave, and presently
I'll be with you.   Exit.

ANTONIO: Hie thee, gentle° Jew.
The Hebrew will turn Christian; he grows kind.   170

BASSANIO:
I like not fair terms and a villain's mind.

ANTONIO:
Come on. In this there can be no dismay;
My ships come home a month before the day.

                                                          Exeunt.

ACT 2, SCENE 1°

[Flourish of cornets.] Enter [the Prince of] Morocco, a tawny Moor all in white, and
three or four followers accordingly,° with Portia, Nerissa, and their train.

MOROCCO:
Mislike me not for my complexion,
The shadowed livery° of the burnished sun,
To whom I am a neighbor and near bred.°
Bring me the fairest creature northward born,
Where Phoebus'° fire scarce thaws the icicles,   5
And let us make incision for your love
To prove whose blood is reddest,° his or mine.

158. estimable: valuable. 161. so: well and good. 162. wrong me not: do not think evil of
me. 167. fearful: to be mistrusted. 169. gentle: gracious, courteous (with a play on "gentile"). ACT 2, SCENE 1. Location: Belmont. Portia's house. s.d. accordingly: similarly (i.e.,
dressed in white and dark-skinned like Morocco). 2. shadowed livery: i.e., dark complexion,
worn as though it were a costume of the sun's servants. 3. near bred: closely related.
5. Phoebus': i.e., the sun's. 7. reddest: (Red blood was regarded as a sign of courage.)

I tell thee, lady, this aspect° of mine
Hath feared° the valiant. By my love I swear,
The best-regarded virgins of our clime   10
Have loved it too. I would not change this hue,
Except to steal your thoughts, my gentle queen.
PORTIA:
In terms of choice I am not solely led
By nice direction° of a maiden's eyes;
Besides, the lott'ry of my destiny   15
Bars me the right of voluntary choosing.
But if my father had not scanted° me,
And hedged me by his wit° to yield myself
His wife who° wins me by that means I told you,
Yourself, renownèd Prince, then stood as fair°   20
As any comer I have looked on yet
For my° affection.
MOROCCO: Even for that I thank you.
Therefore, I pray you, lead me to the caskets
To try my fortune. By this scimitar
That slew the Sophy° and a Persian prince,   25
That won three fields° of Sultan Solyman,°
I would o'erstare° the sternest eyes that look,
Outbrave the heart most daring on the earth,
Pluck the young sucking cubs from the she-bear,
Yea, mock the lion when 'a° roars for prey,   30
To win thee, lady. But alas the while!
If Hercules and Lichas° play at dice
Which is the better man, the greater throw
May turn by fortune from the weaker hand.
So is Alcides beaten by his page,   35
And so may I, blind Fortune leading me,
Miss that which one unworthier may attain,
And die with grieving.
PORTIA: You must take your chance,
And either not attempt to choose at all

---

8. **aspect:** visage.   9. **feared:** frightened.   14. **nice direction:** careful guidance.   17. **scanted:** limited.   18. **wit:** wisdom.   18–19. **yield . . . who:** give myself to be the wife of him who.   20. **then . . . fair:** would then have looked as attractive and stood as fair a chance (with a play on "fair-skinned").   22. **For my:** of gaining my.   25. **Sophy:** Shah of Persia.   26. **fields:** battles.   **Solyman:** a Turkish sultan ruling from 1520 to 1566.   27. **o'erstare:** outstare.   30. **'a:** he.   32. **Lichas:** a page of Hercules (Alcides). See the note for 3.2.55.

Or swear before you choose, if you choose wrong   40
Never to speak to lady afterward
In way of marriage. Therefore be advised.°
MOROCCO:
   Nor will not.° Come, bring me unto my chance.
PORTIA:
   First, forward to the temple.° After dinner
Your hazard shall be made.
MOROCCO: Good fortune then!   45
To make me blest or cursed'st among men.

                              [*Cornets, and*] *exeunt.*

ACT 2, SCENE 2°

Enter [*Launcelot*] *the clown,*° *alone.*

LAUNCELOT: Certainly my conscience will serve° me to run from this Jew
my master. The fiend is at mine elbow and tempts me, saying to me,
"Gobbo, Launcelot Gobbo, good Launcelot," or "Good Gobbo," or
"Good Launcelot Gobbo, use your legs, take the start, run away." My
conscience says, "No, take heed, honest Launcelot, take heed, honest   5
Gobbo," or, as aforesaid, "Honest Launcelot Gobbo, do not run; scorn
running with thy heels."° Well, the most courageous fiend bids me pack.°
"Fia!"° says the fiend. "Away!" says the fiend. "For the heavens,° rouse up a
brave mind," says the fiend, "and run." Well, my conscience, hanging
about the neck of my heart,° says very wisely to me, "My honest friend   10
Launcelot, being an honest man's son," or rather an honest woman's
son — for indeed my father did something smack, something grow to, he
had a kind of taste° — well, my conscience says, "Launcelot, budge not."
"Budge," says the fiend. "Budge not," says my conscience. "Conscience,"
say I, "you counsel well." "Fiend," say I, "you counsel well." To be ruled by 15
my conscience, I should stay with the Jew my master, who, God bless the
mark,° is a kind of devil; and to run away from the Jew, I should be ruled

---

42. **be advised:** take warning, consider.   43. **Nor will not:** i.e., nor indeed will I violate the
oath.   44. **to the temple:** i.e., in order to take the oaths.   ACT 2, SCENE 2. Location: Venice.
A street.   **s.d. clown:** (1) country bumpkin (2) comic type in an Elizabethan acting company.
1. **serve:** permit.   6-7. **with thy heels:** i.e., emphatically (with a pun on the literal sense).
7. **pack:** begone.   **Fia:** i.e., via, away.   8. **For the heavens:** i.e., in heaven's name.
9-10. **hanging . . . heart:** i.e., timidly.   12-13. **something . . . taste:** i.e., had a tendency to
lechery.   16. **God . . . mark:** (An expression by way of apology for introducing something
potentially offensive, as also in *saving your reverence.*)

by the fiend, who, saving your reverence, is the devil himself. Certainly the Jew is the very devil incarnation;° and, in my conscience, my conscience is but a kind of hard conscience to offer to counsel me to stay with the Jew. The fiend gives the more friendly counsel. I will run, fiend. My heels are at your commandment. I will run.

Enter Old Gobbo, with a basket.

GOBBO: Master young man, you,° I pray you, which is the way to master Jew's?

LAUNCELOT: [*aside*]   O heavens, this is my true-begotten father, who, being more than sand-blind,° high-gravel-blind,° knows me not. I will try confusions° with him.

GOBBO: Master young gentleman, I pray you, which is the way to master Jew's?

LAUNCELOT: Turn up on your right hand at the next turning, but at the next turning of all on your left; marry,° at the very next turning, turn of no hand, but turn down indirectly to the Jew's house.

GOBBO: By God's sonties,° 'twill be a hard way to hit. Can you tell me whether one Launcelot, that dwells with him, dwell with him or no?

LAUNCELOT: Talk you of young Master Launcelot? [*Aside.*] Mark me now; now will I raise the waters.° — Talk you of young Master Launcelot?

GOBBO: No master,° sir, but a poor man's son. His father, though I say 't, is an honest exceeding poor man and, God be thanked, well to live.°

LAUNCELOT: Well, let his father be what 'a° will, we talk of young Master Launcelot.

GOBBO: Your worship's friend, and Launcelot,° sir.

LAUNCELOT: But I pray you, ergo,° old man, ergo, I beseech you, talk you of young Master Launcelot?

GOBBO: Of Launcelot, an 't please your mastership.

LAUNCELOT: Ergo, Master Launcelot. Talk not of Master Launcelot, Father,° for the young gentleman, according to Fates and Destinies and

---

19. **incarnation:** (Launcelot means "incarnate.")   23. **you:** (Gobbo uses the formal *you* but switches to the familiar *thou,* line 69, when he accepts Launcelot as his son.)   26. **sand-blind:** dimsighted.   **high-gravel-blind:** blinder than sand-blind. (A term seemingly invented by Launcelot.)   26-27. **try confusions:** (Launcelot's blunder for "try conclusions," i.e., experiment, though his error is comically apt.)   31. **marry:** i.e., by the Virgin Mary, indeed. (A mild interjection.)   33. **sonties:** saints.   36. **raise the waters:** i.e., start tears.   37. **master:** (The title was applied to gentlefolk only.)   38. **well to live:** enjoying a good livelihood. (Perhaps Old Gobbo intends the phrase to mean "in good health," since he protests that he is poor.)   39. **'a:** he.   41. **Your . . . Launcelot:** (Again, Old Gobbo denies that Launcelot is entitled to be called "Master.")   42. **ergo:** therefore. (But Launcelot may use this Latin word with no particular meaning in mind.)   45. **Father:** (1) old man (2) Father.

such odd sayings, the Sisters Three° and such branches of learning, is indeed deceased, or, as you would say in plain terms, gone to heaven.

GOBBO: Marry, God forbid! The boy was the very staff of my age, my very prop. 50

LAUNCELOT: Do I look like a cudgel or a hovel post,° a staff, or a prop? Do you know me, Father?

GOBBO: Alack the day, I know you not, young gentleman. But I pray you, tell me, is my boy, God rest his soul, alive or dead?

LAUNCELOT: Do you not know me, Father? 55

GOBBO: Alack, sir, I am sand-blind. I know you not.

LAUNCELOT: Nay, indeed, if you had your eyes you might fail of the knowing me; it is a wise father that knows his own child.° Well, old man, I will tell you news of your son. [*He kneels.*] Give me your blessing. Truth will come to light; murder cannot be hid long; a man's son may, but in the end 60 truth will out.

GOBBO: Pray you, sir, stand up. I am sure you are not Launcelot, my boy.

LAUNCELOT: Pray you, let's have no more fooling about it, but give me your blessing. I am Launcelot, your boy that was, your son that is, your child that shall be.° 65

GOBBO: I cannot think you are my son.

LAUNCELOT: I know not what I shall think of that; but I am Launcelot, the Jew's man, and I am sure Margery your wife is my mother.

GOBBO: Her name is Margery indeed. I'll be sworn, if thou be Launcelot, thou art mine own flesh and blood. Lord worshiped might he be, what a 70 beard° hast thou got! Thou hast got more hair on thy chin than Dobbin my fill horse° has on his tail.

LAUNCELOT: [*rising*] It should seem then that Dobbin's tail grows backward.° I am sure he had more hair of his tail than I have of° my face when I last saw him. 75

GOBBO: Lord, how art thou changed! How dost thou and thy master agree? I have brought him a present. How 'gree you now?

LAUNCELOT: Well, well; but for mine own part, as I have set up my rest° to run away, so I will not rest° till I have run some ground. My master's a

---

47. **the Sisters Three:** the three Fates. 51. **hovel post:** support for a hovel or open shed. 58. **it is . . . child:** (Reverses the proverb "It is a wise child that knows his own father.") 64–65. **your . . . shall be:** (Echoes the *Gloria* from the Book of Common Prayer: "As it was in the beginning, is now, and ever shall be.") 71. **beard:** (Stage tradition has Old Gobbo mistaking Launcelot's long hair for a beard.) 72. **fill horse:** cart horse. 73–74. **grows backward:** (1) grows inward, shorter (2) grows at the wrong end. 74. **of:** in, on. 78. **set up my rest:** determined, risked all. (A metaphor from the card game *primero*, in which a final wager is made, with a pun also on *rest* as "place of residence.") 79. **not rest:** i.e., not stop running. (More punning on *rest.*)

very° Jew.° Give him a present? Give him a halter!° I am famished in his 80
service; you may tell° every finger I have with my ribs.° Father, I am glad
you are come. Give me° your present to one Master Bassanio, who
indeed gives rare new liveries.° If I serve not him, I will run as far as God
has any ground. O rare fortune! Here comes the man. To him, Father, for
I am a Jew° if I serve the Jew any longer. 85

Enter Bassanio, with [Leonardo and] a follower or two.

BASSANIO: You may do so, but let it be so hasted° that supper be ready at
the farthest° by five of the clock. See these letters delivered, put the liveries to making, and desire Gratiano to come anon to my lodging.

[Exit a Servant.]

LAUNCELOT: To him, Father.
GOBBO: *[advancing]* God bless your worship! 90
BASSANIO: Gramercy.° Wouldst thou aught° with me?
GOBBO: Here's my son, sir, a poor boy —
LAUNCELOT: Not a poor° boy, sir, but the rich Jew's man, that would, sir, as
my father shall specify —
GOBBO: He hath a great infection,° sir, as one would say, to serve — 95
LAUNCELOT: Indeed, the short and the long is, I serve the Jew, and have a
desire, as my father shall specify —
GOBBO: His master and he, saving your worship's reverence, are scarce
cater-cousins° —
LAUNCELOT: To be brief, the very truth is that the Jew, having done me 100
wrong, doth cause me, as my father, being, I hope, an old man, shall
frutify° unto you —
GOBBO: I have here a dish of doves that I would bestow upon your worship, and my suit is —
LAUNCELOT: In very brief, the suit is impertinent° to myself, as your wor- 105
ship shall know by this honest old man, and, though I say it, though old
man, yet poor man, my father.
BASSANIO: One speak for both. What would you?
LAUNCELOT: Serve you, sir.

---

80. **very:** veritable. **Jew:** (1) Jew (2) grasping old usurer. **halter:** hangman's noose. 81. **tell:**
count. **tell . . . ribs:** (Comically reverses the usual saying of counting one's ribs with one's fingers.) 82. **Give me:** give. (*Me* suggests "on my behalf.") 83. **liveries:** uniforms or costumes
for servants. 85. **a Jew:** i.e., a villain (punning on the literal sense in *the Jew.* Compare with
line 80.) 86. **hasted:** hastened, hurried. 87. **farthest:** latest. 91. **Gramercy:** many thanks.
aught: anything. 92–93. **poor . . . poor:** unfortunate . . . impoverished. 95. **infection:**
(Blunder for "affection" or "inclination.") 99. **cater-cousins:** good friends. 102. **frutify:**
(Launcelot may be trying to say "fructify," but he means "certify" or "notify.") 105. **imperti-
nent:** (Blunder for "pertinent.")

GOBBO: That is the very defect° of the matter, sir.   110

BASSANIO:
   I know thee well; thou hast obtained thy suit.
   Shylock thy master spoke with me this day,
   And hath preferred° thee, if it be preferment
   To leave a rich Jew's service to become
   The follower of so poor a gentleman.   115

LAUNCELOT: The old proverb° is very well parted° between my master
   Shylock and you, sir: you have the grace of God, sir, and he hath enough.

BASSANIO:
   Thou speak'st it well. Go, father, with thy son.
   Take leave of thy old master, and inquire
   My lodging out. [*To a Servant.*] Give him a livery   120
   More guarded° than his fellows'. See it done.

LAUNCELOT: Father, in. I cannot get a service, no! I have ne'er a tongue in
   my head, well! [*He looks at his palm.*] If any man in Italy have a fairer table°
   which doth offer to swear upon a book,° I shall have good fortune. Go
   to,° here's a simple° line of life.° Here's a small trifle of wives! Alas, fifteen   125
   wives is nothing. Eleven widows and nine maids is a simple coming-in°
   for one man. And then to scape drowning thrice, and to be in peril of my
   life with the edge of a feather bed!° Here are simple scapes.° Well, if
   Fortune be a woman,° she's a good wench for this gear.° Father, come. I'll
   take my leave of the Jew in the twinkling.   130

                                        Exit clown [*Launcelot, with Old Gobbo*].

BASSANIO: [*giving Leonardo a list*]
   I pray thee, good Leonardo, think on this:
   These things being bought and orderly bestowed,°
   Return in haste, for I do feast° tonight
   My best-esteemed acquaintance. Hie thee, go.

---

110. **defect:** (Blunder for "effect," i.e., "purport.")   113. **preferred:** recommended.   116. **prov-erb:** i.e., "He who has the grace of God has enough."   **parted:** divided.   121. **guarded:** trimmed with braided ornament.   123. **table:** palm of the hand. (Launcelot now reads the lines of his palm.)   124. **book:** i.e., Bible. (The image is of a hand being laid on the Bible to take an oath.)   124–125. **Go to:** (An expression of impatience.)   125. **simple:** unremarkable. (Said ironically.)   **line of life:** curved line at the base of the thumb.   126. **simple coming-in:** modest beginning or income (with sexual suggestion).   128. **feather bed:** (Suggesting mar-riage bed or love bed; Launcelot sees sexual adventure and the dangers of marriage in his palm reading.)   **scapes:** (1) adventures (2) transgressions.   129. **Fortune . . . woman:** (Fortune was personified as a goddess.)   **gear:** matter.   132. **bestowed:** i.e., stowed on board ship.   133. **feast:** give a feast for.

LEONARDO:
My best endeavors shall be done herein.   135

[He starts to leave.]

Enter Gratiano.

GRATIANO: [*to Leonardo*]
Where's your master?
LEONARDO: Yonder, sir, he walks.

Exit Leonardo.

GRATIANO: Signor Bassanio!
BASSANIO: Gratiano!
GRATIANO:
I have a suit to you.
BASSANIO: You have obtained it.
GRATIANO: You must not deny me. I must go with you to Belmont.   140
BASSANIO:
Why, then you must. But hear thee, Gratiano;
Thou art too wild, too rude and bold of voice —
Parts° that become thee happily enough,
And in such eyes as ours appear not faults,
But where thou art not known, why, there they show   145
Something too liberal.° Pray thee, take pain
To allay° with some cold drops of modesty°
Thy skipping spirit, lest through thy wild behavior
I be misconstered° in the place I go to
And lose my hopes.
GRATIANO: Signor Bassanio, hear me: 150
If I do not put on a sober habit,°
Talk with respect and swear but now and then,
Wear prayer books in my pocket, look demurely,
Nay more, while grace is saying,° hood mine eyes
Thus with my hat, and sigh and say "amen,"   155
Use all the observance of civility,
Like one well studied in a sad ostent°
To please his grandam,° never trust me more.
BASSANIO: Well, we shall see your bearing.

---

143. Parts: qualities.   **146. liberal:** free of manner (often with sexual connotation).
147. allay: temper, moderate.   **modesty:** decorum.   **149. misconstered:** misconstrued.
151. habit: demeanor (with a suggestion of "clothes").   **154. saying:** being said.   **157. sad
ostent:** grave appearance.   **158. grandam:** grandmother.

GRATIANO:
> Nay, but I bar tonight. You shall not gauge me 160
> By what we do tonight.

BASSANIO: No, that were pity.
> I would entreat you rather to put on
> Your boldest suit of mirth, for we have friends
> That purpose merriment. But fare you well;
> I have some business. 165

GRATIANO:
> And I must to Lorenzo and the rest,
> But we will visit you at suppertime. *Exeunt.*

ACT 2, SCENE 3°

Enter Jessica and [Launcelot] the clown.

JESSICA:
> I am sorry thou wilt leave my father so.
> Our house is hell, and thou, a merry devil,
> Didst rob it of some taste of tediousness.
> But fare thee well. There is a ducat for thee.
> *[Giving money.]*
> And, Launcelot, soon at supper shalt thou see 5
> Lorenzo, who is thy new master's guest.
> Give him this letter; do it secretly. *[Giving a letter.]*
> And so farewell. I would not have my father
> See me in talk with thee.

LAUNCELOT: Adieu! Tears exhibit° my tongue. Most beautiful pagan, most 10
sweet Jew! If a Christian did not play the knave and get° thee, I am much
deceived. But, adieu! These foolish drops do something drown my manly
spirit. Adieu!

JESSICA: Farewell, good Launcelot. *[Exit Launcelot.]*
> Alack, what heinous sin is it in me 15
> To be ashamed to be my father's child!
> But though I am a daughter to his blood,
> I am not to his manners. O Lorenzo,
> If thou keep promise, I shall end this strife,
> Become a Christian and thy loving wife. *Exit.* 20

ACT 2, SCENE 3. Location: Venice. Shylock's house. 10. exhibit: (Blunder for "inhibit," "restrain.") 11. get: beget.

ACT 2, SCENE 4°

Enter Gratiano, Lorenzo, Salerio, and Solanio.

LORENZO:
Nay, we will slink away in° suppertime,
Disguise us at my lodging, and return
All in an hour.

GRATIANO:
We have not made good preparation.

SALERIO:
We have not spoke us yet of° torchbearers.   5

SOLANIO:
'Tis vile, unless it may be quaintly ordered,°
And better in my mind not undertook.

LORENZO:
'Tis now but four o'clock. We have two hours
To furnish us.

Enter Launcelot [with a letter].

　　　　　　　　Friend Launcelot, what's the news?

LAUNCELOT:　An° it shall please you to break up this,° it shall seem   10
to signify.   [*Giving the letter.*]

LORENZO:
I know the hand. In faith, 'tis a fair hand,
And whiter than the paper it writ on
Is the fair hand that writ.

GRATIANO:　　　　　　　Love news, in faith.

LAUNCELOT:　By your leave, sir.   [*He starts to leave.*]   15

LORENZO:　Whither goest thou?

LAUNCELOT:　Marry, sir, to bid my old master the Jew to sup tonight with
my new master the Christian.

LORENZO:
Hold here, take this. [*He gives money.*] Tell gentle Jessica
I will not fail her. Speak it privately.   20

　　　　　　　　　　　　　　Exit clown [Launcelot].

ACT 2, SCENE 4. **Location:** Venice. A street.　**1. in:** during.　**5. spoke . . . of:** yet bespoken, ordered.　**6. quaintly ordered:** skillfully and tastefully managed.　**10. An:** if.　**break up this:** i.e., open the seal.

Go, gentlemen,
Will you prepare you for this masque tonight?
I am provided of a torchbearer.

SALERIO:
Ay, marry, I'll be gone about it straight.°

SOLANIO:
And so will I.

LORENZO: Meet me and Gratiano.   25
At Gratiano's lodging some hour° hence.

SALERIO: 'Tis good we do so.   *Exit [with Solanio].*

GRATIANO:
Was not that letter from fair Jessica?

LORENZO:
I must needs° tell thee all. She hath directed
How I shall take her from her father's house,   30
What gold and jewels she is furnished with,
What page's suit she hath in readiness.
If e'er the Jew her father come to heaven,
It will be for his gentle° daughter's sake;
And never dare misfortune cross her foot,°   35
Unless she° do it under this excuse,
That she is issue° to a faithless° Jew.
Come, go with me. Peruse this as thou goest.
                                   [He gives Gratiano the letter.]
Fair Jessica shall be my torchbearer.   *Exeunt.*

ACT 2, SCENE 5°

Enter [Shylock the] Jew and [Launcelot,] his man that was, the clown.

SHYLOCK:
Well, thou shalt see, thy eyes shall be thy judge,
The difference of° old Shylock and Bassanio. —
What, Jessica! — Thou shalt not gormandize,°
As thou hast done with me — What, Jessica! —
And sleep and snore, and rend apparel out° —   5

---

24. **straight:** at once.   26. **some hour:** about an hour.   29. **must needs:** must.   34. **gentle:** (With pun on "gentile"?)   35. **foot:** footpath.   36. **she:** i.e., Misfortune.   37. **she is issue:** i.e., Jessica is daughter.   **faithless:** pagan.   ACT 2, SCENE 5. **Location:** Venice. Before Shylock's house.   2. **of:** between.   3. **gormandize:** eat gluttonously.   5. **rend apparel out:** i.e., wear out your clothes.

Why, Jessica, I say!

LAUNCELOT: Why, Jessica!

SHYLOCK:

Who bids thee call? I did not bid thee call.

LAUNCELOT: Your worship was wont to tell me I could do nothing without
bidding.   10

Enter Jessica.

JESSICA: Call you? What is your will?

SHYLOCK:

I am bid forth to supper, Jessica.
There are my keys. But wherefore° should I go?
I am not bid for love — they flatter me —
But yet I'll go in hate, to feed upon   15
The prodigal Christian. Jessica, my girl,
Look to my house. I am right loath° to go.
There is some ill a-brewing towards my rest,
For I did dream of moneybags tonight.°

LAUNCELOT: I beseech you, sir, go. My young master doth expect your   20
reproach.°

SHYLOCK: So do I his.

LAUNCELOT: And they have conspired together. I will not say you shall see
a masque, but if you do, then it was not for nothing that my nose fell
a-bleeding on Black Monday° last at six o'clock i' the morning, falling   25
out that year on Ash Wednesday was four year in th' afternoon.

SHYLOCK:

What, are there masques? Hear you me, Jessica:
Lock up my doors, and when you hear the drum
And the vile squealing of the wry-necked° fife,
Clamber not you up to the casements then,   30
Nor thrust your head into the public street
To gaze on Christian fools with varnished faces,°
But stop my house's ears — I mean my casements.
Let not the sound of shallow foppery enter
My sober house. By Jacob's staff° I swear   35

---

13. **wherefore:** why.   17. **right loath:** reluctant.   19. **tonight:** last night.   21. **reproach:**
(Launcelot's blunder for "approach." Shylock takes it in grim humor.)   25. **Black Monday:**
Easter Monday. (Launcelot's talk of omens is perhaps intentional gibberish, a parody of Shylock's fears.)   29. **wry-necked:** i.e., played with the musician's head awry.   32. **varnished
faces:** i.e., painted masks.   35. **Jacob's staff:** (See Genesis 32:10 and Hebrews 11:21.)

FIGURE 2 *Venetian Masque, from Pietro Bertelli,* Diversarum Nationem *(1594). Shylock condemns the masques as "shallow foppery." Here the celebrants in the street wear masks while the women above look out at the spectacle.*

I have no mind of feasting forth tonight.
But I will go. — Go you before me, sirrah.
Say I will come.
LAUNCELOT: I will go before, sir. [*Aside to Jessica.*] Mistress, look out at window, for all this;

40

There will come a Christian by,
Will be worth a Jewess' eye.   [*Exit.*]

SHYLOCK:
What says that fool of Hagar's offspring,° ha?

JESSICA:
His words were "Farewell, mistress," nothing else.

SHYLOCK:
The patch° is kind enough, but a huge feeder,   45
Snail-slow in profit,° and he sleeps by day
More than the wildcat. Drones hive not with me;
Therefore I part with him, and part with him
To one that I would have him help to waste
His borrowed purse. Well, Jessica, go in.   50
Perhaps I will return immediately.
Do as I bid you. Shut doors after you.
Fast bind, fast find° —
A proverb never stale in thrifty mind.   *Exit.*

JESSICA:
Farewell, and if my fortune be not crossed,   55
I have a father, you a daughter, lost.   *Exit.*

ACT 2, SCENE 6°

Enter the masquers, Gratiano and Salerio.

GRATIANO:
This is the penthouse° under which Lorenzo
Desired us to make stand.

SALERIO: His hour is almost past.

GRATIANO:
And it is marvel he outdwells his hour,°
For lovers ever run before the clock.   5

SALERIO:
O, ten times faster Venus' pigeons° fly
To seal love's bonds new-made than they are wont
To keep obligèd° faith unforfeited.°

43. **Hagar's offspring:** (Hagar, a gentile and Abraham's servant, gave birth to Ishmael; both mother and son were cast out after the birth of Isaac.)   45. **patch:** fool.   46. **profit:** profitable labor.   53. **Fast . . . find:** i.e., something firmly secured or bound will always be easily located.   **ACT 2, SCENE 6. Location:** Before Shylock's house, as in act 2, scene 5.   1. **penthouse:** projecting roof or upper story of a house.   4. **it . . . hour:** i.e., it is surprising that he is late.   6. **Venus' pigeons:** the doves that drew Venus' chariot.   8. **obligèd:** bound by marriage or engagement.   **unforfeited:** unbroken.

GRATIANO:
> That ever holds.° Who riseth from a feast
> With that keen appetite that he sits down?   10
> Where is the horse that doth untread° again
> His tedious measures° with the unbated fire
> That he did pace them first? All things that are
> Are with more spirit chasèd than enjoyed.
> How like a younger° or a prodigal   15
> The scarfèd° bark puts from her native bay,
> Hugged and embracèd by the strumpet° wind!
> How like the prodigal doth she return,
> With overweathered ribs° and ragged sails,
> Lean, rent,° and beggared by the strumpet wind!   20

Enter Lorenzo.

SALERIO:
> Here comes Lorenzo. More of this hereafter.

LORENZO:
> Sweet friends, your patience° for my long abode;°
> Not I, but my affairs, have made you wait.
> When you shall please to pay the thieves for wives,
> I'll watch as long for you then. Approach;   25
> Here dwells my father° Jew. Ho! Who's within?

[Enter] Jessica, above [in boy's clothes].

JESSICA:
> Who are you? Tell me for more certainty,
> Albeit I'll swear that I do know your tongue.

LORENZO:
> Lorenzo, and thy love.

JESSICA:
> Lorenzo, certain, and my love indeed,
> For who love I so much? And now who knows   30
> But you, Lorenzo, whether I am yours?

---

9. **ever holds:** always holds true. 11. **untread:** retrace. 12. **measures:** paces. 15. **younger:** i.e., younger son, as in the parable of the Prodigal Son (Luke 15). (Often emended to *younker,* youth.) 16. **scarfèd:** decorated with flags or streamers. 17. **strumpet:** i.e., inconsistent, variable (likened metaphorically to the harlots with whom the Prodigal Son wasted his fortune). 19. **overweathered ribs:** i.e., weather-beaten and leaking timbers. 20. **rent:** torn. 22. **your patience:** i.e., I beg your patience. **abode:** delay. 26. **father:** i.e., father-in-law.

LORENZO:
Heaven and thy thoughts are witness that thou art.
JESSICA: [*throwing down a casket*]
Here, catch this casket. It is worth the pains.
I am glad 'tis night, you do not look on me,   35
For I am much ashamed of my exchange.°
But love is blind, and lovers cannot see
The pretty° follies that themselves commit,
For if they could, Cupid himself would blush
To see me thus transformèd to a boy.   40
LORENZO:
Descend, for you must be my torchbearer.
JESSICA:
What, must I hold a candle° to my shames?
They in themselves, good sooth, are too too light.°
Why, 'tis an office of discovery,° love,
And I should be obscured.
LORENZO: So are you, sweet,   45
Even in the lovely garnish° of a boy.
But come at once,
For the close° night doth play the runaway,°
And we are stayed° for at Bassanio's feast.
JESSICA:
I will make fast the doors, and gild° myself   50
With some more ducats, and be with you straight.
                                              [*Exit above.*]
GRATIANO:
Now, by my hood,° a gentle° and no Jew.
LORENZO:
Beshrew° me but I love her heartily,
For she is wise, if I can judge of her,
And fair she is, if that mine eyes be true,   55
And true she is, as she hath proved herself;

---

36. **exchange:** change of clothes.   38. **pretty:** ingenious, artful.   42. **hold a candle:** i.e., stand by and witness (with a play on the idea of acting as torchbearer).   43. **light:** immodest (with pun on literal meaning).   44. **'tis . . . discovery:** i.e., torchbearing is intended to shed light on matters.   46. **garnish:** outfit, trimmings.   48. **close:** dark, secretive. **doth . . . runaway:** i.e., is quickly passing.   49. **stayed:** waited.   50. **gild:** adorn. (Literally, cover with gold.)   52. **by my hood:** (An asseveration.)   **gentle:** gracious person (with pun on "gentile," as at 2.4.34).   53. **Beshrew:** i.e., a mischief on. (A mild oath.)

And therefore, like herself, wise, fair, and true,
Shall she be placèd in my constant soul.

Enter Jessica [below].

What, art thou come? On, gentlemen, away!
Our masquing mates by this time for us stay.°   60
        Exit [with Jessica and Salerio; Gratiano is about to follow them].

Enter Antonio.

ANTONIO: Who's there?
GRATIANO: Signor Antonio?
ANTONIO:
  Fie, fie, Gratiano! Where are all the rest?
  'Tis nine o'clock; our friends all stay for you.
  No masque tonight. The wind is come about;   65
  Bassanio presently will go aboard.
  I have sent twenty out to seek for you.
GRATIANO:
  I am glad on 't. I desire no more delight
  Than to be under sail and gone tonight.   *Exeunt.*

ACT 2, SCENE 7°

[Flourish of cornets.] Enter Portia, with [the Prince of] Morocco, and both their trains.

PORTIA:
  Go, draw aside the curtains and discover°
  The several° caskets to this noble prince.
  Now make your choice.   *[The curtains are drawn.]*
MOROCCO:
  The first, of gold, who° this inscription bears,
  "Who chooseth me shall gain what many men desire";   5
  The second, silver, which this promise carries,
  "Who chooseth me shall get as much as he deserves";
  This third, dull° lead, with warning all as blunt,°
  "Who chooseth me must give and hazard all he hath."
  How shall I know if I do choose the right?   10

---

60. **stay:** wait. (Also in line 64.) **ACT 2, SCENE 7. Location:** Belmont. Portia's house.
1. **discover:** reveal. 2. **several:** different, various. 4. **who:** which. 8. **dull:** (1) dull-colored (2) blunt. **all as blunt:** as blunt as lead.

PORTIA:
   The one of them contains my picture, Prince.
   If you choose that, then I am yours withal.°
MOROCCO:
   Some god direct my judgment! Let me see,
   I will survey th' inscriptions back again.
   What says this leaden casket?   15
   "Who chooseth me must give and hazard all he hath."
   Must give — for what? For lead? Hazard for lead?
   This casket threatens. Men that hazard all
   Do it in hope of fair advantages.
   A golden mind stoops not to shows of dross.°   20
   I'll then nor give° nor hazard aught for lead.
   What says the silver with her virgin hue?
   "Who chooseth me shall get as much as he deserves."
   As much as he deserves! Pause there, Morocco,
   And weigh thy value with an even° hand.   25
   If thou be'st rated by thy estimation,°
   Thou dost deserve enough; and yet enough
   May not extend so far as to the lady;
   And yet to be afeard of my deserving
   Were but a weak disabling° of myself.   30
   As much as I deserve? Why, that's the lady.
   I do in birth deserve her, and in fortunes,
   In graces, and in qualities of breeding;
   But more than these, in love I do deserve.
   What if I strayed no farther, but chose here?   35
   Let's see once more this saying graved in gold:
   "Who chooseth me shall gain what many men desire."
   Why, that's the lady; all the world desires her.
   From the four corners of the earth they come
   To kiss this shrine, this mortal breathing° saint.   40
   The Hyrcanian° deserts and the vasty° wilds
   Of wide Arabia are as throughfares now
   For princes to come view fair Portia.
   The watery kingdom, whose ambitious head

---

12. **withal:** with it.   20. **dross:** worthless matter. (Literally, the impurities cast off in the melting down of metals.)   21. **nor give:** neither give.   25. **even:** impartial.   26. **estimation:** valuation.   30. **disabling:** underrating.   40. **mortal breathing:** living.   41. **Hyrcanian:** (Hyrcania was the country south of the Caspian Sea celebrated for its wildness.)   **vasty:** vast.

Spits° in the face of heaven, is no bar   45
To stop the foreign spirits,° but they come,
As o'er a brook, to see fair Portia.
One of these three contains her heavenly picture.
Is 't like° that lead contains her? 'Twere damnation
To think so base° a thought; it were° too gross   50
To rib° her cerecloth° in the obscure grave.°
Or shall I think in silver she's immured,°
Being ten times undervalued to° tried° gold?
O, sinful thought! Never so rich a gem
Was set in worse than gold. They have in England   55
A coin° that bears the figure of an angel
Stamped in gold, but that's insculped upon;°
But here an angel in a golden bed
Lies all within. Deliver me the key.
Here do I choose, and thrive I as I may!   60
PORTIA:
There, take it, Prince; and if my form° lie there,
Then I am yours.   [*He unlocks the golden casket.*]
MOROCCO: O hell! What have we here?
A carrion Death,° within whose empty eye
There is a written scroll! I'll read the writing.
[*He reads.*]
          "All that glisters is not gold;   65
          Often have you heard that told.
          Many a man his life hath sold
          But my outside to behold.
          Gilded tombs do worms infold.
          Had you been as wise as bold,   70
          Young in limbs, in judgment old,
          Your answer had not been inscrolled.°
          Fare you well; your suit is cold."

---

45. **Spits:** (The image is of huge waves breaking at sea.) 46. **spirits:** i.e., men of courage.
49. **like:** likely. 50. **base:** (1) ignoble (2) low in the natural scale, as with lead, a *base* metal.
it were: i.e., lead would be. 51. **rib:** i.e., enclose. **cerecloth:** wax cloth used in wrapping for
burial. **grave:** i.e., lead casket used for burial. 52. **immured:** enclosed, confined.
53. **Being . . . to:** which has only one tenth the value of. **tried:** assayed, purified. 56. **coin:**
i.e., the gold coin known as the *angel,* which bore the device of the archangel Michael treading
on the dragon. 57. **insculped upon:** merely engraved upon the surface. 61. **form:** image.
63. **carrion Death:** death's-head. 72. **inscrolled:** i.e., written on this scroll.

Cold, indeed, and labor lost.
Then, farewell, heat, and welcome, frost! 75
Portia, adieu. I have too grieved a heart
To take a tedious leave. Thus losers part.°

Exit [with his train. Flourish of cornets].

PORTIA:
A gentle riddance. Draw the curtains, go.
Let all of his complexion° choose me so.

[The curtains are closed, and] exeunt.

ACT 2, SCENE 8°

Enter Salerio and Solanio.

SALERIO:
Why, man, I saw Bassanio under sail.
With him is Gratiano gone along,
And in their ship I am sure Lorenzo is not.

SOLANIO:
The villain Jew with outcries raised° the Duke,
Who went with him to search Bassanio's ship. 5

SALERIO:
He came too late. The ship was under sail.
But there the Duke was given to understand
That in a gondola were seen together
Lorenzo and his amorous Jessica.
Besides, Antonio certified the Duke 10
They were not with Bassanio in his ship.

SOLANIO:
I never heard a passion° so confused,
So strange, outrageous, and so variable
As the dog Jew did utter in the streets:
"My daughter! O, my ducats! O, my daughter! 15
Fled with a Christian! O, my Christian ducats!
Justice! The law! My ducats, and my daughter!
A sealèd bag, two sealèd bags of ducats,
Of double ducats, stol'n from me by my daughter!
And jewels, two stones, two rich and precious stones, 20

77. **part:** depart.   79. **complexion:** temperament (not merely skin color).   **ACT 2, SCENE 8.**
Location: Venice. A street.   4. **raised:** roused.   12. **passion:** passionate outburst.

Stol'n by my daughter! Justice! Find the girl!
She hath the stones upon her, and the ducats."

SALERIO:
Why, all the boys in Venice follow him,
Crying his stones,° his daughter, and his ducats.

SOLANIO:
Let good Antonio look he keep his day,°   25
Or he shall pay for this.

SALERIO: Marry, well remembered.
I reasoned° with a Frenchman yesterday,
Who told me, in the narrow seas° that part
The French and English, there miscarrièd
A vessel of our country richly fraught.°   30
I thought upon Antonio when he told me,
And wished in silence that it were not his.

SOLANIO:
You were best to tell Antonio what you hear.
Yet do not suddenly, for it may grieve him.

SALERIO:
A kinder gentleman treads not the earth.   35
I saw Bassanio and Antonio part.
Bassanio told him he would make some speed
Of his return; he answered, "Do not so.
Slubber not business° for my sake, Bassanio,
But stay the very riping of the time;°   40
And for° the Jew's bond which he hath of me,
Let it not enter in your mind of° love.
Be merry, and employ your chiefest thoughts
To courtship and such fair ostents° of love
As shall conveniently become you there."   45
And even there,° his eye being big with tears,
Turning his face, he put his hand behind him,°
And with affection wondrous sensible°
He wrung Bassanio's hand; and so they parted.

24. **stones:** (In the boys' jeering cry, the *two stones* suggest testicles; see line 20.)   25. **look . . . day:** see to it that he repays his loan on time.   27. **reasoned:** talked.   28. **narrow seas:** English Channel.   30. **fraught:** freighted.   39. **Slubber not business:** don't do the business hastily and badly.   40. **But . . . time:** i.e., pursue your business at Belmont until it is brought to completion.   41. **for:** as for.   42. **of:** preoccupied with.   44. **ostents:** expressions, shows.   46. **there:** thereupon, then.   47. **behind him:** (Antonio turns away in tears while extending his hand back to Bassanio.)   48. **affection wondrous sensible:** wondrously sensitive and keen emotion.

SOLANIO:
> I think he only loves the world for him.°   50
> I pray thee, let us go and find him out
> And quicken his embracèd heaviness°
> With some delight or other.

SALERIO: Do we so.   *Exeunt.*

Act 2, Scene 9°

Enter Nerissa and a Servitor.°

NERISSA:
> Quick, quick, I pray thee, draw the curtain straight.°
> The Prince of Aragon hath ta'en his oath,
> And comes to his election° presently.°

> > [*The curtains are drawn back.*]

[*Flourish of cornets.*] *Enter* [*the Prince of*] *Aragon, his train, and Portia.*

PORTIA:
> Behold, there stand the caskets, noble Prince.
> If you choose that wherein I am contained,   5
> Straight shall our nuptial rites be solemnized;
> But if you fail, without more speech, my lord,
> You must be gone from hence immediately.

ARAGON:
> I am enjoined by oath to observe three things:
> First, never to unfold to anyone   10
> Which casket 'twas I chose; next, if I fail
> Of the right casket, never in my life
> To woo a maid in way of marriage;
> Lastly,
> If I do fail in fortune of my choice,   15
> Immediately to leave you and be gone.

PORTIA:
> To these injunctions everyone doth swear
> That comes to hazard for my worthless self.

---

50. **he . . . him:** i.e., Bassanio is all he lives for.   52. **quicken . . . heaviness:** lighten the sorrow he has embraced.   **Act 2, Scene 9. Location:** Belmont. Portia's house.   **s.d. Servitor:** servant.   1. **straight:** at once.   3. **election:** choice.   **presently:** immediately.

ARAGON:
>And so have I addressed me.° Fortune now
>To my heart's hope! Gold, silver, and base lead.   20
>"Who chooseth me must give and hazard all he hath."
>You shall look fairer ere I give or hazard.
>What says the golden chest? Ha, let me see:
>"Who chooseth me shall gain what many men desire."
>What many men desire! That "many" may be meant   25
>By° the fool multitude, that choose by show,
>Not learning more than the fond° eye doth teach,
>Which pries not to th' interior, but like the martlet°
>Builds in° the weather on the outward wall,
>Even in the force and road of casualty.°   30
>I will not choose what many men desire,
>Because I will not jump° with common spirits
>And rank me with the barbarous multitudes.
>Why then, to thee, thou silver treasure-house!
>Tell me once more what title thou dost bear: 35
>"Who chooseth me shall get as much as he deserves."
>And well said too; for who shall go about
>To cozen° fortune, and be honorable
>Without the stamp° of merit? Let none presume
>To wear an undeservèd dignity.   40
>O, that estates, degrees,° and offices
>Were not derived corruptly, and that clear honor
>Were purchased by the merit of the wearer!
>How many then should cover that stand bare?°
>How many be commanded that command?°   45
>How much low peasantry would then be gleaned°
>From the true seed of honor,° and how much honor
>Picked from the chaff and ruin of the times
>To be new-varnished?° Well, but to my choice:
>"Who chooseth me shall get as much as he deserves."   50

---

19. **addressed me:** prepared myself (by this swearing). 25–26. **meant By:** meant for. 27. **fond:** foolish. 28. **martlet:** swift. 29. **in:** exposed to. 30. **force . . . casualty:** power and path of mischance. 32. **jump:** agree. 38. **cozen:** cheat. 39. **stamp:** seal of approval. 41. **estates, degrees:** status, social rank. 44. **cover . . . bare:** i.e., wear hats (of authority) who now stand bareheaded. 45. **How . . . command:** how many then should be servants that are now masters. 46. **gleaned:** culled out. 47. **the true seed of honor:** i.e., persons of noble descent. 49. **new-varnished:** i.e., having the luster of their true nobility restored to them.

I will assume desert. Give me a key for this,
And instantly unlock my fortunes here.

[*He opens the silver casket.*]

PORTIA:

Too long a pause for that which you find there.

ARAGON:

What's here? The portrait of a blinking idiot,
Presenting me a schedule!° I will read it.   55
How much unlike art thou to Portia!
How much unlike my hopes and my deservings!
"Who chooseth me shall have as much as he deserves."
Did I deserve no more than a fool's head?
Is that my prize? Are my deserts no better?   60

PORTIA:

To offend and judge are distinct offices
And of opposèd natures.°

ARAGON: What is here?

[*He reads.*] "The fire seven times tried° this;°
            Seven times tried that judgment is
            That did never choose amiss.   65
            Some there be that shadows° kiss;
            Such have but a shadow's bliss.
            There be fools alive, iwis,°
            Silvered o'er,° and so was this.
            Take what wife you will to bed,   70
            I will ever be your head.°
            So begone; you are sped."°

Still more fool I shall appear
By the time I linger here.°
With one fool's head I came to woo,   75
But I go away with two.

---

55. **schedule:** written paper.   61–62. **To offend . . . natures:** i.e., you have no right, having
submitted your case to judgment, to attempt to judge your own case; or, it is not for me to say,
since I've been the indirect cause of your discomfiture.   63. **tried:** tested, purified (?).   **this:**
i.e., the wise sayings on the scroll (that have often been proved right by hard experience); or, the
silver.   66. **shadows:** illusions.   68. **iwis:** certainly.   69. **Silvered o'er:** i.e., with silver hair
and so apparently wise.   71. **I . . . head:** i.e., you will always have a fool's head, be a fool.
72. **sped:** done for.   73–74. **Still . . . here:** i.e., I shall seem all the greater fool for wasting any
more time here.

Sweet, adieu. I'll keep my oath,
Patiently to bear my wroth.°

[*Exeunt Aragon and train.*]

PORTIA:
Thus hath the candle signed the moth.
O, these deliberate° fools! When they do choose,   80
They have the wisdom by their wit to lose.

NERISSA:
The ancient saying is no heresy:
Hanging and wiving goes by destiny.

PORTIA: Come, draw the curtain, Nerissa.

[*The curtains are closed.*]

Enter Messenger.

MESSENGER:
Where is my lady?

PORTIA: Here. What would my lord?°   85

MESSENGER:
Madam, there is alighted at your gate
A young Venetian, one that comes before
To signify th' approaching of his lord,
From whom he bringeth sensible regreets,°
To wit, besides commends° and courteous breath,°   90
Gifts of rich value. Yet° I have not seen
So likely° an ambassador of love.
A day in April never came so sweet,
To show how costly° summer was at hand,
As this fore-spurrer° comes before his lord.   95

PORTIA:
No more, I pray thee. I am half afeard
Thou wilt say anon he is some kin to thee,
Thou spend'st such high-day° wit in praising him.
Come, come, Nerissa, for I long to see
Quick Cupid's post° that comes so mannerly.   100

NERISSA:
Bassanio, Lord Love, if thy will it be!   *Exeunt.*

---

78. **wroth:** sorrow, unhappy lot (a variant of *ruth*); or, anger.   80. **deliberate:** reasoning, calculating.   85. **my lord:** (A jesting response to "my lady.")   89. **sensible regreets:** tangible gifts, greetings.   90. **commends:** greetings.   **breath:** speech.   91. **Yet:** heretofore.   92. **likely:** promising.   94. **costly:** lavish, rich.   95. **fore-spurrer:** herald, harbinger.   98. **high-day:** holiday (i.e., extravagant).   100. **post:** messenger.

ACT 3, SCENE 1°

[Enter] Solanio and Salerio.

SOLANIO: Now, what news on the Rialto?

SALERIO: Why, yet it lives there unchecked° that Antonio hath a ship of
rich lading wrecked on the narrow seas° — the Goodwins,° I think they
call the place, a very dangerous flat,° and fatal, where the carcasses of
many a tall° ship lie buried, as they say, if my gossip Report° be an honest   5
woman of her word.

SOLANIO: I would she were as lying a gossip in that as ever knapped° gin-
ger or made her neighbors believe she wept for the death of a third husband. But it is true, without any slips of prolixity° or crossing the plain
highway of talk,° that the good Antonio, the honest Antonio — O, that I   10
had a title good enough to keep his name company! —

SALERIO: Come, the full stop.°

SOLANIO: Ha, what sayest thou? Why, the end is, he hath lost a ship.

SALERIO: I would it might prove the end of his losses.

SOLANIO: Let me say "amen" betimes,° lest the devil cross° my prayer, for   15
here he comes in the likeness of a Jew.

Enter Shylock.

How now, Shylock, what news among the merchants?

SHYLOCK: You knew, none so well, none so well as you, of my daughter's
flight.

SALERIO: That's certain. I for my part knew the tailor that made the wings   20
she flew withal.°

SOLANIO: And Shylock for his own part knew the bird was fledge,° and
then it is the complexion° of them all to leave the dam.°

SHYLOCK: She is damned for it.

SALERIO: That's certain, if the devil may be her judge.   25

ACT 3, SCENE 1. Location: Venice. A street.   **2. yet . . . unchecked:** i.e., a rumor is spreading
undenied.   **3. the narrow seas:** the English Channel, as at 2.8.28.   **Goodwins:** Goodwin
Sands, off the Kentish coast near the Thames estuary.   **4. flat:** shoal, sandbank.   **5. tall:** gallant.   **gossip Report:** i.e., Dame Rumor.   **7. knapped:** nibbled.   **9. slips of prolixity:** lapses
into long-windedness; or, long-winded lies. *Slips* may be the cuttings or offshoots of tedious-
ness.   **9–10. crossing . . . talk:** deviating from honest, plain speech.   **12. Come . . . stop:** fin-
ish your sentence; rein in your tongue as a horse is checked in its manage.   **15. betimes:** while
there is yet time.   **cross:** thwart; make the sign of the cross following.   **20–21. the wings . . .
withal:** i.e., the disguise she escaped in; with a play on *wings* or ornamented shoulder flaps sewn
on garments.   **22. fledge:** ready to fly.   **23. complexion:** natural disposition, as at 2.7.79.
dam: mother.

SHYLOCK: My own flesh and blood to rebel!

SOLANIO: Out upon it, old carrion! Rebels it at these years?°

SHYLOCK: I say my daughter is my flesh and my blood.

SALERIO: There is more difference between thy flesh and hers than between jet° and ivory, more between your bloods than there is between 30 red wine and Rhenish.° But tell us, do you hear whether Antonio have had any loss at sea or no?

SHYLOCK: There I have another bad match!° A bankrupt, a prodigal, who dare scarce show his head on the Rialto; a beggar, that was used to come so smug upon the mart!° Let him look to his bond. He was wont to call 35 me usurer. Let him look to his bond. He was wont to lend money for a Christian courtesy. Let him look to his bond.

SALERIO: Why, I am sure, if he forfeit, thou wilt not take his flesh. What's that good for?

SHYLOCK: To bait° fish withal.° If it will feed nothing else, it will feed my 40 revenge. He hath disgraced me, and hindered me half a million, laughed at my losses, mocked at my gains, scorned my nation, thwarted my bargains, cooled my friends, healed mine enemies; and what's his reason? I am a Jew. Hath not a Jew eyes? Hath not a Jew hands, organs, dimensions, senses, affections, passions? Fed with the same food, hurt with the 45 same weapons, subject to the same diseases, healed by the same means, warmed and cooled by the same winter and summer, as a Christian is? If you prick us, do we not bleed? If you tickle us, do we not laugh? If you poison us, do we not die? And if you wrong us, shall we not revenge? If we are like you in the rest, we will resemble you in that. If a Jew wrong a 50 Christian, what is his humility?° Revenge. If a Christian wrong a Jew, what should his sufferance° be by Christian example? Why, revenge. The villainy you teach me I will execute, and it shall go hard but I will better the instruction.

Enter a Man from Antonio.

MAN: Gentlemen, my master Antonio is at his house and desires to speak 55 with you both.

---

27. Rebels . . . years: (Solanio pretends to interpret Shylock's cry about the rebellion of his own flesh and blood as referring to his own carnal desires, his own erection.) 30. jet: a black, hard mineral, here contrasted with the whiteness of ivory and Jessica's fair complexion. 31. Rhenish: i.e., a German white wine from the Rhine valley. Salerio seems to prefer the white wine as more refined than the red. 33. match: bargain. 35. mart: marketplace, Rialto. 40. To bait: to act as bait for. withal: with. 51. what is his humility: how does the Christian humble himself? 52. his sufferance: the Jew's patient endurance.

SALERIO: We have been up and down to seek him.

Enter Tubal.

SOLANIO: Here comes another of the tribe. A third cannot be matched,° unless the devil himself turn Jew.

Exeunt gentlemen [Solanio, Salerio, with Man].

SHYLOCK: How now, Tubal, what news from Genoa? Hast thou found my 60 daughter?

TUBAL: I often came where I did hear of her, but cannot find her.

SHYLOCK: Why, there, there, there, there! A diamond gone, cost me two thousand ducats in Frankfort! The curse° never fell upon our nation till now; I never felt it till now. Two thousand ducats in that, and other pre- 65 cious, precious jewels. I would my daughter were dead at my foot, and the jewels in her ear! Would she were hearsed° at my foot, and the ducats in her coffin! No news of them? Why, so — and I know not what's spent in the search. Why, thou loss upon loss! The thief gone with so much, and so much to find the thief, and no satisfaction, no revenge! Nor no ill 70 luck stirring but what lights o' my shoulders, no sighs but o' my breathing, no tears but o' my shedding.

TUBAL: Yes, other men have ill luck too. Antonio, as I heard in Genoa —

SHYLOCK: What, what, what? Ill luck, ill luck?

TUBAL: Hath an argosy cast away,° coming from Tripolis. 75

SHYLOCK: I thank God, I thank God. Is it true, is it true?

TUBAL: I spoke with some of the sailors that escaped the wreck.

SHYLOCK: I thank thee, good Tubal. Good news, good news! Ha, ha! Heard in Genoa?

TUBAL: Your daughter spent in Genoa, as I heard, one night fourscore 80 ducats.

SHYLOCK: Thou stick'st a dagger in me. I shall never see my gold again. Fourscore ducats at a sitting? Fourscore ducats?

TUBAL: There came divers of Antonio's creditors in my company to Venice that swear he cannot choose but break.° 85

SHYLOCK: I am very glad of it. I'll plague him, I'll torture him. I am glad of it.

TUBAL: One of them showed me a ring that he had of your daughter for a monkey.

---

58. matched: i.e., found to match them.   64. The curse: God's curse (such as the plagues visited upon Egypt in Exodus 7-12).   67. hearsed: coffined.   75. cast away: shipwrecked.   85. break: go bankrupt.

SHYLOCK: Out upon her! Thou torturest me, Tubal. It was my turquoise; I   90
had it of Leah° when I was a bachelor. I would not have given it for a
wilderness of monkeys.

TUBAL: But Antonio is certainly undone.

SHYLOCK: Nay, that's true, that's very true. Go, Tubal, fee° me an officer;°
bespeak° him a fortnight before. I will have the heart of him if he forfeit,   95
for were he out of Venice I can make what merchandise I will.° Go,
Tubal, and meet me at our synagogue. Go, good Tubal; at our synagogue,
Tubal.   *Exeunt [separately].*

ACT 3, SCENE 2°

Enter Bassanio, Portia, Gratiano, [Nerissa,] and all their trains.

PORTIA:
I pray you, tarry. Pause a day or two
Before you hazard, for in choosing° wrong
I lose your company. Therefore forbear awhile.
There's something tells me — but it is not love —
I would not lose you; and you know yourself   5
Hate counsels not in such a quality.°
But lest you should not understand me well —
And yet a maiden hath no tongue but thought —
I would detain you here some month or two
Before you venture for me. I could teach you   10
How to choose right, but then I am forsworn.
So will I never be. So° may you miss° me.
But if you do, you'll make me wish a sin,
That I had been forsworn. Beshrew your eyes,
They have o'erlooked° me and divided me!   15
One half of me is yours, the other half yours —
Mine own, I would say; but if mine, then yours,
And so all yours. O, these naughty° times
Puts bars° between the owners and their rights!
And so, though yours, not yours.° Prove it so,   20

---

91. **Leah:** Shylock's wife. 94. **fee:** hire. **officer:** bailiff. 95. **bespeak:** engage.
96. **make . . . I will:** drive whatever bargains I please. ACT 3, SCENE 2. Location: Belmont.
Portia's house. 2. **in choosing:** in your choosing. 6. **quality:** way, manner. 12. **So . . . So:**
that . . . therefore. **miss:** i.e., fail to win. 15. **o'erlooked:** bewitched. 18. **naughty:** worthless, wicked. 19. **bars:** barriers. 20. **though yours, not yours:** (I am) yours by right but not
by actual possession.

Let Fortune go to hell for it, not I.°
I speak too long, but 'tis to peise° the time,
To eke° it and to draw it out in length,
To stay you from election.°

BASSANIO: Let me choose,
For as I am, I live upon the rack.   25

PORTIA:
Upon the rack, Bassanio? Then confess
What treason° there is mingled with your love.

BASSANIO:
None but that ugly treason of mistrust,°
Which makes me fear° th' enjoying of my love.
There may as well be amity and life   30
'Tween snow and fire, as treason and my love.

PORTIA:
Ay, but I fear you speak upon the rack,
Where men enforcèd do speak anything.

BASSANIO:
Promise me life, and I'll confess the truth.

PORTIA:
Well then, confess and live.

BASSANIO: "Confess and love"   35
Had been the very sum of my confession.
O happy torment, when my torturer
Doth teach me answers for deliverance!
But let me to my fortune and the caskets.°

PORTIA:
Away, then! I am locked in one of them.
If you do love me, you will find me out.   40
Nerissa and the rest, stand all aloof.°
Let music sound while he doth make his choice;
Then, if he lose, he makes a swanlike° end,
Fading in music. That the comparison   45
May stand more proper, my eye shall be the stream
And watery deathbed for him. He may win;

20–21. Prove . . . not I: i.e., if it turn out thus (that you are cheated of what is justly yours, i.e., of me), let Fortune be blamed for it, not I, for I will not be forsworn.   **22. peise:** retard (by hanging on of weights).   **23. eke:** eke out, augment.   **24. election:** choice.   **26–27. confess What treason:** (The rack was used to force traitors to confess.)   **28. mistrust:** misapprehension.   **29. fear:** fearful about.   **39. fortune . . . caskets:** (Presumably the curtains are drawn at about this point, as in the previous "casket" scenes, revealing the three caskets.)   **42. aloof:** apart, at a distance.   **44. swanlike:** (Swans were believed to sing when they came to die.)

And what is music then? Then music is
Even as the flourish° when true subjects bow
To a new-crownèd monarch. Such it is   50
As are those dulcet sounds in break of day
That creep into the dreaming bridegroom's ear
And summon him to marriage. Now he goes,
With no less presence,° but with much more love,
Than young Alcides° when he did redeem   55
The virgin tribute paid by howling° Troy
To the sea monster. I stand for sacrifice;°
The rest aloof are the Dardanian° wives,
With blearèd° visages, come forth to view
The issue° of th' exploit. Go, Hercules!   60
Live thou,° I live. With much, much more dismay
I view the fight than thou that mak'st the fray.

A song, the whilst Bassanio comments on the caskets to himself.

Tell me where is fancy° bred,
Or° in the heart or in the head?
How begot, how nourishèd?   65
    Reply, reply.
It is engenderèd in the eyes,°
With gazing fed, and fancy dies
In the cradle° where it lies.
    Let us all ring fancy's knell.   70
    I'll begin it — Ding, dong, bell.
ALL: Ding, dong, bell.

BASSANIO:
So may the outward shows be least themselves;°
The world is still° deceived with ornament.
In law, what plea so tainted and corrupt   75

---

49. **flourish**: sounding of trumpets.   54. **presence**: noble bearing.   55. **Alcides**: Hercules
(called *Alcides*, as at 2.1.32–35, because he was the grandson of Alcaeus) rescued Hesione,
daughter of the Trojan king Laomedon, from a monster to which, by command of Neptune,
she was about to be sacrificed. Hercules was rewarded, however, not with the lady's love, but
with a famous pair of horses.   56. **howling**: lamenting.   57. **stand for sacrifice**: represent the
sacrificial victim.   58. **Dardanian**: Trojan.   59. **blearèd**: tear-stained.   60. **issue**: outcome.
61. **Live thou**: if you live.   63. **fancy**: love.   64. **Or**: either.   67. **eyes**: (Love entered the
heart especially through the eyes.)   69. **In the cradle**: i.e., in its infancy, in the eyes.   73. **be
least themselves**: least represent the inner reality.   74. **still**: ever.

But, being seasoned with a gracious voice,
Obscures the show of evil? In religion,
What damnèd error but some sober brow
Will bless it and approve° it with a text,
Hiding the grossness with fair ornament?   80
There is no vice so simple° but assumes
Some mark of virtue on his° outward parts.
How many cowards, whose hearts are all as false
As stairs° of sand, wear yet upon their chins
The beards of Hercules and frowning Mars,   85
Who, inward searched,° have livers° white as milk?
And these assume but valor's excrement°
To render them redoubted.° Look on beauty,
And you shall see 'tis purchased by the weight,°
Which therein works a miracle in nature,   90
Making them lightest° that wear most of it.
So are those crispèd,° snaky, golden locks,
Which maketh such wanton gambols with the wind
Upon supposèd fairness,° often known
To be the dowry of a second head,   95
The skull that bred them in the sepulcher.°
Thus ornament is but the guilèd° shore
To a most dangerous sea, the beauteous scarf
Veiling an Indian° beauty; in a word,
The seeming truth which cunning times put on   100
To entrap the wisest. Therefore, thou gaudy gold,
Hard food for Midas,° I will none of thee;
Nor none of thee, thou pale and common drudge
'Tween man and man.° But thou, thou meager lead,
Which rather threaten'st than dost promise aught,   105

79. **approve:** confirm. 81. **simple:** unadulterated. 82. **his:** its. 84. **stairs:** steps.
86. **searched:** surgically probed.   **livers:** (The liver was thought to be the seat of courage; for it
to be deserted by the blood would be the condition of cowardice.) 87. **excrement:** outgrowth,
such as a beard (as in this case). 88. **redoubted:** feared. 89. **purchased by the weight:**
bought (as cosmetics) at so much per ounce. 91. **lightest:** most lascivious (with pun on the
sense of "least heavy"). 92. **crispèd:** curly. 94. **Upon supposèd fairness:** i.e., on a woman
supposed beautiful and fair-haired. 95–96. **To . . . sepulcher:** i.e., to be a wig of hair taken
from a woman now dead. 97. **guilèd:** treacherous. 99. **Indian:** i.e., swarthy, not fair.
102. **Midas:** the Phrygian king whose touch turned everything to gold, including his food.
103–104. **pale . . . man:** i.e., silver, used in commerce.

Thy paleness moves me more than eloquence;
And here choose I. Joy be the consequence!

PORTIA: [*aside*]
How all the other passions fleet to air,
As° doubtful thoughts, and rash-embraced despair,
And shuddering fear, and green-eyed jealousy!　　　　　　　110
O love, be moderate, allay thy ecstasy,
In measure rain° thy joy, scant° this excess!
I feel too much thy blessing. Make it less,
For fear I surfeit.

BASSANIO: [*opening the leaden casket*]
　　　　　　　　What find I here?
Fair Portia's counterfeit!° What demigod°　　　　　　　115
Hath come so near creation? Move these eyes?
Or whether,° riding on the balls of mine,
Seem they in motion? Here are severed lips,
Parted with sugar breath; so sweet a bar°
Should sunder such sweet friends.° Here in her hairs　　　120
The painter plays the spider, and hath woven
A golden mesh t' entrap the hearts of men
Faster° than gnats in cobwebs. But her eyes —
How could he see to do them? Having made one,
Methinks it should have power to steal both his　　　　　125
And leave itself unfurnished.° Yet look how far°
The substance of my praise doth wrong this shadow°
In underprizing it,° so far° this shadow
Doth limp behind the substance.° Here's the scroll,
The continent° and summary of my fortune.　　　　　　　130
[*He reads.*] "You that choose not by the view
　　　　　Chance as fair,° and choose as true.
　　　　　Since this fortune falls to you,
　　　　　Be content and seek no new.

---

109. **As:** such as. 112. **rain:** rain down, or perhaps "rein." **scant:** lessen. 115. **counterfeit:** portrait. **demigod:** i.e., the painter as creator. 117. **Or whether:** or. 119. **so sweet a bar:** i.e., Portia's breath. 120. **sweet friends:** i.e., her lips. 123. **Faster:** more tightly. 126. **unfurnished:** i.e., without a companion. **look how far:** however far. 127. **shadow:** painting, semblance. 128. **underprizing it:** failing to do it justice. **so far:** to a similar extent. 129. **the substance:** the subject, i.e., Portia. 130. **continent:** container. 132. **Chance as fair:** hazard as fortunately.

> If you be well pleased with this,   135
> And hold your fortune for your bliss,
> Turn you where your lady is
> And claim her with a loving kiss."

A gentle scroll. Fair lady, by your leave,
I come by note,° to give and to receive.   140
Like one of two contending in a prize,°
That thinks he hath done well in people's eyes,
Hearing applause and universal shout,
Giddy in spirit, still gazing in a doubt
Whether those peals of praise be his° or no,   145
So, thrice-fair lady, stand I, even so,
As doubtful whether what I see be true,
Until confirmed, signed, ratified by you.

PORTIA:
You see me, Lord Bassanio, where I stand,
Such as I am. Though for myself alone   150
I would not be ambitious in my wish
To wish myself much better, yet for you
I would be trebled twenty times myself,
A thousand times more fair, ten thousand times more rich,
That only to stand high in your account°   155
I might in virtues, beauties, livings,° friends,
Exceed account.° But the full sum of me
Is sum of something,° which, to term in gross,
Is an unlessoned girl, unschooled, unpracticèd;
Happy in this, she is not yet so old   160
But she may learn; happier than this,
She is not bred so dull but she can learn;
Happiest of all is that her gentle spirit
Commits itself to yours to be directed
As from her lord, her governor, her king.   165
Myself and what is mine to you and yours
Is now converted. But now° I was the lord

---

140. by note: by a bill of dues (i.e., the scroll). The commercial metaphor continues in *confirmed, signed, ratified* (line 148), *account* (155), *sum* (157), *term in gross* (158), etc. **141. prize:** competition. **145. his:** for him. **155. account:** estimation. **156. livings:** possessions. **157. account:** calculation (playing on *account*, estimation, in line 155). **157–158. But ... something:** i.e., but the full sum of my worth can only be the sum of whatever I am. **158. term in gross:** denote in full. **167. But now:** a moment ago.

Of this fair mansion, master of my servants,
Queen o'er myself; and even now, but now,
This house, these servants, and this same myself   170
Are yours, my lord's. I give them with this ring,
Which when you part from, lose, or give away,
Let it presage the ruin of your love
And be my vantage to exclaim on° you.

[*She puts a ring on his finger.*]

BASSANIO:

Madam, you have bereft me of all words.   175
Only my blood speaks to you in my veins,
And there is such confusion in my powers°
As, after some oration fairly spoke
By a belovèd prince, there doth appear
Among the buzzing pleasèd multitude,   180
Where every something being blent together
Turns to a wild of nothing save of joy
Expressed and not expressed.° But when this ring
Parts from this finger, then parts life from hence.
O, then be bold to say Bassanio's dead!   185

NERISSA:

My lord and lady, it is now our time,
That° have stood by and seen our wishes prosper,
To cry, "good joy." Good joy, my lord and lady!

GRATIANO:

My lord Bassanio and my gentle lady,
I wish you all the joy that you can wish —   190
For I am sure you can wish none from me.°
And when your honors mean to solemnize
The bargain of your faith, I do beseech you
Even at that time I may be married too.

BASSANIO:

With all my heart, so° thou canst get a wife.   195

---

174. **vantage to exclaim on:** opportunity to reproach. 177. **powers:** faculties.
181–183. **Where . . . expressed:** i.e., in which every individual utterance, being blended and confused, turns into a hubbub of joy in which individual utterances cannot be distinguished.
187. **That:** we who. 191. **For . . . me:** i.e., I'm sure I can't wish you any more joy than you could wish for yourselves, or, I'm sure your wishes for happiness cannot take away from my happiness. 195. **so:** provided.

GRATIANO:
> I thank your lordship, you have got me one.
> My eyes, my lord, can look as swift as yours.
> You saw the mistress, I beheld the maid;°
> You loved, I loved; for intermission°
> No more pertains to me, my lord, than you.   200
> Your fortune stood upon the caskets there,
> And so did mine too, as the matter falls;°
> For wooing here until I sweat again,°
> And swearing till my very roof° was dry
> With oaths of love, at last, if promise last,°   205
> I got a promise of this fair one here
> To have her love, provided that your fortune
> Achieved her mistress.

PORTIA: Is this true, Nerissa?

NERISSA:
> Madam, it is, so° you stand pleased withal.

BASSANIO:
> And do you, Gratiano, mean good faith?   210

GRATIANO: Yes, faith, my lord.

BASSANIO:
> Our feast shall be much honored in your marriage.

GRATIANO: We'll play° with them the first boy for a thousand ducats.

NERISSA: What, and stake down?°

GRATIANO: No, we shall ne'er win at that sport, and stake down.   215

Enter Lorenzo, Jessica, and Salerio, a messenger from Venice.

> But who comes here? Lorenzo and his infidel?
> What, and my old Venetian friend Salerio?

BASSANIO:
> Lorenzo and Salerio, welcome hither,
> If that the youth of my new interest° here
> Have power to bid you welcome. — By your leave,   220

---

198. **maid:** (Nerissa is a lady-in-waiting, not a house servant.)   199. **intermission:** delay (in loving).   202. **falls:** falls out, happens.   203. **sweat again:** sweated repeatedly.   204. **roof:** roof of my mouth.   205. **if promise last:** i.e., if Nerissa's promise should last, hold out (with a play on *last* and *at last,* "finally").   209. **so:** provided.   213. **play:** wager.   214. **stake down:** cash placed in advance. (But Gratiano, in his reply, turns the phrase into a bawdy joke; *stake down* to him suggests a non-erect phallus.)   219. **youth . . . interest:** i.e., newness of my household authority.

I bid my very° friends and countrymen,
Sweet Portia, welcome.

PORTIA: So do I, my lord.
They are entirely welcome.

LORENZO:
I thank your honor. For my part, my lord,
My purpose was not to have seen you here,   225
But meeting with Salerio by the way,
He did intreat me, past all saying nay,
To come with him along.

SALERIO: I did, my lord,
And I have reason for it. Signor Antonio
Commends him° to you.   [*He gives Bassanio a letter.*]

BASSANIO: Ere I ope his letter,   230
I pray you tell me how my good friend doth.

SALERIO:
Not sick, my lord, unless it be in mind,
Nor well, unless in mind. His letter there
Will show you his estate.° [*Bassanio*] open[*s*] *the letter.*

GRATIANO: [*indicating Jessica*]
Nerissa, cheer yond stranger,° bid her welcome.   235
Your hand, Salerio. What's the news from Venice?
How doth that royal merchant,° good Antonio?
I know he will be glad of our success.
We are the Jasons; we have won the fleece.°

SALERIO:
I would you had won the fleece that he hath lost.   240

PORTIA:
There are some shrewd° contents in yond same paper
That steals the color from Bassanio's cheek —
Some dear friend dead, else nothing in the world
Could turn so much the constitution
Of any constant° man. What, worse and worse?   245
With leave, Bassanio; I am half yourself,

---

221. **very:** true.   230. **Commends him:** desires to be remembered.   234. **estate:** condition.
235. **stranger:** alien.   237. **royal merchant:** i.e., chief among merchants.   239. **Jasons . . .**
fleece: (Compare with 1.1.171.)   241. **shrewd:** cursed, grievous.   245. **constant:** settled, not
swayed by passion.

And I must freely have the half of anything
That this same paper brings you.
BASSANIO: O sweet Portia,
Here are a few of the unpleasant'st words
That ever blotted paper! Gentle lady,   250
When I did first impart my love to you,
I freely told you all the wealth I had
Ran in my veins, I was a gentleman;
And then I told you true. And yet, dear lady,
Rating myself at nothing, you shall see   255
How much I was a braggart. When I told you
My state° was nothing, I should then have told you
That I was worse than nothing; for indeed
I have engaged myself to a dear friend,
Engaged my friend to his mere° enemy,   260
To feed my means. Here is a letter, lady,
The paper as the body of my friend,
And every word in it a gaping wound
Issuing lifeblood. But is it true, Salerio?
Hath all his ventures failed? What, not one hit?°   265
From Tripolis, from Mexico, and England,
From Lisbon, Barbary, and India,
And not one vessel scape the dreadful touch
Of merchant°-marring rocks?
SALERIO: Not one, my lord.
Besides, it should appear that if he had   270
The present° money to discharge° the Jew
He° would not take it. Never did I know
A creature that did bear the shape of man
So keen° and greedy to confound° a man.
He plies the Duke at morning and at night,   275
And doth impeach the freedom of the state°
If they deny him justice. Twenty merchants,
The Duke himself, and the magnificoes°

257. state: estate. 260. mere: absolute. 265. hit: success. 269. merchant: merchant ship.
271. present: available. discharge: pay off. 272. He: i.e., Shylock. 274. keen: cruel.
confound: destroy. 276. impeach . . . state: i.e., call in question the ability of Venice to
defend legally the freedom of commerce of its citizens. 278. magnificoes: chief men of
Venice.

Of greatest port° have all persuaded° with him,
But none can drive him from the envious° plea   280
Of forfeiture, of justice, and his bond.

JESSICA:
When I was with him I have heard him swear
To Tubal and to Chus,° his countrymen,
That he would rather have Antonio's flesh
Than twenty times the value of the sum   285
That he did owe him; and I know, my lord,
If law, authority, and power deny not,
It will go hard with poor Antonio.

PORTIA: [*to Bassanio*]
Is it your dear friend that is thus in trouble?

BASSANIO:
The dearest friend to me, the kindest man,   290
The best-conditioned° and unwearied spirit
In doing courtesies, and one in whom
The ancient Roman honor more appears
Than any that draws breath in Italy.

PORTIA: What sum owes he the Jew?   295

BASSANIO:
For me, three thousand ducats.

PORTIA: What, no more?
Pay him six thousand, and deface° the bond;
Double six thousand, and then treble that,
Before a friend of this description
Shall lose a hair through Bassanio's fault.   300
First go with me to church and call me wife,
And then away to Venice to your friend;
For never shall you lie by Portia's side
With an unquiet soul. You shall have gold
To pay the petty debt twenty times over.   305
When it is paid, bring your true friend along.
My maid Nerissa and myself meantime
Will live as maids and widows. Come, away!
For you shall hence upon your wedding day.

---

279. port: dignity. **persuaded:** argued. **280. envious:** malicious. **283. Chus:** the Bishops' Bible spelling of *Cush*, son of Ham and grandson of Noah. Tubal was son of Japheth and grandson of Noah (Genesis 10:2, 6). **291. best-conditioned:** best natured. **297. deface:** erase.

Bid your friends welcome, show a merry cheer;° 310
Since you are dear bought, I will love you dear.°
But let me hear the letter of your friend.

BASSANIO: [*reads*]   "Sweet Bassanio, my ships have all miscarried, my creditors grow cruel, my estate is very low, my bond to the Jew is forfeit; and
since in paying it, it is impossible I should live, all debts are cleared 315
between you and I if I might but see you at my death. Nothwithstanding,
use your pleasure. If your love do not persuade you to come, let not my
letter."

PORTIA:
O love, dispatch all business, and begone!

BASSANIO:
Since I have your good leave to go away, 320
I will make haste; but till I come again
No bed shall e'er be guilty of my stay,
Nor rest be interposer twixt us twain.   *Exeunt.*

ACT 3, SCENE 3°

Enter [Shylock] the Jew and Solanio and Antonio and the Jailer.

SHYLOCK:
Jailer, look to him. Tell not me of mercy.
This is the fool that lent out money gratis.
Jailer, look to him.

ANTONIO: Hear me yet, good Shylock.

SHYLOCK:
I'll have my bond. Speak not against my bond.
I have sworn an oath that I will have my bond. 5
Thou calledst me dog before thou hadst a cause,
But since I am a dog, beware my fangs.
The Duke shall grant me justice. I do wonder,
Thou naughty° jailer, that thou art so fond°
To come abroad° with him at his request. 10

ANTONIO: I pray thee, hear me speak.

SHYLOCK:
I'll have my bond. I will not hear thee speak.
I'll have my bond, and therefore speak no more.

---

310. **cheer:** countenance.   311. **dear . . . dear:** expensively . . . dearly.   ACT 3, SCENE 3.
Location: Venice. A street.   9. **naughty:** worthless, wicked.   **fond:** foolish.   10. **abroad:**
outside.

I'll not be made a soft and dull-eyed° fool,
To shake the head, relent, and sigh, and yield   15
To Christian intercessors. Follow not.
I'll have no speaking. I will have my bond.   *Exit Jew.*

SOLANIO:
It is the most impenetrable cur
That ever kept° with men.

ANTONIO: Let him alone.
I'll follow him no more with bootless° prayers.   20
He seeks my life. His reason well I know:
I oft delivered from his forfeitures
Many that have at times made moan to me;
Therefore he hates me.

SOLANIO: I am sure the Duke
Will never grant this forfeiture to hold.   25

ANTONIO:
The Duke cannot deny the course of law;
For the commodity° that strangers° have
With us in Venice, if it be denied,
Will much impeach the justice of the state,
Since that° the trade and profit of the city   30
Consisteth of all nations. Therefore go.
These griefs and losses have so bated° me
That I shall hardly spare a pound of flesh
Tomorrow to my bloody creditor.
Well, jailer, on. Pray God Bassanio come   35
To see me pay his debt, and then I care not.   *Exeunt.*

ACT 3, SCENE 4°

Enter Portia, Nerissa, Lorenzo, Jessica, and [Balthasar,] a man of Portia's.

LORENZO:
Madam, although I speak it in your presence,
You have a noble and a true conceit°
Of godlike amity, which appears most strongly
In bearing thus the absence of your lord.

---

14. **dull-eyed:** easily duped. 19. **kept:** associated, dwelt. 20. **bootless:** unavailing.
27. **commodity:** facilities or privileges for trading. **strangers:** noncitizens, including Jews.
30. **Since that:** since. 32. **bated:** reduced. ACT 3, SCENE 4. **Location:** Belmont. Portia's
house. 2. **conceit:** understanding.

But if you knew to whom you show this honor,   5
How true a gentleman you send relief,
How dear a lover° of my lord your husband,
I know you would be prouder of the work
Than customary bounty can enforce you.°

PORTIA:

I never did repent for doing good,   10
Nor shall not now; for in companions
That do converse and waste° the time together,
Whose souls do bear an equal yoke of love,
There must be needs° a like proportion
Of lineaments,° of manners, and of spirit;   15
Which makes me think that this Antonio,
Being the bosom lover° of my lord,
Must needs be like my lord. If it be so,
How little is the cost I have bestowed
In purchasing the semblance of my soul°   20
From out the state of hellish cruelty!°
This comes too near the praising of myself;
Therefore no more of it. Hear other things:
Lorenzo, I commit into your hands
The husbandry and manage° of my house   25
Until my lord's return. For mine own part,
I have toward heaven breathed a secret vow
To live in prayer and contemplation,
Only attended by Nerissa here,
Until her husband and my lord's return.   30
There is a monastery two miles off,
And there we will abide. I do desire you
Not to deny this imposition,°
The which my love and some necessity
Now lays upon you.

LORENZO: Madam, with all my heart,   35
I shall obey you in all fair commands.

---

7. **lover:** friend.   9. **Than . . . you:** than ordinary benevolence can make you.   12. **waste:** spend.   14. **must be needs:** must be.   15. **lineaments:** physical features.   17. **bosom lover:** dear friend.   20. **the semblance of my soul:** i.e., Antonio, so like my Bassanio.   21. **of hellish cruelty:** to which hellish cruelty has reduced him.   25. **husbandry and manage:** care and management.   33. **deny this imposition:** refuse this charge imposed.

PORTIA:
My people do already know my mind,
And will acknowledge you and Jessica
In place of Lord Bassanio and myself.
So fare you well till we shall meet again.   40

LORENZO:
Fair thoughts and happy hours attend on you!

JESSICA:
I wish your ladyship all heart's content.

PORTIA:
I thank you for your wish and am well pleased
To wish it back on you. Fare you well, Jessica.

                                        Exeunt [Jessica and Lorenzo].

Now, Balthasar,   45
As I have ever found thee honest-true,
So let me find thee still. Take this same letter,

                                        [*giving a letter*]

And use thou all th' endeavor of a man
In speed to Padua. See thou render this
Into my cousin's hands, Doctor Bellario;   50
And look what° notes and garments he doth give thee,
Bring them, I pray thee, with imagined° speed
Unto the traject,° to the common° ferry
Which trades° to Venice. Waste no time in words,
But get thee gone. I shall be there before thee.   55

BALTHASAR:
Madam, I go with all convenient° speed.   [*Exit.*]

PORTIA:
Come on, Nerissa, I have work in hand
That you yet know not of. We'll see our husbands
Before they think of us.

NERISSA: Shall they see us?

PORTIA:
They shall, Nerissa, but in such a habit°   60
That they shall think we are accomplishèd°
With that° we lack. I'll hold thee any wager,
When we are both accoutred like young men

51. **look what:** whatever.   52. **imagined:** all imaginable.   53. **traject:** ferry. (Italian *traghetto*.)
common: public.   54. **trades:** plies back and forth.   56. **convenient:** due, proper.
60. **habit:** apparel, garb.   61. **accomplishèd:** supplied.   62. **that:** that which (with a bawdy
suggestion).

I'll prove the prettier fellow of the two,
And wear my dagger with the braver grace, 65
And speak between the change of man and boy
With a reed voice, and turn two mincing steps
Into a manly stride, and speak of frays
Like a fine bragging youth, and tell quaint° lies,
How honorable ladies sought my love, 70
Which I denying, they fell sick and died —
I could not do withal!° Then I'll repent,
And wish, for all that, that I had not killed them;
And twenty of these puny° lies I'll tell,
That men shall swear I have discontinued school 75
Above° a twelvemonth.° I have within my mind
A thousand raw tricks of these bragging Jacks,°
Which I will practice.
NERISSA: Why, shall we turn to° men?
PORTIA: Fie, what a question's that,
If thou wert near a lewd interpreter! 80
But come, I'll tell thee all my whole device
When I am in my coach, which stays for us
At the park gate; and therefore haste away,
For we must measure° twenty miles today.   *Exeunt.*

ACT 3, SCENE 5°

Enter [Launcelot the] clown and Jessica.

LAUNCELOT: Yes, truly, for look you, the sins of the father are to be laid
upon the children; therefore, I promise you, I fear you.° I was always
plain with you, and so now I speak my agitation° of the matter. Therefore
be o' good cheer, for truly I think you are damned. There is but one hope
in it that can do you any good, and that is but a kind of bastard° hope, 5
neither.°
JESSICA: And what hope is that, I pray thee?

---

69. **quaint:** elaborate, clever.   72. **do withal:** help it.   74. **puny:** childish.   75–76. **I . . .
twelvemonth:** i.e., that I am no mere schoolboy.   76. **Above:** more than.   77. **Jacks:** fellows.
78. **turn to:** turn into. (But Portia sees the occasion for a bawdy quibble on the idea of "turning
toward, lying next to.")   84. **measure:** traverse.   **ACT 3, SCENE 5. Location:** Belmont. Outside Portia's house.   2. **fear you:** fear for you.   3. **agitation:** (A blunder for "cogitation"?)
5. **bastard:** i.e., unfounded (but also anticipating the usual meaning in line 8).   6. **neither:**
i.e., to be sure.

LAUNCELOT: Marry, you may partly hope that your father got° you not, that you are not the Jew's daughter.

JESSICA: That were a kind of bastard hope, indeed! So the sins of my mother should be visited upon me. 10

LAUNCELOT: Truly, then, I fear you are damned both by father and mother. Thus when I shun Scylla,° your father, I fall into Charybdis,° your mother. Well, you are gone° both ways.

JESSICA: I shall be saved by my husband.° He hath made me a Christian. 15

LAUNCELOT: Truly, the more to blame he! We were Christians enough° before, e'en as many as could well live one by another.° This making of Christians will raise the price of hogs. If we grow all to be pork eaters, we shall not shortly have a rasher° on the coals for money.°

Enter Lorenzo.

JESSICA: I'll tell my husband, Launcelot, what you say. Here he comes. 20

LORENZO: I shall grow jealous of you shortly, Launcelot, if you thus get my wife into corners.

JESSICA: Nay, you need not fear us, Lorenzo. Launcelot and I are out.° He tells me flatly there's no mercy for me in heaven because I am a Jew's daughter; and he says you are no good member of the commonwealth, 25 for in converting Jews to Christians you raise the price of pork.

LORENZO: I shall answer that better to the commonwealth than you can the getting up of the Negro's belly. The Moor° is with child by you, Launcelot.

LAUNCELOT: It is much that the Moor should be more than reason;° but 30 if she be less than an honest woman, she is indeed more than I took her for.°

LORENZO: How every fool can play upon the word! I think the best grace° of wit will shortly turn into silence, and discourse grow commendable in none only but parrots. Go in, sirrah, bid them prepare for dinner. 35

LAUNCELOT: That is done, sir. They have all stomachs.°

---

8. **got:** begot. 13. **Scylla, Charybdis:** twin dangers of the *Odyssey*, 12.255, a monster and a whirlpool guarding the straits presumably between Italy and Sicily. (*Fall into* plays on the idea of "whirlpool" as the female sexual anatomy.) 14. **gone:** done for. 15. **I . . . husband:** (Compare 1 Corinthians 7:14: "the unbelieving wife is sanctified by the husband.) 16. **We . . . enough:** there were enough of us Christians. 17. **one by another:** (1) as neighbors (2) off one another. 19. **rasher:** i.e., of bacon. **for money:** even for ready money, at any price. 23. **are out:** have fallen out. 28. **The Moor:** (Launcelot has evidently impregnated some woman of the household, who, being of African heritage, is referred to as both "Negro" and "Moor.") 30. **more than reason:** larger than is reasonable, than she should be (with wordplay on "Moor," "more," continued in line 31). 31–32. **if she . . . for:** i.e., to describe her as something less than perfectly chaste would be to pay her a compliment, in my view. 33. **best grace:** highest quality. 36. **stomachs:** appetites.

LORENZO: Goodly Lord, what a wit-snapper are you! Then bid them prepare dinner.
LAUNCELOT: That is done too, sir, only "cover"° is the word.
LORENZO: Will you cover° then, sir?   40
LAUNCELOT: Not so, sir, neither. I know my duty.
LORENZO: Yet more quarreling with occasion!° Wilt thou show the whole
  wealth of thy wit in an instant? I pray thee, understand a plain man in his
  plain meaning: go to thy fellows, bid them cover the table, serve in the
  meat,° and we will come in to dinner.   45
LAUNCELOT: For° the table,° sir, it shall be served in; for the meat, sir, it
  shall be covered;° for your coming in to dinner, sir, why, let it be as
  humors and conceits° shall govern.   *Exit [Launcelot the] clown.*
LORENZO:
  O dear discretion,° how his words are suited!°
  The fool hath planted in his memory
  An army of good words; and I do know   50
  A many° fools, that stand in better place,°
  Garnished° like him, that for a tricksy word
  Defy the matter.° How cheer'st thou,° Jessica?
  And now, good sweet, say thy opinion: 55
  How dost thou like the Lord Bassanio's wife?
JESSICA:
  Past all expressing. It is very meet°
  The Lord Bassanio live an upright life,
  For, having such a blessing in his lady,
  He finds the joys of heaven here on earth;   60
  And if on earth he do not merit it,
  In reason° he should never come to heaven.
  Why, if two gods should play some heavenly match
  And on the wager lay° two earthly women,
  And Portia one, there must be something else°   65

39, 40. cover: spread the table for the meal. (But in line 41 Launcelot uses the word to
mean "put on one's hat.")   **42. Yet . . . occasion:** i.e., still quibbling at every opportunity.
45. meat: food.   **46. For:** as for.   **table:** (Here Launcelot quibblingly uses the word to
mean the food itself.)   **47. covered:** (Here used in the sense of providing a cover for each separate
dish.)   **47–48. humors and conceits:** whims and fancies.   **49. O dear discretion:** O what
precious discrimination.   **suited:** suited to the occasion.   **52. A many:** many.   **better place:**
higher social station.   **53. Garnished:** i.e., furnished with words, or with garments.
53–54. that . . . matter: who for the sake of ingenious wordplay disdain to speak sensibly.
54. How cheer'st thou: i.e., what cheer, how are you doing.   **57. meet:** fitting.   **62. In reason:** it stands to reason. (Jessica jokes that for Bassanio to receive unmerited bliss on earth —
unmerited because no person can earn bliss through his or her own deserving — is to run the
risk of eternal damnation.)   **64. lay:** stake.   **65. else:** more.

Pawned° with the other, for the poor rude world
Hath not her fellow.

LORENZO: Even such a husband
Hast thou of me as she is for a wife.

JESSICA:
Nay, but ask my opinion too of that!

LORENZO:
I will anon. First let us go to dinner.   70

JESSICA:
Nay, let me praise you while I have a stomach.°

LORENZO:
No, pray thee, let it serve for table talk;
Then, howsoe'er thou speak'st, 'mong other things
I shall digest° it.

JESSICA: Well, I'll set you forth.°   *Exeunt.*

ACT 4, SCENE 1°

Enter the Duke, the Magnificoes, Antonio, Bassanio, [Salerio,] and Gratiano [with others. The judges take their places].

DUKE: What, is Antonio here?

ANTONIO: Ready, so please Your Grace.

DUKE:
I am sorry for thee. Thou art come to answer°
A stony adversary, an inhuman wretch
Uncapable of pity, void and empty   5
From any dram° of mercy.

ANTONIO: I have heard
Your Grace hath ta'en great pains to qualify°
His rigorous course; but since he stands obdurate
And that no lawful means can carry me
Out of his envy's° reach, I do oppose   10
My patience to his fury and am armed
To suffer with a quietness of spirit
The very tyranny° and rage of his.

66. **Pawned:** staked, wagered.   71. **stomach:** (1) appetite (2) inclination.   74. **digest:** (1) ponder, analyze (2) "swallow," put up with (with a play also on the gastronomic sense).   **set you forth:** (1) serve you up, as at a feast (2) set forth your praises.   **ACT 4, SCENE 1. Location:** Venice. A court of justice. Benches, etc., are provided for the justices.   3. **answer:** defend yourself against. (A legal term.)   6. **dram:** sixty grains apothecaries' weight, a tiny quantity   7. **qualify:** moderate.   10. **envy's:** malice's.   13. **tyranny:** cruelty.

DUKE:
Go one, and call the Jew into the court.
SALERIO:
He is ready at the door. He comes, my lord.   15

Enter Shylock.

DUKE:
Make room, and let him stand before our° face.
Shylock, the world thinks, and I think so too,
That thou but leadest this fashion° of thy malice
To the last hour of act,° and then 'tis thought
Thou'lt show thy mercy and remorse° more strange°   20
Than is thy strange° apparent° cruelty;
And where thou now exacts the penalty,
Which is a pound of this poor merchant's flesh,
Thou wilt not only loose° the forfeiture,
But, touched with human gentleness and love,   25
Forgive a moiety° of the principal,
Glancing an eye of pity on his losses
That have of late so huddled on his back —
Enough to press a royal merchant down
And pluck commiseration of his state   30
From brassy° bosoms and rough hearts of flint,
From stubborn Turks and Tartars never trained
To offices of tender courtesy.
We all expect a gentle answer, Jew.
SHYLOCK:
I have possessed° Your Grace of what I purpose,   35
And by our holy Sabbath have I sworn
To have the due and forfeit of my bond.
If you deny it, let the danger° light
Upon your charter and your city's freedom!°
You'll ask me why I rather choose to have   40
A weight of carrion flesh than to receive
Three thousand ducats. I'll not answer that,
But say it is my humor.° Is it answered?

16. **our:** (The royal plural.)   18. **thou . . . fashion:** you only maintain this pretense or form.
19. **the last . . . act:** the brink of action.   20. **remorse:** pity.   **strange:** remarkable.
21. **strange:** unnatural, foreign.   **apparent:** (1) manifest, overt (2) seeming.   24. **loose:** release,
waive.   26. **moiety:** part, portion.   31. **brassy:** unfeeling, hard like brass.   35. **possessed:**
informed.   38. **danger:** injury.   39. **Upon . . . freedom:** (See 3.2.276.)   43. **humor:** whim.

What if my house be troubled with a rat
And I be pleased to give ten thousand ducats   45
To have it baned?° What, are you answered yet?
Some men there are love° not a gaping pig,°
Some that are mad if they behold a cat,
And others, when the bagpipe sings i' the nose,
Cannot contain their urine; for affection,°   50
Mistress of passion, sways it to the mood
Of what it likes or loathes. Now, for your answer:
As there is no firm reason to be rendered
Why he° cannot abide a gaping pig,
Why he a harmless necessary° cat,   55
Why he a woolen° bagpipe, but of force
Must yield to such inevitable shame
As to offend, himself being offended,
So can I give no reason, nor I will not,
More than a lodged° hate and a certain° loathing   60
I bear Antonio, that I follow thus
A losing° suit against him. Are you answered?

BASSANIO:
This is no answer, thou unfeeling man,
To excuse the current° of thy cruelty.

SHYLOCK:
I am not bound to please thee with my answers.   65

BASSANIO:
Do all men kill the things they do not love?

SHYLOCK:
Hates any man the thing he would not kill?

BASSANIO:
Every offense is not a hate at first.

SHYLOCK:
What, wouldst thou have a serpent sting thee twice?

ANTONIO:
I pray you, think° you question° with the Jew.   70
You may as well go stand upon the beach

---

46. **baned:** killed, especially by poison or ratsbane.   47. **love:** who love.   **gaping pig:** pig roasted whole with its mouth open.   50. **affection:** feeling, inclination.   54, 55, 56. **he, he, he:** one person, another, yet another.   55. **necessary:** i.e., useful for catching rats and mice. 56. **woolen:** i.e., with flannel-covered bag.   60. **lodged:** settled, steadfast.   **certain:** unwavering, fixed.   62. **losing:** unprofitable.   64. **current:** flow, tendency.   70. **think:** bear in mind. **question:** argue.

And bid the main flood° bate his° usual height;
You may as well use question with° the wolf
Why he hath made the ewe bleat for the lamb;
You may as well forbid the mountain pines   75
To wag their high tops and to make no noise
When they are fretten° with the gusts of heaven;
You may as well do anything most hard
As seek to soften that — than which what's harder? —
His Jewish heart. Therefore, I do beseech you,   80
Make no more offers, use no farther means,
But with all brief and plain conveniency°
Let me have judgment, and the Jew his will.
BASSANIO: [*to Shylock*]
For thy three thousand ducats here is six.
SHYLOCK:
If every ducat in six thousand ducats   85
Were in six parts, and every part a ducat,
I would not draw° them. I would have my bond.
DUKE:
How shalt thou hope for mercy, rendering none?
SHYLOCK:
What judgment shall I dread, doing no wrong?
You have among you many a purchased slave,   90
Which, like your asses and your dogs and mules,
You use in abject and in slavish parts,°
Because you bought them. Shall I say to you,
"Let them be free, marry them to your heirs!
Why sweat they under burdens? Let their beds   95
Be made as soft as yours, and let their palates
Be seasoned with such viands"?° You will answer
"The slaves are ours." So do I answer you:
The pound of flesh which I demand of him
Is dearly bought, is mine, and I will have it.   100
If you deny me, fie upon your law!
There is no force in the decrees of Venice.
I stand for judgment. Answer: shall I have it?

---

72. **main flood:** sea at high tide. **bate his:** abate its. 73. **use question with:** interrogate.
77. **fretten:** fretted, i.e., disturbed, ruffled. 82. **conveniency:** propriety. 87. **draw:** receive.
92. **parts:** duties, capacities. 97. **viands:** food.

DUKE:
> Upon° my power I may dismiss this court,
> Unless Bellario, a learnèd doctor,   105
> Whom I have sent for to determine this,°
> Come here today.

SALERIO: My lord, here stays without°
> A messenger with letters from the doctor,
> New come from Padua.

DUKE:
> Bring us the letters. Call the messenger.   [*Exit one.*]   110

BASSANIO:
> Good cheer, Antonio. What, man, courage yet!
> The Jew shall have my flesh, blood, bones, and all,
> Ere thou shalt lose for me one drop of blood.

ANTONIO:
> I am a tainted wether° of the flock,
> Meetest° for death. The weakest kind of fruit   115
> Drops earliest to the ground, and so let me.
> You cannot better be employed, Bassanio,
> Than to live still and write mine epitaph.

Enter Nerissa [dressed like a lawyer's clerk].

DUKE:
> Came you from Padua, from Bellario?

NERISSA:
> From both, my lord. Bellario greets Your Grace.   120
> [*She presents a letter. Shylock whets his knife on his shoe.*]

BASSANIO:
> Why dost thou whet thy knife so earnestly?

SHYLOCK:
> To cut the forefeiture from that bankrupt there.

GRATIANO:
> Not on thy sole, but on thy soul, harsh Jew,
> Thou mak'st thy knife keen; but no metal can,
> No, not the hangman's° ax, bear half the keenness°   125
> Of thy sharp envy.° Can no prayers pierce thee?

---

104. **Upon**: in accordance with.   106. **determine this**: resolve this legal dispute.   107. **stays without**: waits outside.   114. **wether**: ram, especially a castrated ram.   115. **Meetest**: fittest. 125. **hangman's**: executioner's.   **keenness**: (1) sharpness (2) savagery.   126. **envy**: malice.

SHYLOCK:
No, none that thou hast wit enough to make.

GRATIANO:
O, be thou damned, inexecrable° dog,
And for thy life° let justice be accused!
Thou almost mak'st me waver in my faith   130
To hold opinion with Pythagoras°
That souls of animals infuse themselves
Into the trunks of men. Thy currish spirit
Governed a wolf who, hanged for human slaughter,°
Even from the gallows did his fell° soul fleet,°   135
And, whilst thou layest in thy unhallowed dam,°
Infused itself in thee; for thy desires
Are wolvish, bloody, starved, and ravenous.

SHYLOCK:
Till thou canst rail° the seal from off my bond,
Thou but offend'st° thy lungs to speak so loud.   140
Repair thy wit, good youth, or it will fall
To cureless° ruin. I stand here for law.

DUKE:
This letter from Bellario doth commend
A young and learnèd doctor to our court.
Where is he?

NERISSA: He attendeth here hard by   145
To know your answer, whether you'll admit him.

DUKE:
With all my heart. Some three or four of you
Go give him courteous conduct to this place.

                                        [*Exeunt some.*]

Meantime the court shall hear Bellario's letter.
[*He reads.*]° "Your Grace shall understand that at the receipt of your letter   150
I am very sick; but in the instant that your messenger came, in loving
visitation was with me a young doctor of Rome. His name is Balthasar.

---

128. **inexecrable**: that cannot be overly execrated or detested.   129. **for thy life**: i.e., because
you are allowed to live.   131. **Pythagoras**: ancient Greek philosopher who argued for the
transmigration of souls.   134. **hanged for human slaughter**: (A possible allusion to the Elizabethan practice of trying and punishing animals for various crimes.)   135. **fell**: fierce, cruel.
fleet: flit, i.e., pass from the body.   136. **dam**: mother. (Usually used of animals.)   139. **rail**:
revile, use abusive language.   140. **offend'st**: injurest.   142. **cureless**: incurable.   150. *[He
reads.]*: (In many modern editions, the reading of the letter is assigned to a clerk, but the original text gives no such indication.)

I acquainted him with the cause in controversy between the Jew
and Antonio the merchant. We turned o'er many books together. He
is furnished with my opinion, which bettered with his own learning,   155
the greatness whereof I cannot enough commend, comes with him,° at
my importunity,° to fill up Your Grace's request in my stead. I beseech
you, let his lack of years be no impediment to let him lack° a reverend
estimation, for I never knew so young a body with so old a head. I leave
him to your gracious acceptance, whose trial° shall better publish° his   160
commendation."

Enter Portia for° Balthasar [dressed like a doctor of laws, escorted].

You hear the learn'd Bellario, what he writes;
And here, I take it, is the doctor come.
Give me your hand. Come you from old Bellario?

PORTIA:
I did, my lord.

DUKE: You are welcome. Take your place.   165

[Portia takes her place.]

Are you acquainted with the difference°
That holds this present question in the court?

PORTIA:
I am informèd throughly° of the cause.°
Which is the merchant here, and which the Jew?

DUKE:
Antonio and old Shylock, both stand forth.   170

[Antonio and Shylock stand forth.]

PORTIA:
Is your name Shylock?

SHYLOCK: Shylock is my name.

PORTIA:
Of a strange nature is the suit you follow,
Yet in such rule° that the Venetian law
Cannot impugn° you as you do proceed. —
You stand within his danger,° do you not?   175

---

156. **comes with him:** i.e., my opinion is brought by him.   157. **importunity:** insistence.
158. **to let him lack:** such as would deprive him of.   160. **trial:** testing, performance.   **publish:** make known.   **s.d. for:** i.e., disguised as.   166. **difference:** argument.   168. **throughly:**
thoroughly.   **cause:** case.   173. **rule:** order.   174. **impugn:** find fault with.   175. **within his
danger:** in his power.

ANTONIO:
  Ay, so he says.
PORTIA: Do you confess the bond?
ANTONIO:
  I do.
PORTIA: Then must the Jew be merciful.
SHYLOCK:
  On what compulsion must I? Tell me that.
PORTIA:
  The quality of mercy is not strained.°
  It droppeth as the gentle rain from heaven   180
  Upon the place beneath. It is twice blest:°
  It blesseth him that gives and him that takes.
  'Tis mightiest in the mightiest; it becomes
  The thronèd monarch better than his crown.
  His scepter shows the force of temporal power,   185
  The attribute to° awe and majesty,
  Wherein doth sit the dread and fear of kings.
  But mercy is above this sceptered sway;
  It is enthronèd in the hearts of kings;
  It is an attribute to God himself;   190
  And earthly power doth then show likest God's
  When mercy seasons justice. Therefore, Jew,
  Though justice be thy plea, consider this,
  That in the course of justice none of us
  Should see salvation. We do pray for mercy,   195
  And that same prayer doth teach us all to render
  The deeds of mercy. I have spoke thus much
  To mitigate the justice of thy plea,°
  Which if thou follow, this strict court of Venice
  Must needs give sentence 'gainst the merchant there.   200
SHYLOCK:
  My deeds upon my head!° I crave the law,
  The penalty and forfeit of my bond.
PORTIA:
  Is he not able to discharge the money?

---

179. **strained:** forced, constrained.   181. **is twice blest:** grants a double blessing.
186. **attribute to:** symbol of.   198. **To . . . plea:** i.e., to moderate your plea for strict justice.
201. **My . . . head:** (Compare the cry of the crowd at Jesus' crucifixion: "His blood be on us,
and on our children," Matthew 27:25.)

FIGURE 3 *Mercy Overcoming Revenge, from Richard Day,* A Book of Christian Prayers *(1578). The text accompanying this image reads "Mercy beareth with infirmities, cruelty seeketh revenge."*

BASSANIO:

    Yes, here I tender it for him in the court,

    Yea, twice the sum. If that will not suffice,   205

    I will be bound to pay it ten times o'er,

    On forfeit of my hands, my head, my heart.

    If this will not suffice, it must appear

    That malice bears down truth.° And I beseech you,

    Wrest once° the law to your authority.   210

    To do a great right, do a little wrong,

    And curb this cruel devil of his will.

PORTIA:

    It must not be. There is no power in Venice

    Can alter a decree establishèd.

    'Twill be recorded for a precedent,   215

    And many an error by the same example

    Will rush into the state. It cannot be.

SHYLOCK:

    A Daniel° come to judgment! Yea, a Daniel!

    O wise young judge, how I do honor thee!

PORTIA:

    I pray you, let me look upon the bond.   220

SHYLOCK: [*giving the bond*]

    Here 'tis, most reverend doctor, here it is.

PORTIA:

    Shylock, there's thrice thy money offered thee.

SHYLOCK:

    An oath, an oath! I have an oath in heaven.

    Shall I lay perjury upon my soul?

    No, not for Venice.

PORTIA: Why, this bond is forfeit,   225

    And lawfully by this the Jew may claim

    A pound of flesh, to be by him cut off

    Nearest the merchant's heart. — Be merciful.

    Take thrice thy money; bid me tear the bond.

SHYLOCK:

    When it is paid according to the tenor.°   230

    It doth appear you are a worthy judge.

---

209. **bears down truth:** overwhelms righteousness.   210. **Wrest once:** for once, forcibly subject.   218. **Daniel:** (In the Apocrypha's story of Susannah and the Elders, Daniel is the young man who rescues Susannah from her false accusers.)   230. **tenor:** conditions.

You know the law. Your exposition
Hath been most sound. I charge you by the law,
Whereof you are a well-deserving pillar,
Proceed to judgment. By my soul I swear   235
There is no power in the tongue of man
To alter me. I stay here on° my bond.

ANTONIO:
Most heartily I do beseech the court
To give the judgment.

PORTIA: Why then, thus it is:
You must prepare your bosom for his knife.   240

SHYLOCK:
O noble judge! O excellent young man!

PORTIA:
For the intent and purpose of the law
Hath full relation to° the penalty
Which here appeareth due upon the bond.

SHYLOCK:
'Tis very true. O wise and upright judge!   245
How much more elder art thou than thy looks!

PORTIA:
Therefore lay bare your bosom.

SHYLOCK: Ay, his breast.
So says the bond, doth it not, noble judge?
"Nearest his heart," those are the very words.

PORTIA:
It is so. Are there balance° here   250
To weigh the flesh?

SHYLOCK: I have them ready.

PORTIA:
Have by some surgeon, Shylock, on your charge,°
To stop his wounds, lest he do bleed to death.

SHYLOCK:
Is it so nominated in the bond?

PORTIA:
It is not so expressed, but what of that?   255
'Twere good you do so much for charity.

---

237. **stay here on:** remain committed to, insist upon.   243. **Hath . . . to:** is fully in accord with.   250. **balance:** scales.   252. **on your charge:** at your personal expense.

SHYLOCK:

I cannot find it. 'Tis not in the bond.

PORTIA:

You, merchant, have you anything to say?

ANTONIO:

But little. I am armed° and well prepared.

Give my your hand, Bassanio; fare you well!   260

Grieve not that I am fall'n to this for you,

For herein Fortune shows herself more kind

Than is her custom. It is still her use°

To let the wretched man outlive his wealth

To view with hollow eye and wrinkled brow   265

An age of poverty; from which ling'ring penance

Of such misery doth she cut me off.

Commend me to your honorable wife.

Tell her the process° of Antonio's end,

Say how I loved you, speak me fair° in death;   270

And, when the tale is told, bid her be judge

Whether Bassanio had not once a love.°

Repent but you° that you shall lose your friend.

And he repents not that he pays your debt.

For if the Jew do cut but deep enough,   275

I'll pay it instantly with all my heart.°

BASSANIO:

Antonio, I am married to a wife

Which° is as dear to me as life itself;

But life itself, my wife, and all the world

Are not with me esteemed above thy life.   280

I would lose all, ay, sacrifice them all

Here to this devil, to deliver you.

PORTIA:

Your wife would give you little thanks for that,

If she were by° to hear you make the offer.

GRATIANO:

I have a wife who, I protest, I love;   285

I would she were in heaven, so she could

Entreat some power to change this currish Jew.

---

259. **armed**: i.e., fortified in spirit.   263. **still her use**: i.e., commonly Fortune's practice.
269. **process**: story, manner.   270. **speak me fair**: speak well of me.   272. **a love**: a friend's
love.   273. **Repent but you**: grieve only.   276. **with . . . heart**: (1) wholeheartedly (2) literally,
with my heart's blood.   278. **Which**: who.   284. **by**: nearby.

NERISSA:
'Tis well you offer it behind her back;
The wish would make else an unquiet house.

SHYLOCK:
These be the Christian husbands. I have a daughter;   290
Would any of the stock of Barabbas°
Had been her husband rather than a Christian! —
We trifle° time. I pray thee, pursue° sentence.

PORTIA:
A pound of that same merchant's flesh is thine.
The court awards it, and the law doth give it.   295

SHYLOCK: Most rightful judge!

PORTIA:
And you must cut this flesh from off his breast.
The law allows it, and the court awards it.

SHYLOCK:
Most learnèd judge! A sentence! Come, prepare.

PORTIA:
Tarry a little; there is something else.   300
This bond doth give thee here no jot of blood;
The words expressly are "a pound of flesh."
Take then thy bond, take thou thy pound of flesh,
But in the cutting it if thou dost shed
One drop of Christian blood, thy lands and goods   305
Are by the laws of Venice confiscate
Unto the state of Venice.

GRATIANO:
O upright judge! Mark, Jew. O learnèd judge!

SHYLOCK:
Is that the law?

PORTIA: Thyself shalt see the act;
For, as thou urgest justice, be assured   310
Thou shalt have justice, more than thou desir'st.

GRATIANO:
O learnèd judge! Mark, Jew, a learnèd judge!

---

291. **Barabbas:** a thief whom Pontius Pilate set free instead of Christ in response to the people's demand (see Mark 15); also, the villainous protagonist of Marlowe's *The Jew of Malta*. 293. **trifle:** waste. **pursue:** proceed with.

SHYLOCK:

    I take this offer, then. Pay the bond thrice
    And let the Christian go.

BASSANIO: Here is the money.

PORTIA: Soft!°   315

    The Jew shall have all justice.° Soft, no haste.
    He shall have nothing but the penalty.

GRATIANO:

    O Jew! An upright judge, a learnèd judge!

PORTIA:

    Therefore prepare thee to cut off the flesh.
    Shed thou no blood, nor cut thou less nor more   320
    But just a pound of flesh. If thou tak'st more
    Or less than a just pound, be it but so much
    As makes it light or heavy in the substance°
    Or the division° of the twentieth part
    Of one poor scruple,° nay, if the scale do turn   325
    But in the estimation of a hair,
    Thou diest, and all thy goods are confiscate.

GRATIANO:

    A second Daniel, a Daniel, Jew!
    Now, infidel, I have you on the hip.°

PORTIA:

    Why doth the Jew pause? Take thy forfeiture.   330

SHYLOCK:

    Give me my principal, and let me go.

BASSANIO:

    I have it ready for thee. Here it is.

PORTIA:

    He hath refused it in the open court.
    He shall have merely justice and his bond.

GRATIANO:

    A Daniel, still say I, a second Daniel!   335
    I thank thee, Jew, for teaching me that word.

SHYLOCK:

    Shall I not have barely my principal?

---

315. **Soft:** i.e., not so fast.   316. **all justice:** precisely what the law provides.   323. **substance:** mass or gross weight.   324. **division:** fraction.   325. **scruple:** twenty grains apothecaries' weight, a small quantity.   329. **on the hip:** i.e., at a disadvantage. (A phrase from wrestling.)

FIGURE 4 *Daniel and Susannah, by Hans Holbein, from* The Images of the Old Testament *(1549). Daniel, the young boy standing on the left with a rod in his hand, judges one of the two elders who sought to slander the innocent Susannah and condemn her to death.*

PORTIA:
   Thou shalt have nothing but the forfeiture,
   To be so taken at thy peril, Jew.
SHYLOCK:
   Why, then the devil give him good of it!   340
   I'll stay no longer question.°   [*He starts to go.*]
PORTIA: Tarry, Jew!
   The law hath yet another hold on you.
   It is enacted in the laws of Venice,
   If it be proved against an alien
   That by direct or indirect attempts   345
   He seek the life of any citizen,
   The party 'gainst the which he doth contrive
   Shall seize one half his goods; the other half
   Comes to the privy coffer° of the state,
   And the offender's life lies in° the mercy   350

341. **I'll . . . question:** I'll stay no further pursuing of the case.   349. **privy coffer:** private treasury.   350. **lies in:** lies at.

Of the Duke only, 'gainst all other voice.°
In which predicament, I say, thou stand'st;
For it appears, by manifest proceeding,
That indirectly and directly too
Thou hast contrived against the very life   355
Of the defendant; and thou hast incurred
The danger formerly by me rehearsed.°
Down therefore, and beg mercy of the Duke.
GRATIANO:
Beg that thou mayst have leave to hang thyself!
And yet, thy wealth being forfeit to the state,   360
Thou hast not left the value of a cord;
Therefore thou must be hanged at the state's charge.°
DUKE:
That thou shalt see the difference of our spirit,
I pardon thee thy life before thou ask it.
For° half thy wealth, it is Antonio's;   365
The other half comes to the general state,
Which humbleness may drive unto a fine.°
PORTIA:
Ay, for the state, not for Antonio.°
SHYLOCK:
Nay, take my life and all! Pardon not that!
You take my house when you do take the prop   370
That doth sustain my house. You take my life
When you do take the means whereby I live.
PORTIA:
What mercy can you render him, Antonio?
GRATIANO:
A halter° gratis! Nothing else, for God's sake.
ANTONIO:
So please my lord the Duke and all the court   375
To quit° the fine for one half of his goods,
I am content, so° he will let me have
The other half in use,° to render it,

351. 'gainst . . . voice: without appeal.   357. danger . . . rehearsed: penalty already cited by
me.   362. charge: expense.   365. For: as for.   367. Which . . . fine: i.e., which penitence on
your part may persuade me to reduce to a fine.   368. Ay . . . Antonio: i.e., yes, the state's half
may be reduced to a fine, but not Antonio's half.   374. halter: hangman's noose.   376. quit:
remit, relinquish, or perhaps settle for. (That is, Antonio may ask the court to forgive even the
fine imposed in lieu of a heavier penalty.)   377. so: provided that.   378. in use: in trust, or
possibly, to be used as a source of income.

Upon his death, unto the gentleman
That lately stole his daughter.   380
Two things provided more: that for this favor
He presently° become a Christian;
The other, that he do record a gift
Here in the court of all he dies possessed°
Unto his son Lorenzo and his daughter.   385

DUKE:
He shall do this, or else I do recant
The pardon that I late° pronouncèd here.

PORTIA:
Art thou contented, Jew? What dost thou say?

SHYLOCK:
I am content.

PORTIA: Clerk, draw a deed of gift.

SHYLOCK:
I pray you, give me leave to go from hence;   390
I am not well. Send the deed after me,
And I will sign it.

DUKE: Get thee gone, but do it.

GRATIANO:
In christening shalt thou have two godfathers.
Had I been judge, thou shouldst have had ten more,°
To bring thee to the gallows, not the font.   395

Exit [Shylock].

DUKE: [*to Portia*]
Sir, I entreat you home with me to dinner.

PORTIA:
I humbly do desire Your Grace of pardon.
I must away this night toward Padua,
And it is meet I presently set forth.

DUKE:
I am sorry that your leisure serves you not.   400
Antonio, gratify° this gentleman,
For in my mind you are much bound to him.

Exeunt Duke and his train.

---

382. **presently:** at once. 384. **of . . . possessed:** i.e., what remains of the portion not placed under Antonio's trust (which will also go to Lorenzo and Jessica). 387. **late:** lately. 394. **ten more:** i.e., to make up a jury of twelve. (Jurors were colloquially termed *godfathers*.) 401. **gratify:** reward.

BASSANIO: [*to Portia*]
Most worthy gentleman, I and my friend
Have by your wisdom been this day acquitted
Of grievous penalties, in lieu whereof,° 405
Three thousand ducats due unto the Jew
We freely cope° your courteous pains withal.

[*He offers money.*]

ANTONIO:
And stand indebted over and above
In love and service to you evermore.

PORTIA:
He is well paid that is well satisfied, 410
And I, delivering you, am satisfied,
And therein do account myself well paid.
My mind was never yet more mercenary.
I pray you, know me when we meet again.°
I wish you well, and so I take my leave. 415

[*She starts to leave.*]

BASSANIO:
Dear sir, of force° I must attempt° you further.
Take some remembrance of us as a tribute,
Not as fee. Grant me two things, I pray you:
Not to deny me, and to pardon me.°

PORTIA:
You press me far, and therefore I will yield. 420
Give me your gloves,° I'll wear them for your sake.
And, for your love,° I'll take this ring from you.
Do not draw back your hand; I'll take no more,
And you in love shall not deny me this.

BASSANIO:
This ring, good sir? Alas, it is a trifle! 425
I will not shame myself to give you this.

PORTIA:
I will have nothing else but only this;
And now, methinks, I have a mind to it.

405. **in lieu whereof:** in return for which. 407. **cope:** requite. 414. **know . . . again:** i.e., consider our acquaintance well established. (But punning on *know* in the sense of "recognize" and "have sexual relations with" — meanings that are hidden from Bassanio by Portia's disguise.) 416. **of force:** of necessity. **attempt:** urge. 419. **pardon me:** i.e., pardon my presumption in pressing the matter. 421. **gloves:** (Perhaps Bassanio removes his gloves, thereby revealing the ring that "Balthasar" asks of him.) 422. **for your love:** i.e., for friendship's sake — a polite phrase, but with ironic double meaning as applied to husband and wife.

BASSANIO:

    There's more depends on this than on the value.

    The dearest° ring in Venice will I give you,   430

    And find it out by proclamation.

    Only for this, I pray you, pardon me.

PORTIA:

    I see, sir, you are liberal° in offers.

    You taught me first to beg, and now, methinks,

    You teach me how a beggar should be answered.   435

BASSANIO:

    Good sir, this ring was given me by my wife,

    And when she put it on she made me vow

    That I should neither sell nor give nor lose it.

PORTIA:

    That 'scuse serves many men to save their gifts.

    An if your wife be not a madwoman,   440

    And know how well I have deserved this ring,

    She would not hold out° enemy forever

    For giving it to me. Well, peace be with you!

                        Exeunt [Portia and Nerissa].

ANTONIO:

    My lord Bassanio, let him have the ring.

    Let his deservings and my love withal   445

    Be valued 'gainst your wife's commandement.°

BASSANIO:

    Go, Gratiano, run and overtake him;

    Give him the ring, and bring him, if thou canst,

    Unto Antonio's house. Away, make haste!

                        Exit Gratiano [with the ring].

    Come, you and I will thither presently,   450

    And in the morning early will we both

    Fly toward Belmont. Come, Antonio.   *Exeunt.*

---

430. **dearest:** most expensive.   433. **liberal:** generous.   442. **would . . . out:** i.e., would not remain.   446. **commandement:** (Pronounced in four syllables.)

ACT 4, SCENE 2°

Enter [Portia and] Nerissa [still disguised].

PORTIA: *[giving a deed to Nerissa]*
Inquire the Jew's house out; give him this deed°
And let him sign it. We'll away tonight
And be a day before our husbands home.
This deed will be well welcome to Lorenzo.

Enter Gratiano.

GRATIANO: Fair sir, you are well o'erta'en.°   5
My lord Bassanio upon more advice°
Hath sent you here this ring and doth entreat
Your company at dinner.   *[He gives a ring.]*
PORTIA: That cannot be.
His ring I do accept most thankfully,
And so, I pray you, tell him. Furthermore,   10
I pray you, show my youth old Shylock's house.
GRATIANO:
That will I do.
NERISSA: Sir, I would speak with you.
[Aside to Portia.] I'll see if I can get my husband's ring,
Which I did make him swear to keep forever.
PORTIA: *[aside to Nerissa]*
Thou mayst, I warrant. We shall have old° swearing   15
That they did give the rings away to men;
But we'll outface° them, and outswear them too. —
Away, make haste! Thou know'st where I will tarry.
NERISSA:
Come, good sir, will you show me to this house?
                    [Exeunt, Portia separately from the others.]

ACT 4, SCENE 2. Location: Venice. A street.   1. **this deed:** i.e., the deed of gift.   5. **you . . . o'erta'en:** I'm happy to have caught up with you.   6. **advice:** consideration.   15. **old:** plenty of.   17. **outface:** boldly contradict.

ACT 5, SCENE 1°

Enter Lorenzo and Jessica.

LORENZO:

    The moon shines bright. In such a night as this,
    When the sweet wind did gently kiss the trees
    And they did make no noise, in such a night
    Troilus° methinks mounted the Trojan walls
    And sighed his soul toward the Grecian tents   5
    Where Cressid lay that night.

JESSICA: In such a night

    Did Thisbe° fearfully o'ertrip the dew,
    And saw the lion's shadow ere himself,
    And ran dismayed away.

LORENZO: In such a night

    Stood Dido° with a willow° in her hand   10
    Upon the wild sea banks, and waft° her love
    To come again to Carthage.

JESSICA: In such a night

    Medea° gathered the enchanted herbs
    That did renew old Aeson.

LORENZO: In such a night

    Did Jessica steal° from the wealthy Jew   15
    And with an unthrift° love did run from Venice
    As far as Belmont.

JESSICA: In such a night

    Did young Lorenzo swear he loved her well,
    Stealing her soul with many vows of faith,
    And ne'er a true one.

LORENZO: In such a night   20

    Did pretty Jessica, like a little shrew,
    Slander her love, and he forgave it her.

ACT 5, SCENE 1. Location: Belmont. Outside Portia's house. **4. Troilus:** Trojan prince betrayed by his beloved, Cressida, after she had been transferred to the Greek camp. **7. Thisbe:** beloved of Pyramus who, arranging to meet him by night, was frightened by a lion and fled; the tragic misunderstanding of her absence led to the suicides of both lovers. (See *A Midsummer Night's Dream*, act 5.) **10. Dido:** Queen of Carthage, deserted by Aeneas. **willow:** (A symbol of forsaken love.) **11. waft:** wafted, beckoned. **13. Medea:** famous sorceress of Colchis who, after falling in love with Jason and helping him to gain the Golden Fleece, used her magic to restore youth to Aeson, Jason's father. **15. steal:** (1) escape (2) rob. **16. unthrift:** prodigal.

JESSICA:
>   I would out-night° you, did nobody come.
>   But hark, I hear the footing° of a man.
>
>   *Enter [Stephano,] a messenger.*

LORENZO:
>   Who comes so fast in silence of the night?   25
STEPHANO: A friend.
LORENZO:
>   A friend? What friend? Your name, I pray you, friend?
STEPHANO:
>   Stephano is my name, and I bring word
>   My mistress will before the break of day
>   Be here at Belmont. She doth stray about   30
>   By holy crosses,° where she kneels and prays
>   For happy wedlock hours.
LORENZO: Who comes with her?
STEPHANO:
>   None but a holy hermit and her maid.
>   I pray you, is my master yet returned?
LORENZO:
>   He is not, nor we have not heard from him.   35
>   But go we in, I pray thee, Jessica,
>   And ceremoniously let us prepare
>   Some welcome for the mistress of the house.
>
>   *Enter [Launcelot, the] clown.*

LAUNCELOT: Sola,° sola! Wo ha, ho! Sola, sola!
LORENZO: Who calls?   40
LAUNCELOT: Sola! Did you see Master Lorenzo? Master Lorenzo, sola, sola!
LORENZO: Leave holloing, man! Here.
LAUNCELOT: Sola! Where, where?
LORENZO: Here.
LAUNCELOT: Tell him there's a post° come from my master, with his horn°   45

---

23. **out-night:** i.e., outdo in the verbal games we've been playing.   24. **footing:** footsteps.
31. **holy crosses:** wayside shrines.   39. **Sola:** (Imitation of a post horn.)   45. **post:** courier.
horn: (Launcelot jestingly compares the courier's post horn to a cornucopia.)

full of good news: my master will be here ere morning.   [*Exit.*]

LORENZO:

Sweet soul, let's in, and there expect° their coming.
And yet no matter. Why should we go in?
My friend Stephano, signify,° I pray you,
Within the house, your mistress is at hand,   50
And bring your music forth into the air.

[*Exit Stephano.*]

How sweet the moonlight sleeps upon this bank!
Here will we sit and let the sounds of music
Creep in our ears. Soft stillness and the night
Become° the touches° of sweet harmony.   55
Sit, Jessica. [*They sit.*] Look how the floor of heaven
Is thick inlaid with patens° of bright gold.
There's not the smallest orb which thou behold'st
But in his motion like an angel sings,
Still choiring° to the young-eyed° cherubins.   60
Such harmony is in immortal souls,
But whilst this muddy vesture of decay°
Doth grossly close it in,° we cannot hear it.°

[*Enter musicians.*]

Come, ho, and wake Diana° with a hymn!
With sweetest touches pierce your mistress' ear   65
And draw her home with music.   *Play music.*

JESSICA:

I am never merry when I hear sweet music.

LORENZO:

The reason is, your spirits are attentive.°
For do but note a wild and wanton herd,
Or race° of youthful and unhandled colts,   70
Fetching mad bounds, bellowing and neighing loud,
Which is the hot condition of their blood;

47. **expect:** await. 49. **signify:** make known. 55. **Become:** suit. **touches:** strains, notes (produced by the fingering of an instrument). 57. **patens:** thin, circular plates of metal. 60. **Still choiring:** continually singing. **young-eyed:** eternally clear-sighted. In Ezekiel 10:12, the bodies and wings of cherubim are "full of eyes round about." 62. **muddy . . . decay:** i.e., mortal flesh. 63. **close it in:** i.e., enclose the soul. **hear it:** i.e., hear the music of the spheres. 64. **Diana:** (Here, goddess of the moon; compare with 1.2.78.) 68. **spirits are attentive:** (The spirits would be in motion within the body in merriment, whereas in sadness they would be drawn to the heart and, as it were, busy listening.) 70. **race:** herd.

If they but hear perchance a trumpet sound,
Or any air of music touch their ears,
You shall perceive them make a mutual° stand,   75
Their savage eyes turned to a modest gaze
By the sweet power of music. Therefore the poet°
Did feign that Orpheus° drew° trees, stones, and floods,
Since naught so stockish,° hard, and full of rage
But music for the time doth change his nature.   80
The man that hath no music in himself,
Nor is not moved with concord of sweet sounds,
Is fit for treasons, stratagems, and spoils;°
The motions of his spirit are dull as night
And his affections dark as Erebus.°   85
Let no such man be trusted. Mark the music.

Enter Portia and Nerissa.

PORTIA:
That light we see is burning in my hall.
How far that little candle throws his beams!
So shines a good deed in a naughty° world.
NERISSA:
When the moon shone, we did not see the candle.   90
PORTIA:
So doth the greater glory dim the less.
A substitute shines brightly as a king
Until a king be by, and then his° state
Empties itself, as doth an inland brook
Into the main of waters.° Music! Hark!   95
NERISSA:
It is your music, madam, of the house.
PORTIA:
Nothing is good, I see, without respect.°
Methinks it sounds much sweeter than by day.
NERISSA:

75. **mutual:** common or simultaneous.   77. **poet:** perhaps Ovid, with whom the story of Orpheus was a favorite theme.   78. **Orpheus:** legendary musician. **drew:** attracted, charmed.   79. **stockish:** unfeeling.   83. **spoils:** acts of pillage.   85. **Erebus:** a place of primeval darkness on the way to Hades.   89. **naughty:** wicked.   93. **his:** i.e., the substitute's.   95. **main of waters:** sea.   97. **respect:** comparison, context.

Silence bestows that virtue on it, madam.

PORTIA:

The crow doth sing as sweetly as the lark
When neither is attended;° and I think   100
The nightingale, if she should sing by day,
When every goose is cackling, would be thought
No better a musician than the wren.
How many things by season° seasoned are   105
To their right praise and true perfection!
Peace, ho! The moon sleeps with Endymion°
And would not be awaked.   [*The music ceases.*]

LORENZO: That is the voice,
Or I am much deceived, of Portia.

PORTIA:

He knows me as the blind man knows the cuckoo,   110
By the bad voice.

LORENZO: Dear lady, welcome home.

PORTIA:

We have been praying for our husbands' welfare,
Which speed,° we hope, the better for our words.
Are they returned?

LORENZO: Madam, they are not yet;
But there is come a messenger before,   115
To signify their coming.

PORTIA: Go in, Nerissa.
Give order to my servants that they take
No note at all of our being absent hence;
Nor you, Lorenzo; Jessica, nor you. [*A tucket° sounds.*]

LORENZO:

Your husband is at hand. I hear his trumpet.   120
We are no telltales, madam, fear you not.

PORTIA:

This night, methinks, is but the daylight sick;
It looks a little paler. 'Tis a day
Such as the day is when the sun is hid.

---

101. **neither is attended:** i.e., either is alone. 105. **season:** fit occasion (but playing on the idea of seasoning, spices). 107. **Endymion:** a shepherd loved by the moon goddess, who caused him to sleep a perennial sleep in a cave on Mount Latmos where she could visit him. 113. **speed:** prosper (with a suggestion too of speedy return). 119. **s.d. tucket:** flourish on a trumpet.

Enter Bassanio, Antonio, Gratiano, and their followers.

BASSANIO:
We should hold day with the Antipodes,   125
If you would walk in absence of the sun.°

PORTIA:
Let me give light, but let me not be light;°
For a light wife doth make a heavy° husband,
And never be Bassanio so for me.
But God sort° all! You are welcome home, my lord.   130

BASSANIO:
I thank you, madam. Give welcome to my friend.
This is the man, this is Antonio,
To whom I am so infinitely bound.

PORTIA:
You should in all sense° be much bound to him,
For, as I hear, he was much bound° for you.   135

ANTONIO:
No more than I am well acquitted of.°

PORTIA:
Sir, you are very welcome to our house.
It must appear in other ways than words;
Therefore I scant this breathing courtesy.°

GRATIANO: [*to Nerissa*]
By yonder moon I swear you do me wrong!   140
In faith, I gave it to the judge's clerk.
Would he were gelt° that had it, for my part,°
Since you do take it, love, so much at heart.

PORTIA:
A quarrel, ho, already? What's the matter?

GRATIANO:
About a hoop of gold, a paltry ring   145
That she did give me, whose posy° was

125–26. **We . . . sun:** i.e., if you, Portia, like a second sun, would always walk about during the sun's absence, we should never have night but would enjoy daylight even when the Antipodes, those who dwell on the opposite side of the globe, enjoy daylight.   127. **be light:** be wanton, unchaste.   128. **heavy:** sad (with wordplay on the antithesis of *light* and *heavy*).   130. **sort:** decide, dispose.   134. **in all sense:** in every way, with every reason.   134–35. **bound . . . bound:** Portia plays on (1) obligated (2) indebted and imprisoned.   136. **acquitted of:** freed from.   139. **scant . . . courtesy:** make brief these empty (i.e., merely verbal) compliments.   142. **gelt:** gelded.   **for my part:** as far as I'm concerned.   146. **posy:** a motto on a ring.

For all the world like cutler's poetry
Upon a knife, "Love me, and leave me not."

NERISSA:

What talk you of the posy or the value?
You swore to me, when I did give it you,   150
That you would wear it till your hour of death
And that it should lie with you in your grave.
Though not for me, yet for your vehement oaths
You should have been respective° and have kept it.
Gave it a judge's clerk! No, God's my judge,   155
The clerk will ne'er wear hair on 's face that had it.

GRATIANO:

He will, an if° he live to be a man.

NERISSA:

Ay, if a woman live to be a man.

GRATIANO:

Now, by this hand, I gave it to a youth,
A kind of boy, a little scrubbèd° boy   160
No higher than thyself, the judge's clerk,
A prating° boy, that begged it as a fee.
I could not for my heart deny it him.

PORTIA:

You were to blame — I must be plain with you —
To part so slightly with your wife's first gift,   165
A thing stuck on with oaths upon your finger,
And so riveted with faith unto your flesh.
I gave my love a ring and made him swear
Never to part with it; and here he stands.
I dare be sworn for him he would not leave it,   170
Nor pluck it from his finger, for the wealth
That the world masters.° Now, in faith, Gratiano,
You give your wife too unkind a cause of grief.
An° 'twere to me, I should be mad° at it.

BASSANIO: [aside]

Why, I were best to cut my left hand off   175
And swear I lost the ring defending it.

GRATIANO:

My lord Bassanio gave his ring away
Unto the judge that begged it and indeed
Deserved it too; and then the boy, his clerk,
That took some pains in writing, he begged mine;   180
And neither man nor master would take aught°
But the two rings.

PORTIA: What ring gave you, my lord?
Not that, I hope, which you received of me.

BASSANIO:
If I could add a lie unto a fault,
I would deny it; but you see my finger   185
Hath not the ring upon it. It is gone.

PORTIA:
Even so void is your false heart of truth.
By heaven, I will ne'er come in your bed
Until I see the ring!

NERISSA: [to Gratiano]   Nor I in yours
Till I again see mine.

BASSANIO: Sweet Portia,   190
If you did know to whom I gave the ring,
If you did know for whom I gave the ring,
And would conceive for what I gave the ring,
And how unwillingly I left the ring,
When naught would be accepted but the ring,   195
You would abate the strength of your displeasure.

PORTIA:
If you had known the virtue° of the ring,
Or half her worthiness that gave the ring,
Or your own honor to contain° the ring,
You would not then have parted with the ring.   200
What man is there so much unreasonable,
If you had pleased to have defended it
With any terms of zeal, wanted the modesty°
To urge° the thing held as a ceremony?°
Nerissa teaches me what to believe: 205
I'll die for 't but some woman had the ring.

BASSANIO:

181. aught: anything. 197. virtue: power. 199. contain: retain. 203. wanted the modesty: who would have been so lacking in consideration as. 204. urge: insist upon receiving.
ceremony: something sacred. 208. civil doctor: i.e., doctor of civil law.

No, by my honor, madam! By my soul,
No woman had it, but a civil doctor,°
Which did refuse three thousand ducats of me
And begged the ring, the which I did deny him   210
And suffered him to go displeased away —
Even he that had held up the very life
Of my dear friend. What should I say, sweet lady?
I was enforced to send it after him.
I was beset with shame and courtesy.   215
My honor would not let ingratitude
So much besmear it. Pardon me, good lady!
For by these blessèd candles of the night,°
Had you been there, I think you would have begged
The ring of me to give the worthy doctor.   220

PORTIA:
Let not that doctor e'er come near my house.
Since he hath got the jewel that I loved,
And that which you did swear to keep for me,
I will become as liberal° as you:
I'll not deny him anything I have,   225
No, not my body nor my husband's bed.
Know° him I shall, I am well sure of it.
Lie not a night from° home. Watch me like Argus;°
If you do not, if I be left alone,
Now, by mine honor, which is yet mine own,   230
I'll have that doctor for my bedfellow.

NERISSA:
And I his clerk; therefore be well advised°
How you do leave me to mine own protection.

GRATIANO:
Well, do you so. Let not me take° him, then!
For if I do, I'll mar the young clerk's pen.°   235

ANTONIO:
I am th' unhappy subject of these quarrels.

PORTIA:
Sir, grieve not you; you are welcome notwithstanding.

BASSANIO:

218. blessèd . . . night: i.e., stars.   224. liberal: generous (sexually as well as otherwise).
227. Know: (With the suggestion of carnal knowledge.)   228. from: away from.   Argus:
mythological monster with a hundred eyes.   232. be well advised: take care.   234. take:
apprehend.   235. pen: (With sexual double meaning.)

Portia, forgive me this enforcèd wrong,
And in the hearing of these many friends
I swear to thee, even by thine own fair eyes   240
Wherein I see myself —

PORTIA: Mark you but that!
In both my eyes he doubly sees himself;
In each eye, one. Swear by your double° self,
And there's an oath of credit.°

BASSANIO: Nay, but hear me.
Pardon this fault, and by my soul I swear   245
I never more will break an oath with thee.

ANTONIO:
I once did lend my body for his wealth,°
Which, but for him that had your husband's ring,
Had quite miscarried. I dare be bound again,
My soul upon the forfeit,° that your lord   250
Will nevermore break faith advisedly.°

PORTIA:
Then you shall be his surety.° Give him this,
And bid him keep it better than the other.

[*She gives the ring to Antonio, who gives it to Bassanio.*]

ANTONIO:
Here, Lord Bassanio. Swear to keep this ring.

BASSANIO:
By heaven, it is the same I gave the doctor!   255

PORTIA:
I had it of him. Pardon me, Bassanio,
For by this ring the doctor lay with me.

NERISSA:
And pardon me, my gentle Gratiano,
For that same scrubbèd boy, the doctor's clerk,
In lieu of° this last night did lie with me.   260

                                          [*Presenting her ring.*]

GRATIANO:
Why, this is like the mending of highways

---

243. **double:** i.e., deceitful.   244. **of credit:** worthy to be believed. (Said ironically.)
247. **wealth:** welfare.   250. **My . . . forfeit:** at the risk of eternal damnation.   251. **advisedly:**
intentionally.   252. **surety:** guarantor.   260. **In lieu of:** in return for.   262. **In . . . enough:**
i.e., before repair is necessary.   263. **cuckolds:** husbands whose wives are unfaithful.

In summer, where the ways are fair enough.°
What, are we cuckolds° ere we have deserved it?

PORTIA:
Speak not so grossly.° You are all amazed.°
Here is a letter; read it at your leisure.   265

[She gives a letter.]

It comes from Padua, from Bellario.
There you shall find that Portia was the doctor,
Nerissa there her clerk. Lorenzo here
Shall witness I set forth as soon as you,
And even but now returned; I have not yet   270
Entered my house. Antonio, you are welcome,
And I have better news in store for you
Than you expect. Unseal this letter soon.

[She gives him a letter.]

There you shall find three of your argosies
Are richly come to harbor suddenly.   275
You shall not know by what strange accident
I chancèd on this letter.

ANTONIO: I am dumb.°

BASSANIO: [*to Portia*]
Were you the doctor and I knew you not?

GRATIANO: [*to Nerissa*]
Were you the clerk that is to make me cuckold?

NERISSA:
Ay, but the clerk that never means to do it,   280
Unless he live until he be a man.

BASSANIO:
Sweet doctor, you shall be my bedfellow.
When I am absent, then lie with my wife.

ANTONIO:
Sweet lady, you have given me life and living;
For here I read for certain that my ships   285
Are safely come to road.°

PORTIA: How now, Lorenzo?
My clerk hath some good comforts too for you.

NERISSA:

---

264. **grossly:** stupidly, licentiously. **amazed:** bewildered. 277. **dumb:** at a loss for words.
286. **road:** anchorage.

Ay, and I'll give them him without a fee.

[She gives a deed.]

There do I give to you and Jessica,
From the rich Jew, a special deed of gift,   290
After his death, of all he dies possessed of.

LORENZO:

Fair ladies, you drop manna° in the way
Of starvèd people.

PORTIA: It is almost morning,
And yet I am sure you are not satisfied
Of these events at full. Let us go in;   295
And charge us there upon inter'gatories,°
And we will answer all things faithfully.

GRATIANO:

Let it be so. The first inter'gatory
That my Nerissa shall be sworn on is
Whether till the next night she had rather stay°   300
Or go to bed now, being two hours to day.
But were the day come, I should wish it dark
Till I were couching with the doctor's clerk.
Well, while I live I'll fear no other thing
So sore as keeping safe Nerissa's ring.°   *Exeunt.* 305

FINIS

---

292. **manna:** the food from heaven that was miraculously supplied to the Israelites in the wilderness (Exodus 16). 296. **charge . . . inter'gatories:** require ourselves to answer all things under oath. 300. **stay:** wait. 305. **ring:** (With sexual suggestion.)

TEXTUAL NOTES FOR THE MERCHANT OF VENICE

Copy text: the First Quarto of 1600 [Q]. The act and scene divisions are absent from the Quarto; the Folio provides act divisions only.

ACT 1, SCENE 1. s.d. [and elsewhere] Salerio, and Solanio: *Salaryno,* and *Salanio.* 7. s.p. [and elsewhere] Salerio: *Salarino.* 19. Peering: Piring. 27. docked: docks. 85. jaundice: *Jaundies.* 112. tongue: togue. 113. Is: It is. 127. off: of. 150. back: bake.

ACT 1, SCENE 2. 34. Palatine: Palentine [and at line 44]. 41. Bon: *Boune.* 45. throstle: Trassell. 88. s.d. Enter a Servingman: [after line 120 in Q].

ACT 1, SCENE 3. 22. s.p. [and elsewhere] Shylock: *Jew.* 69. compromised: compremyzd. 75. peeled: pyld. 103. spit: spet [also at lines 117 and 122]. 118. day, another time: day another time,.

ACT 2, SCENE 1. s.d. Morocco: *Morochus.* 25. Sophy . . . prince: Sophy, and a Persian Prince. 31. thee: the. 35. page: rage.

ACT 2, SCENE 2. 1. s.p. [and elsewhere] Launcelot: *Clowne.* 3. [and elsewhere in this scene] Gobbo: *Jobbe.* 33. By: Be. 60. murder: muder. 75. last: lost. 136. s.d. Exit Leonardo: [after line 164 in Q]. 139. a suit: sute.

ACT 2, SCENE 3. 11. did: doe.

ACT 2, SCENE 4. 8. o'clock: of clocke. 9. s.d.: [after line 9 in Q]. 14. Love news: Loue, newes. 20. s.d. Exit clown: [after line 23 in Q]. 39. s.d. Exeunt: *Exit.*

ACT 2, SCENE 5. 42. Jewess': Jewes.

ACT 2, SCENE 6. 26. Who's: whose [also at line 61]. 35. night, you: night you. 59. gentlemen: gentleman.

ACT 2, SCENE 7. 18. threatens. Men: threatens men. 45. Spits: Spets. 69. tombs: *timber.*

ACT 2, SCENE 8. 8. gondola: Gondylo. 39. slubber: slumber.

ACT 2, SCENE 9. 48. chaff: chaft. 49. varnished: varnist. 64. judgment: *iudement.* 73.: [Q provides a s.p.: *Arrag.*]

ACT 3, SCENE 1. 16. s.d. Enter Shylock: [after line 22 in Q]. 37. courtesy: cursie. 54. s.p. Man: [not in Q]. 59. s.d.: [Q repeats the s.d. *"Enter* Tubal"]. 80. Heard: heere. 90. turquoise: Turkies.

ACT 3, SCENE 2. 23. eke: ech. 61. live. With: liue with. 67. eyes: *eye.* 81. vice: voyce. 84. stairs: stayers. 101. Therefore: Therefore then. 110. shuddering: shyddring. 199. loved; for intermission: lou'd for intermission. 204. roof: rough. 215. s.d.: [after line 219 in Q]. 313. s.p. Bassanio: [not in Q].

ACT 3, SCENE 3. s.d. Solanio: *Salerio.* 24. s.p. Solanio: *Sal.*

ACT 3, SCENE 4. 23. Hear other things: heere other things. 49. Padua: Mantua. 50. cousin's: cosin. 53. traject: Tranect. 80. near: nere. 81. my: my my.

ACT 3, SCENE 5. 17. e'en: in. 20. comes: come. 61. merit it: meane it, it. 68. a wife: wife. 74. s.d. Exeunt: *Exit.*

ACT 4, SCENE 1. 30. his state: this states. 31. flint: flints. 35. s.p. [and elsewhere in this scene] Shylock: *Jewe.* 50. urine; for affection: vrine for affection. 51. Mistress: Maisters. 73. You may as well: well. 74. Why he hath made the: the. bleat: bleake. 75. pines: of Pines. 100. is: as. 113. lose: loose. 136. whilst: whilest. 225: No, not: Not not. 230. tenor: tenure. 267. off: of. 319. off: of. 393. s.p. Gratiano: *Shy.* 402. s.d. Exeunt: *Exit.*

ACT 5, SCENE 1. 26. s.p. Stephano: *Messen.* [also at lines 28 and 33]. 41. Lorenzo: *Lorenzo,* &. 47. Sweet soul: [assigned in Q to Launcelot]. 49. Stephano: Stephen. 85. Erebus: *Terebus.* 107. ho: how. 150. give it: giue. 231. my: mine.

PART TWO

—✶—

Cultural Contexts

CHAPTER I

Venice

—————————————————⤖—————————————————

The republic of Venice inspired numerous early modern English representations of its virtues and vices. The English saw themselves in relation to
this maritime city-state, identifying with its reputed political independence,
navigational expertise, and international trade. Hence Shakespeare's tendency in his plays to transmit a slice of contemporary London life to historically, geographically, and imaginatively distant locales is, in the case of *The
Merchant of Venice*, reinforced by popular associations of England and
Venice. Both are characterized as virgins: the isle of England, "This precious
stone set in the silver sea / Which serves it in the office of a wall / Or as a
moat defensive to a house" (*Richard II* 2.1.46–48) is akin to its virgin queen,
Elizabeth; Venice, in its independence from foreign domination, is popu-
larly likened to a virgin queen. Lewes Lewkenor, in the introduction to his
1599 translation of Gasparo Contarini's 1543 *De Magistratibus et Republica
Venetorum* ("Of the Magistrates and the Republic of Venice"), makes the
connection between the two polities explicit: "[T]he rest of the whole world
honoreth her [i.e. Venice] with the name of a Virgin, a name though in all
places most sacred and venerable, yet in no place more dearely and religiously to be reverenced, than with us [i.e., the English], who have thence
derived our blessednesse" (quoted in McPherson 33). Because England owes
its own fortunate status, according to Lewkenor, to its remarkable virgin

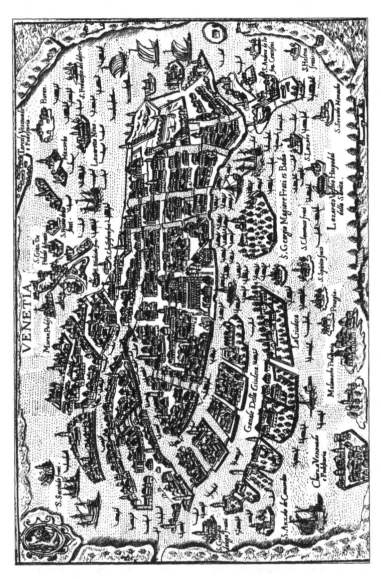

FIGURE 5   *The City-State of Venice, from Pietro Bertelli,* Theatrum Urbium Italicarum *(1599).*

queen, it is best suited to appreciate the polity of Venice, which is also characterized as a virgin. However, while English authors glorified Venice's preeminence in wealth and trade, they also condemned its avarice, which, they argued, contributed to its economic decline by the end of the sixteenth century. The republic's independence from the Roman Catholic pope's authority heightened hopes among the English for the development of Protestantism there, even while they noted and condemned Venetian Catholicism. A consideration of several early modern English stereotypes about Venice suggests a range of positive and negative attitudes that a contemporary audience might have brought to the play.

English Ideas of Venice and Italians

WEALTH

The English association of wealth with Italians appears in a number of contexts from the Middle Ages forward. Visitors to Venice frequently mentioned the opulence and affluence of the republic. Thomas Coryate, in *Coryats Crudities*, praises at length the ravishing splendor of its architecture, which he argues is without rival. In *The History of Italy*, William Thomas, on the other hand, calls this wealth into question in his discussion of the Venetian accumulation of and attitudes toward money. He explains that their revenue derives from the sales tax that is levied on all merchandise sold there: "For there is not a grain of corn, a spoonful of wine, a corn of salt, egg, bird, beast, foul, or fish bought or sold, that payeth not a certain custom" (p. 132). He associates their zealous collecting of taxes with the profit they make on usury charged by the Jews, and he suggests their numerous tariffs make life harder for Christians there than in the Turkish empire. Thomas acknowledges that the character of the individual "gentleman Venetian" may be excellent, but also repeats the charges of strangers who accuse the Venetian of greed and stinginess, and of not only doing business with Jews but even lending money to the state at the rate of 10 to 15 percent. Shakespeare seems to polarize these attitudes by ascribing Venetian greed to the Jewish Shylock and characterizing Antonio as wealthy yet generous. However, the Christians of *The Merchant of Venice* demonstrate an interest in acquiring wealth; are Bassanio and Lorenzo substantively different from Shylock in their attitudes toward money?

By the beginning of the seventeenth century, Dudley Carleton could catalogue the factors contributing to the decline of Venetian trade, and hence, wealth. The Venetians had abandoned "merchandising," had educated their sons to be gentlemen rather than merchants, and had allowed the size of their fleet to dwindle; their high taxes had turned trade to other ports and competition from English merchants had cut into their profits. How might these conditions influence an understanding of Antonio's increasing marginalization in the play? What sense do we have of the viability of Venice's economy at the play's end?

WOMEN

The women of Venice were represented as alluringly beautiful; whether as chaste wives or licentious prostitutes, they, like the goods traded in their city, attracted men from all over the world. Thomas remarks on the sumptuousness of Venetian wives' clothing and Coryate gives a detailed account of their artful cosmetics and coiffure. According to Fynes Moryson, Italian "women of honor" are "much sooner inflamed with love, be it lawful or unlawful, than the women of other nations" (p. 167), perhaps as a result of being constrained to remain at home or to go out only when chaperoned by an older female servant. The play engages these views in a number of ways, representing Jessica and Portia as beautiful and desirable women who, in spite of being strictly confined by their fathers, are pursued by — and also pursue — attractive suitors. Moryson also notes that husbands remain fearful of cuckoldry (their wives' being unfaithful) in spite of the confinement of their wives. Coryate reports that "toleration of such licentious wantons [prostitutes] in so glorious, so potent, so renowned a city" (p. 144) exists, in part, to prevent husbands from being "capricornified," on the assumption that men will not seduce other men's wives if prostitutes are readily available. Prostitution provides another benefit to the state: revenues from the trade subsidize the cost of maintaining "a dozen of their galleys" (p. 145). Prostitutes enjoy great freedom of movement, unlike their "honorable" sisters, and according to Moryson, frequently dress in men's clothing throughout Italy. While Portia and Jessica are clearly not prostitutes, how would an early modern audience view their cross-dressing and freedom of movement, as well as Portia and Nerissa's pretended infidelity, in the context of these assumptions about the behavior of Venetian women?

LAW

Venetian law, like other aspects of the republic, is both praised and criticized by English authors. Thomas explains that although their lawyers studied the civil law on the Continent, in practice, the particular laws of Venice were "so many, that in manner they suffice of themselves" (p. 133). However, cases were apparently settled according to the conscience of the judges, and not necessarily according to the laws. In contrast to capital cases tried under English common law, Venetian defendants were not permitted to speak on their own behalf, although they were allotted advocates to plead for them. Disputants often either hid or resolved their conflicts in secret for fear of banishment or other punishment inflicted by a republic afraid of civil unrest. Thomas praises the structure of the Venetian legal system, claiming that "there can be no better order of justice in a commonwealth than theirs" (p. 134), but laments that in practice advocates have corrupted the system to the disadvantage of poor litigants. Law plays a pivotal but complex role in *The Merchant of Venice*. On the one hand, law is necessary to maintain the economic viability of the republic; from this perspective, it makes a central contribution to the character and function of the state. On the other hand, Portia's rhetoric in the trial scene suggests that Shylock, rather than the Venetian state that is helpless to halt his suit, embodies the letter of the law, which is here contrasted to a superior mercy. Yet, by what means does Portia win the case?

Shylock's status as an alien resonates with contemporary authors' accounts of the stranger's legal position in Venice. While Thomas gives the impression of a somewhat arbitrary system for the citizens of Venice, his discussion of the rights of strangers suggests that the law affords them a large measure of protection. They are free to speak ill of Venice, so long as they foment no action; by contrast, in England there were laws punishing sedition (speech tending toward or advocating the overthrow of government). They enjoy a great deal of legal and religious freedom, as long as their practices remain private. In England, royal proclamations and common and statute law served both to restrict and protect aliens. While the statute law concerning aliens was fairly restrictive, they could avoid the required penalties in a number of ways. Immigrants to England could also apply to become denizens, which gave them the right to hold, but not purchase, land, avoid the payment of customs charged aliens, and engage in litigation. Naturalization, which required an act of Parliament, was rarer, but bestowed the privileges of full English citizenship on a stranger (Yungblut 78–79). Conditions during the reign of Elizabeth led to a mixed policy on aliens: on the one hand, England sought to shelter persecuted Protestants as well as attract foreign experts in technologies that were locally weak; on the other hand, fear of Catholics and radical Protestant aliens and popular disaffection with foreign economic competition worked to limit immigration (85–86, 90). A proclamation from the beginning of Elizabeth's reign (p. 159) shows concern for protecting the rights of aliens and insuring their equitable treatment under the law. In his *Reports*, Sir Edward Coke reviews the rights and restrictions pertaining to the alien as either friend or foe (p. 161). Jews fall into the category of perpetual enemies, according to Coke, who have no recourse to the law, though evidence exists to the contrary. This mixture of perspectives on the legal rights of aliens also registers in the play. In what ways does Shakespeare's Venice protect its alien residents? How does Shylock's status as an alien affect the outcome of the trial? Why is Shylock considered an alien?

Nation, Race, and Religion

Venice's status as a cosmopolitan center of trade and culture attracted the attention of a number of English authors. Coryate admires the daily international pageant on the Piazza San Marco:

> This part of the piazza is worthy to be celebrated for that famous concourse and meeting of so many distinct and sundry nations twice a day. . . . There you may see many Polonians, Slavonians, Persians, Grecians, Turks, Jews, Christians of all the famousest regions of Christendom, and each nation distinguished from another by their proper and peculiar habits. A singular show, and by many degrees the worthiest of all the European countries. (p. 139)

Trade draws people of numerous nations, races, and religions to Venice, rendering it "Europe's predominant site of contact with the world of racial and religious Others" (Boose, "'Getting'" 39). Stereotypes about the "intrinsic character" of these diverse peoples circulated widely in early modern England and on the continent in the emerging genre of travel literature. Travel writers embarked upon voyages with various purposes: some made pilgrimages to holy sites, others explored sea routes or new countries in order to expand trade or extend national territory, while many sought entertainment and education. However, the texts produced by travelers share some common features. In all accounts the author presents a subjective understanding, or misunderstanding, of the culture observed: "accounts of travel are never objective, they inevitably reveal . . . culture-specific and individual patterns of perception and knowledge" (Korte 5–6). Travelogues are constructed texts, written and organized after the events they describe, evaluating experiences in terms of the writer's own identity and familiar contexts; they also often incorporate writings and opinions of other travelers in lieu of first-hand observation (Korte 10–11, 20, 22). Hence, the accuracy of travel writings, such as those by Thomas, Coryate, Moryson, George Best, and John Leo included in this volume, was limited at best, although they reveal a good deal about the cultural assumptions of the writers. Strikingly, authors from different countries often voice the same opinions about the national characteristics under consideration; the Italian writer Giovanni

Botero, the English Fynes Moryson, and Portia all allege a tendency to drunkenness among Germans. All the stereotypes that Portia articulates or that foreign characters embody in the play can be found in contemporary travelogues, from George Sandys's observation in *A Relation of a Journey* (1615) that the Neopolitan "gentry delight much in great horses, where upon they prance continually throughout the streets" (4.257) to Botero's statement in *Relations, of the Most Famous Kingdoms* (1608) that "the Spanish nation by nature is proud, yet base, and such as careth not though they be hated, so they may be feared" (I2v).

Racial stereotypes are closely related to national characterizations in early modern travel writing. The term *nation*, which entered the English language in the thirteenth century, carried the sense often associated with race, "a blood related group"; however, by the early seventeenth century, it was "used to describe the inhabitants of a country, regardless of that population's ethnonational composition" (Connor 38). Hence, the terms *nation* and *race* were often used interchangeably; race was not consistently linked to skin color, but could indicate national and religious "others" such as the Irish or the Jews (Boose, "'Getting'" 36). John Leo, like George Best, situates his consideration of differences of nation/race and skin color within a religious context: the genealogies recorded in Genesis. Although skin color is not explicitly mentioned in the biblical narrative, both authors draw on traditions that infer it; Best goes so far as to argue that a dark complexion is the punishment inflicted on Chus, the offspring of Noah's disobedient son Cham. Blackness and racial distinctiveness are also associated with Jews in the early modern period; Shylock has a countryman named Chus (3.2.283), and in *The Messiah of the Christians and the Jews,* Sebastian Munster comments on the Jews' "peculiar color of face, different from the form and figure of other men; . . . for you are black and uncomely, and not white as other men" (p. 181). Andrew Willet's exploration of the alleged racial distinctiveness of Jews (as opposed to that of national character) suggests how Shylock might be considered an "alien" in the play, but also raises the question of how easily "racially" different Jews could be incorporated into Christian society by conversion.

Religious difference is often delineated in terms of national difference. The anonymous seventeenth-century text *A Discovery of the Great Subtlety and Wonderful Wisdom of the Italians* argues that Italians have conspired, through a practice of various deceptions, to appropriate the wealth of other countries; however, the true culprit here is the Church of Rome, which the anonymous author accuses of seeking religious and political domination of Europe. While he directs his charges against Italy in general, he makes special mention of "The quick spirited Venetians, . . . being a people of Italy very subtle" (p. 152). In *The Three Ladies of London,* Robert Wilson presents the Italian merchant Mercadorus as a double-dealing scoundrel, covetous of wealth and careless of his religion. While sojourning in Turkey where he has come to buy more merchandise, he chooses to convert to Islam, which would allow him debt forgiveness, rather than repay what he owes to Gerontus, the Jewish moneylender. The Venetian association with trade also appears in the context of religion. William Bedell, the Protestant chaplain to the English ambassador, agrees with the Jews who denounce the Christians of Venice "as merchants of God's word" because the preacher requests a collection in the middle of sermons. Although the English had hopes of fostering Protestantism in Venice, Bedell alleges that their enslavement to idolatrous images is worse than that of pagans, even while he admires their capacity for training children and maintaining their churches. However, Thomas suggests Venice is a place of religious freedom, where if one is "a Jew, a Turk, or believeth in the devil (so thou spread not thine opinions abroad) thou are free from all controllment" (p. 137). While Shakespeare downplays intra-Christian discord by contrasting the Venetian Christians with the Jews Shylock and, to a lesser extent, Jessica, an early modern English audience would have been aware of the national and religious differences that separated it from both the Jews and the Venetian Christians.

Travel writers also used the discourse of religious difference by itself to derogate "others," although they occasionally allowed for more positive perspectives. John Leo, born a Muslim, or Moor, in Granada, was captured and brought to Rome, where he was converted by the pope; he then represented Islam as heretical and calculating in its attempts to attract adherents. English representations of Jews also convey negative associations with their religion. Moryson speaks of Jews of the European diaspora as living openly in some countries while "lurking" in others from which they have been banished. While he articulates the common Christian condemnation of the Jews' rejection of Jesus as the messiah, he also notes that the restriction of their livelihood by their host countries results in their reliance on usury and pawnbroking, which in turn profits their rulers. This calls into question ideas of intrinsic Jewish "greed" and provides a different context for thinking about Shylock's attachment to "the prop / That doth sustain [his] house" (4.1.370–71). The Venetian Jewish community provided early modern English travelers with a rare opportunity to meet with practicing Jews. Coryate expresses mixed views upon observing Venetian Jewish life: he is surprised by the attractiveness of Jews even while he "lament[s] their religion" (p. 141). A visit to the synagogue impressed upon him the beauty and rich apparel of the Jewish women; he also notes that they observe a number of commandments, including avoiding idolatry, resting on the Sabbath, and circumcising their sons. Coryate complains about their adherence to Judaism, and blames the Italian practice of confiscating a convert's property for the low rate of conversions there. This practice resonates in *The Merchant of Venice* with the dual threat of conversion and the state confiscation of Shylock's property at the end of act 4. Coryate also relates an exchange with a learned rabbi, possibly Leo of Modena, in which they debate the validity of Christianity. While the rabbi offers a diplomatic and positive assessment of Jesus, although denying his divinity, Coryate offers an abrasive reply, accusing Jews of being "Christ's irreconcilable enemies" (p. 143) even while he seeks to convert them to Christianity. The core of the argument pivots on a disagreement over the interpretation of the Bible: Coryate alleges that its prophecies point to Jesus as the messiah, and the rabbi insists that Christians misinterpret the Hebrew scriptures. Coryate is oddly surprised at the Jews' increasing anger at his "sharply tax[ing] their superstitious ceremonies . . . [and] reprehend[ing of] their religion" (p. 143–44); he beats a hasty and somewhat comical retreat in the face of a growing crowd of irritated Jews. Antonio seems similarly oblivious to the effect on Shylock of his hostile attitudes toward Judaism; although he characterizes the Jew as his enemy and as a devilish misreader of scripture (1.3.89 ff), Antonio simultaneously hopes for Shylock's transformation through conversion: "The Hebrew will turn Christian; he grows kind" (1.3.1).

→ **WILLIAM THOMAS**

From The History of Italy *1549*

William Thomas (d. 1554) was probably educated at Oxford; in 1544 he was obliged to flee the country, and he spent the next five years chiefly in Italy. While abroad he published a defense of Henry VIII in Italian and an Italian-English dictionary, which included rules of grammar. He returned to England in 1549 and published that same year *The History of Italy*. His defense of Henry and his knowledge of modern languages resulted in his appointment as one of the clerks of Edward VI's Privy Council in 1550. In this role he served the king as an informal tutor on principles of government and recorded, but did not publish, some of his ideas on political subjects. Upon Mary's accession to the throne, he sided with her opponents; he was eventually arrested and executed as a traitor. His *History of Italy* was reportedly burned at that time, but it was published again several times early in Elizabeth's reign. The following excerpt

William Thomas, *The History of Italy: A Book Exceeding Profitable to Be Read: Because It Intreateth of the Estate of Many and Divers Commonweals, How They Have Been, and Now Be Governed* (London, 1549), 76v–77r, 81r–85v.

presents his account of Venice and Venetians. While the Christian citizens of Venice who populate Shakespeare's play tend to be characterized by their generosity, even prodigality, Thomas's account reflects the opposite view. He characterizes the very state as focused on profit, to the extent of being in league with Jewish usurers, whom he represents as contributing to the wealth of the republic; in *The Merchant of Venice,* by contrast, the Duke is philosophically, morally, and financially opposed to Shylock. Thomas describes some Venetians who might bear a resemblance to Antonio and Bassanio, respectively: the "old fatherly men, as wise, as honest, as faithful, as honorable, and as virtuous, as in any place can be found . . . [and] some of the young men, as gentle, as liberal, as valiant, as well learned, as full of good qualities, as may be" (p. 134). But he sees these as exceptions to the majority of citizens who are reported to be proud, stingy, lustful, cruel, and as greedy as their Jewish neighbors, whom they emulate in profiting from usury on loans made to the state.

The republic of Venice was famous in the period for its laws, as Thomas avers, "But this is clear, there can be no better order of justice in a commonwealth than theirs" (p. 134), and as Shakespeare represents throughout the play, particularly in Portia's refusal to "wrest the law" to her authority (4.1.210–17). This respect for law extends even to noncitizens, as Antonio himself states: the Duke dare not "deny the course of law [to aliens because it would] impeach the justice of the state" (3.3.26–29). Thomas similarly notes that the commonwealth is a very tolerant place, allowing a variety of religious observances and offering strangers a great deal of freedom. However, he also criticizes the corruption of the justices in the execution of law, and his account of the Venetian policy of religious freedom stands in tension with Shylock's forced conversion at the end of the trial.

From *The History of Italy*

OF REVENUE

As I have been credibly informed by some gentlemen Venetians, that have had to do therein, they levy of their subjects little less than 4 millions of gold by the year, which (after our old reckoning) amounts to the sum of ten hundred thousand pounds sterling. A thing rather to be wondered at than believed, considering they raise it not upon lands, but upon customs after so extreme a sort, that it would make any honest heart sorrowful to hear it. For there is not a grain of corn, a spoonful of wine, a corn[1] of salt, egg, bird, beast, foul, or fish bought or sold, that payeth[2] not a certain custom. And in Venice especially the customer's part in many things is more than the

---

[1] corn: grain.　[2] payeth: pays.

owner's. And if anything be taken by the way uncustomed,[3] be it merchandise or other, never so great or small, it is forfeited. For those customers keep such a sort of prowlers to search all things as they come to and fro, that I think Cerberus[4] was never so greedy at the gates of hell as they be in the channels about Venice. And though they in searching a boat, find no forfeiture, yet would they not depart without drinking money.[5] And many times the meanest laborer or craftsman throughout all their dominion, payeth a rate for the poll[6] by the month. Insomuch that a Candiote[7] my friend (one that had dwelled in Constantinople) swore to me by his faith, the Christians lived a great deal better under the Turk, than under the Venetians. It is almost incredible, what gain the Venetians receive by the usury of the Jews, both privately and in common. For in every city the Jews keep open shops of usury, taking gages of ordinary for fifteen in the hundred by the year:[8] and if at the year's end, the gage not be redeemed, it is forfeit, or at the least done away to a great disadvantage: by reason whereof the Jews are out of measure wealthy in those parts.

Of Laws

Their advocates (as we should say are men of law) study principally civil laws, and besides that the statutes and customs of the city: which are so many, that in manner they suffice of themselves. But he that substantially considereth the manner of their proceedings, shall plainly see that all matters are determined by the judge's consciences, and not by the civil, nor yet their own laws. For in every office there be diverse judges, and that part[9] that has the most ballots, prevails ever: be it in matter of debt, of title of land, upon life and death, or otherwise. And in every trial of theft, murder, or such other, the party himself is never suffered to speak. But there be certain advocates, waged[10] of the common revenue, which with no less study plead in their defense, than the *Avogadori*[11] in the contrary.... This order they observe in Venice only. For out of Venice the gentlemen Venetian, that is *Potestate*[12] of the city, town, or place, has absolute power to judge upon all matters himself alone: howbeit every of them, has a council of learned men,

---

[3] **uncustomed:** not charged duty tax.   [4] **Cerberus:** dog in classical mythology who guarded the gates of Hades (hell).   [5] **drinking money:** the customs agents extort money for personal use, i.e., drinking.   [6] **rate for the poll:** amount per person.   [7] **Candiote:** an inhabitant of Candia, otherwise known as Crete.   [8] **taking gages ... by the year:** the borrower offers an object of value to the lender in exchange for a sum of money; 15 percent interest is charged on the loan, and if the borrower fails to repay the loan with interest on the agreed upon date, she or he forfeits the object (which is usually more valuable than the loan plus interest).   [9] **part:** party, one of the sides litigating.   [10] **waged:** salaried.   [11] *Avogadori:* lawyers.   [12] *Potestate:* magistrate, presumably in one of the surrounding towns over which Venice exercised rule.

to advise him what the law commandeth. Besides that, every five years there be certain inquisitors, called *Sindici,* sent forth to reform extortions, and all other things that they find amiss, throughout their whole dominion. . . . And because they fear, lest civil sedition might be the destruction of their commonwealth, as of diverse other it has been, therefore they have provided an order, that when any two gentlemen happen to fall out, either they do so dissemble it, that their malice never appeareth to the world, or else they agree within themselves.[13] For if it come to the *Signoria's*[14] knowledge, it cannot be chosen, but he that is most faulty receiveth a great rebuke, and many times in those cases diverse are banished, or sharply punished. As for their other laws, though I were sufficiently expert in them, yet partly for briefness, and partly because they are not so much necessary to my purpose, I pass them over. But this is clear, there can be no better order of justice in a commonwealth than theirs, if it were duly observed. Howbeit corruption (by the advocate's means) is so crept in amongst the judges that poor men many times lack no delays in the process of their matters.

Of Customs in Their Living

To speak of the gentleman Venetians' private life and customs I wote[15] not whether it be best to follow the common report, or to dissemble the matter.[16] And yet it seems to me that I cannot do more indifferently[17] than recite what is used to be said on both sides. If any man would say, there were no worthy men amongst the Venetians, he should greatly err. For (as I believe) there be some, and especially of those old fatherly men, as wise, as honest, as faithful, as honorable, and as virtuous, as in any place can be found. Likewise some of the young men, as gentle, as liberal, as valiant, as well learned, as full of good qualities, as may be. But to speak of the greater number, strangers use to[18] report, that the gentleman Venetian is proud, disdainful, covetous, a great niggard,[19] a more lecher, spare of living, tyrant to his tenant, finally never satisfied with hoarding up of money. For though (say they) he has eight, nine, or ten thousand ducats of yearly revenue, yet would he keep no more persons in his house, but his wife and children, with two or three women servants, and one man, or two, at the most, to row his *Gondola.* He would go to the market himself, and spend so miserably,[20] that many a mean man shall fare better than he of his ten thousand ducats a year. If he spend three or four pounds in his house he considers it a wonderful

---

[13] within themselves: privately. [14] *Signoria:* body of lords of the Venetian republic. [15] wote: know. [16] to dissemble the matter: to conceal the truth of the matter. [17] indifferently: impartially. [18] use to: often. [19] niggard: miser. [20] miserably: frugally.

charge.[21] Besides all this he has two or three Jews, that chop and change[22] with him daily: by whose usury he gaineth out of measure.[23] And yet would rather see a poor man starve than relieve him with a penny. It is true, he would have his wife go gay and sumptuously appareled[24] and on his woman besides, if he be a lover (as in manner they all be) he would stick for no cost. To the marriage of his daughter thirty, forty, or fifty thousand ducats is no marvel. Finally his great triumph, when St. Mark has need (for under that name is comprehended their commonwealth), to be able to disburse a huge sum of money in loan to receive yearly till he be repaid ten, twelve, or fifteen of the hundred.

This kind of prest[25] the *Signoria* useth to take (borrowing of all them that are able to lend) when they happen to have wars. And they that may, do the more willingly lend: because they are not only well paid again with the usury, but also the more honored and favored as long as their money is out of their hands.

This is their trade, sayeth[26] the stranger. But the Venetian to the contrary defends himself in this way.

Admit (sayeth he) that this report were true. If I be proud, I have good cause, for I am a prince and no subject. If I be spare of living, it is because my commonwealth alloweth no pomp and measure is wholesome. If I keep few servants, it is because I need no more. If I buy my meat myself, it is because I would eat that that[27] I love, and that (having little ado) I would exercise myself withal. As for my tenant, he liveth by me, and I am no tyrant for husbanding[28] my own. If I gain, I gain upon my money, and hide not my talent in the ground.[29] If I love, I hate not: if she be fair, I am the more worthy. If I spend little, I have the more in my purse. If I spend largely with my daughter, it is because I would bestow her on a gentleman Venetian, to increase the nobility of my own blood, and by means of such alliance to achieve more ability to rule and reign in my commonwealth. Besides that, my money (if her husband die) is hers and no other man's else. If my wife go gay, it is to please my eye and to satisfy her. In keeping my money to lend to St. Mark, it is both a help to my commonwealth and a profit to myself.

And thus defends the Venetian it, that in manner[30] all the world lays to his charge.

But surely many of them trade and bring up their children in so much liberty, that . . . by the time he comes to twenty years of age, he knows as

---

[21] **wonderful charge**: great expense. [22] **chop and change**: barter and exchange. [23] **out of measure**: inordinately. [24] **go gay and sumptuously appareled**: be finely and richly dressed. [25] **prest**: loan; especially one made to the sovereign in an emergency *(OED)*. [26] **sayeth**: says. [27] **that that**: that which. [28] **husbanding**: using in a thrifty manner. [29] **hide not . . . ground**: see the parable of the talents in Matthew 25:14–30. [30] **it, that in manner**: that which.

FIGURE 6 *A "Sumptuously Appareled" Venetian Bride, from Pietro Bertelli,* Diversarum Nationem *(1594).*

much lewdness as is possible to be imagined. For his greatest exercise is to go amongst his companions, to this good woman's[31] house and that. Of which in Venice are many thousands of ordinary[32] less than honest.

THE LIBERTY OF STRANGERS

All men, especially strangers, have so much liberty there, that though they speak very ill by the Venetians, so[33] they attempt nothing in effect against their [i.e., the Venetian's] estate, no man shall control them for it. And in their *Carnivale* time (which we call Shrovetide)[34] you shall see maskers[35] disguise themselves in the Venetians' habit, and come unto their own noses[36] in derision of their customs, their habits, and misery.[37]

Further, he that dwelleth in Venice may reckon himself exempt from subjection. For no man there marketh another's doings, or that meddleth with an other man's living. If thou be a papist,[38] there shalt thou want[39] no kind of superstition to feed upon. If thou be a gospeler,[40] no man shall ask why thou comest not to church. If thou be a Jew, a Turk, or believeth in the devil (so thou spread not thine opinions abroad) thou are free from all controllment. To live married or unmarried, no man shall ask thee why. For eating of flesh in thine own house, what day so ever it be, it makes no matter.[41] And generally of all other things, so thou offend no man privately, no man shall offend thee: which undoubtedly is one principal cause which draws so many strangers there.

[31] **good woman:** prostitute.   [32] **of ordinary:** commonly.   [33] **so:** as long as.   [34] *Carnivale* . . . Shrovetide: time of celebration before Ash Wednesday and the beginning of Lent. [35] **maskers:** people in masquerade.   [36] **come unto their own noses:** right in front of them. [37] **misery:** miserliness.   [38] **papist:** Catholic.   [39] **want:** lack.   [40] **gospeler:** Protestant, who would not attend the Catholic service performed in Venetian churches.   [41] **For eating of flesh . . . no matter:** meat was usually not eaten by Catholics on Fridays; Protestant reformers rejected this practice.

→ THOMAS CORYATE

From Coryats Crudities   *1611*

Thomas Coryate (1577?–1617) studied but did not take a degree at Oxford; eventually he became a kind of jester at the court of James I, where his learning and wit brought him a small income, including a pension from Henry, Prince of Wales. Coryate probably acquired some property after his father's death, which

Thomas Coryate, *Coryats Crudities: Hastily Gobbled Up in Five Months Travels in France, Savoy, Italy, . . . Switzerland &c.* (London, 1611), 171–73, 175–77, 184, 230–37, 261, 264–65.

allowed him to embark on a tour through Europe in 1608. He visited forty-five cities in five months and produced a journal of his travels, which he tried to publish on his return. He requested assistance from numerous courtiers and poets, many of whom wrote poems commending the project, and eventually the book was published in 1611 as *Coryats Crudities: Hastily Gobbled Up in Five Months Travels in France, Savoy, Italy, . . . Switzerland &c.* The following year he renewed his travels, this time setting out to see the East. He visited Turkey, Greece, Asia Minor, Egypt, Palestine, Persia, and India, learning Persian, Turkish, and Hindustani in the course of four years of travel; several of his letters describing his travels were published. He died in Agra, India, while visiting English merchants there in 1616. *Coryats Crudities* seems to have sold well and is considered the first English handbook on continental travel. The excerpt that follows recounts Coryate's impressions of Venice during his stay there. Coryate's description of Venice's brilliance and global allure, drawing merchants from all over the world, resembles more closely the play's representation of Portia in Belmont than the city itself. Though they are not merchants, Portia's suitors parade through like the international "show" typically seen in St. Mark's Place: "There you may see many Polonians, Slavonians, Persians, Grecians, Turks, Jews, Christians of . . . proper and peculiar habits. A singular show, and by many degrees the worthiest of all the European countries" (p. 139). Both Bassanio and the Prince of Morocco describe Portia as a precious object or site to which the whole world is drawn (1.1.166–71; 2.7.38–47), suggesting a global attraction that aligns Portia uncomfortably to another renowned woman: the Venetian courtesan. "For so infinite are the allurements of the famous Calypsos [i.e., prostitutes], that the fame of them hath drawn many to Venice from some of the remotest parts of Christendom, to contemplate their beauties, and enjoy their pleasing dalliances" (p. 145). While Portia and Nerissa act in accordance with contemporary standards of wifely chastity, their joking confessions of sexual infidelity allege that they traded sex for the return of their rings (5.1.236–60). Coryate's description of the Venetian husband's fear of cuckoldry, situated in the context of his discussion of Venetian prostitution, mirrors Bassanio's and Gratiano's dismay when they imagine themselves "capricornified," that is, given horns or cuckolded.

Coryate's encounters with Venetian Jews draw both positive and negative responses. The piety of the Jews excels that even of Christians in some cases, writes Coryate, and he is struck by their physical attractiveness, particularly that of the richly adorned Jewish women he sees in the synagogue. Jessica's beauty enhanced with her father's wealth exercises a similar influence on Lorenzo in the play. The transfer of this Jewish wealth to Christians, particularly that acquired by usury, is an issue in both texts. Coryate condemns the practice of confiscating the wealth of Jewish converts, a policy the English had also practiced in medieval times. Upon converting, Jessica conveys a considerable portion of her father's wealth to her husband, whose property it becomes under law. Although Shylock is permitted half of his remaining estate at his conversion, the remainder of it transfers to his Christian son-in-law upon his death.

Though he disapproves of confiscating converts' property, Coryate does condemn Judaism on theological grounds, arguing, as Antonio does (1.3.89–93), for the superiority of Christian over Jewish interpretation of the Bible. Jewish resistance to Christian doctrine and hostility to proselytizing becomes a menacing force in Coryate's account; his response to these sentiments may underlie the representation of Shylock's murderous malice in the play. While Shakespeare does not explicitly engage theological differences between Christians and Jews, most of his audience would have been familiar with the kinds of arguments Coryate directs against Judaism in this account.

From *Coryats Crudities*

The fairest place of all the city (which is indeed of that admirable and incomparable beauty, that I think no place whatsoever, either in Christendom or Paganism may compare with it) is the piazza, that is, the marketplace of St. Mark, or (as our English merchants commorant[1] in Venice do call it) the place of St. Mark. . . . Truly such is the stupendious (to use a strange epitheton for so strange and rare a place as this) glory of it, that at my first entrance thereof, it did even amaze or rather ravish my senses. For here is the greatest magnificence of architecture to be seen, that any place under the sun doth yield. Here you may both see all manner of fashions of attire, and hear all the languages of Christendom, besides those that are spoken by the barbarous ethnics; the frequency of people being so great twice a day, betwixt six of the clock in the morning and eleven, and again betwixt five in the afternoon and eight, that (as an elegant writer said of it) a man may very properly call it rather *orbis* than *urbis forum*, that is, a marketplace of the world, not of the city. . . .

This part of the piazza is worthy to be celebrated for that famous concourse and meeting of so many distinct and sundry nations twice a day . . . as I have before mentioned, where also the Venetian long-gowned gentlemen do meet together in great troupes. For you shall not see as much as one Venetian there of the patrician rank without his black gown and tippet.[2] There you may see many Polonians, Slavonians, Persians, Grecians, Turks, Jews, Christians of all the famousest regions of Christendom, and each nation distinguished from another by their proper and peculiar habits. A singular show, and by many degrees the worthiest of all the European countries. . . .

---

[1] **commorant:** dwelling. [2] **tippet:** a long narrow slip of cloth worn, either attached to and forming part of the hood, or loose, as a scarf *(OED)*.

FIGURE 7 *The Plaza of St. Mark in Venice, from Pietro Bertelli,* Diversarum Nationem *(1594).*

There you . . . have a synopsis, that is, a general view of little Christendom (for so do many entitle this city of Venice) or rather of the Jerusalem of Christendom. For so me thinks may a man not improperly call this glorious city of Venice: not in respect of the religion thereof, or the situation, but of the sumptuousness of their buildings, for which we read Jerusalem in former times was famoused above all the Eastern cities of the world. . . .

I was at the place where the whole fraternity of the Jews dwelleth together, which is called the ghetto, being an island: for it is enclosed round about with water. It is thought there are of them in all betwixt five and six thousand. They are distinguished and discerned from the Christians by their habits on their heads; for some of them do wear hats and those red . . . [for] those Jews that are born in the western parts of the world, as in Italy, etc. But the eastern Jews being otherwise called the Levantine Jews, which are born in Jerusalem, Alexandria, Constantinople, etc., wear turbans upon their heads as the Turks do, but the difference is this: the Turks wear white, the Jews yellow. By that word turban I understand a roll of fine linen wrapped together upon their heads, which serveth them instead of hats, whereof many have been often worn by the Turks in London. They have diverse synagogues in their ghetto, at the least seven, where all of them, both men, women, and children do meet together upon their Sabbath, which is Saturday, to the end to do their devotion, and serve God in their kind, each company having a several[3] synagogue. . . .

I observed some few of these Jews, especially some of the Levantines, to be such goodly and proper men, that then I said to myself our English proverb: To look like a Jew (whereby is meant sometimes a weather beaten warp-faced fellow, sometimes a frantic and lunatic person, sometimes one discontented) is not true. For indeed I noted some of them to be most elegant and sweet featured persons, which gave me occasion the more to lament their religion.

In the room wherein they celebrate their divine service, no women sit, but have a loft or gallery proper to themselves only, where I saw many Jewish women, whereof some were as beautiful as ever I saw, and so gorgeous in their apparel, jewels, chains of gold, and rings adorned with precious stones, that some of our English Countesses do scarce exceed them, having marvelous long trains like Princesses that are borne up by waiting women serving for the same purpose. An argument to prove that many of the Jews are very rich. . . . They are very religious in two things only, and no more, in that they worship no images, and that they keep their Sabbath so strictly, that upon that day they will neither buy nor sell, nor do any secular, profane,

---

[3] **several:** separate.

or irreligious exercise (I would to God our Christians would imitate the Jews herein), no not so much as dress their victuals,[4] which is always done the day before, but dedicate and consecrate themselves wholly to the strict worship of God. Their circumcision they observe as duly as they did any time betwixt Abraham (in whose time it was first instituted) and the incarnation of Christ, for they . . . circumcise every male child when he is eight days old, with a stony knife. But I had not the opportunity to see it. Likewise they keep many of those ancient feasts that were instituted by Moses. Amongst the rest of the Feast of Tabernacles[5] is very ceremoniously observed by them. From swine's flesh they abstain as their ancient forefathers were wont to do, in which the Turks do imitate them at this day. Truly it is a most lamentable case for a Christian to consider the damnable estate of these miserable Jews, in that they reject the true Messiah and Savior of their souls, hoping to be saved rather by the observation of those Mosaical ceremonies (the date whereof was fully expired at Christ's incarnation)[6] than by the merits of the Savior of the world, without whom all mankind shall perish. And as pitiful it is to see that few of them living in Italy are converted to the Christian religion. For this I understand is the main impediment to their conversion: all their goods are confiscated as soon as they embrace Christianity. And this I heard is the reason: because whereas many of them do raise their fortunes by usury, insomuch that they do sometime not only shear, but also flay many a poor Christian's estate by their griping extortion, it is therefore decreed by the Pope, and other free Princes in whose territories they live, that they shall make a restitution of all their ill gotten goods, and so disclog their souls and consciences, when they are admitted by the holy baptism into the bosom of Christ's Church. Seeing then when their goods are taken from them at their conversion, they are left even naked, and destitute of their means of maintenance, there are fewer Jews converted to Christianity in Italy, than in any country of Christendom. Whereas in Germany, Poland, and other places the Jews that are converted (which doth often happen, as Emanuel Tremellius was converted in Germany) do enjoy their estates as they did before.

But now I will make relation of that which I promised in my treatise of Padua, I mean my discourse with the Jews about their religion. For when as walking in the court of the ghetto, I casually met with a certain learned Jewish Rabbi that spoke good Latin.[7] I insinuated my self after some few terms of compliment into conference with him, and asked him his opinion of

---

[4] **dress their victuals:** cook their food. [5] **Feast of Tabernacles:** Sukkot, a festival celebrated at harvest time. [6] **the date . . . incarnation:** Christian doctrine holds that Jesus fulfilled the Mosaic law so that it no longer needs to be followed. [7] **Jewish Rabbi . . . Latin:** possibly Leon of Modena, according to Adelman's speculation.

Christ, and why he did not receive him for his Messiah; he made me the same answer that the Turk did at Lyons, of whom I have before spoken, that Christ forsooth was a great prophet, and in that respect as highly to be esteemed as any prophet amongst the Jews that ever lived before him; but derogated altogether[8] from his divinity, and would not acknowledge him for the Messiah and Savior of the world, because he came so contemptibly,[9] and not with that pomp and majesty that beseemed the redeemer of mankind. I replied that we Christians do, and will even to the effusion of our vital blood confess him to be the true and only Messiah of the world, seeing he confirmed his doctrine while he was here on earth, with such an innumerable multitude of divine miracles, which did most infallibly testify his divinity; and that they themselves, who are Christ's irreconcilable enemies, could not produce any authority either out of Moses, the Prophets, or any other authentic author to strengthen their opinion concerning the temporal kingdom of the Messiah, seeing it was foretold to be spiritual; and told him, that Christ did as a spiritual king reign over his subjects in conquering their spiritual enemies the flesh, the world, and the devil. Withal I added that the predictions and sacred oracles both of Moses, and all the holy Prophets of God, aimed altogether at Christ as their only mark, in regard he was the full consummation of the law and the Prophets, and I urged a place of Isaiah [17:14] unto him concerning the name *Emmanuel,* and a virgin's conceiving and bearing of a son; and at last descended to the persuasion of him to abandon and renounce his Jewish religion and to undertake the Christian faith, without the which he should be eternally damned. He again replied that we Christians do misinterpret the Prophets, and very perversely wrest them to our own sense, and for his own part he had confidently resolved to live and die in his Jewish faith, hoping to be saved by the observations of Moses's Law. In the end he seemed to be somewhat exasperated against me, because I sharply taxed their superstitious ceremonies. For many of them are such refractory people that they cannot endure to hear any terms of reconciliation to the Church of Christ, in regard that esteem him but for a carpenter's son, and a silly poor wretch that once rode upon an ass, and most unworthy to be the Messiah whom they expect to come with most pompous magnificence and imperial royalty, like a peerless monarch, guarded by many legions of the gallantest worthies, and most eminent personages of the whole world, to conquer not only their old country Judea and all those opulent and flourishing kingdoms, which heretofore belonged to the four ancient monarchies (such is their insupportable pride) but also all the nations generally under the scope of heaven, and make . . . all other Princes whatsoever dwelling in

---

[8] **derogated altogether:** took away. [9] **contemptibly:** in a low and poor manner.

the remotest parts of the habitable world his tributary vassals. Thus hath God justly infatuated their understandings, and given them the spirit of slumber (as Saint Paul spoke out[10] of the prophet Isaiah) eyes that they should not see, and ears that they should not hear unto this day [Romans 11:8]. But to shut up this narration of my conflict with the Jewish rabbi, after there had passed many vehement speeches to and fro betwixt us, it happened that some forty or fifty Jews more flocked about me, and some of them began very insolently to swagger with me, because I durst[11] reprehend their religion. Whereupon fearing least they would have offered me some violence, I withdrew myself by little and little towards the bridge at the entrance into the ghetto, with an intent to flee from them, but by good fortune our noble ambassador Sir Henry Wotton[12] passing under the bridge in his gondola at that very time, espied me somewhat earnestly bickering with them, and so incontinently[13] sent unto me out of his boat one of his principal gentlemen, Master Belford his secretary, who conveyed me safely from the unchristian miscreants, which perhaps would have given me just occasion to forswear any more coming to the ghetto. . . .

But since I have taken occasion to mention some notable particulars of their women, I will insist farther upon that matter, and make relation of their courtesans also, as being a thing incident and very proper to this discourse, especially because the name of a courtesan of Venice is famous over all Christendom. . . .

The woman that professeth this trade is called in the Italian tongue *Cortezana*, which word is derived from the Italian word *cortesia* that signifies courtesy, because these kinds of women are said to receive courtesies of their favorites. . . . As for the number of these Venetian courtesans it is very great. . . . A most ungodly thing without doubt that there should be a toleration of such licentious wantons in so glorious, so potent, so renowned a city. For me thinks that the Venetians should be daily afraid lest their winking at such uncleanness should be an occasion to draw down upon them God's curses and vengeance from heaven, and to consume their city with fire and brimstone, as in times past he did Sodom and Gomorrha. But they not fearing any such thing do grant large dispensation and indulgence unto them, and that for these two causes. First, *ad vitanda maiora mala*.[14] For they think that the chastity of their wives would be the sooner assaulted, and

---

[10] **spoke out**: quoted. [11] **durst**: dared to. [12] **Henry Wotton**: (1568–1639) English ambassador to Venice from 1604 to 1612; he served two other subsequent tours in Venice. If Modena is the rabbi here, he and Wotton knew each other, so it is likely that Coryate is comically exaggerating his danger. [13] **incontinently**: immediately. [14] *ad vitanda maiora mala*: that a greater evil be avoided.

so consequently they should be capricornified[15] (which of all the indignities in the world the Venetian cannot patiently endure), were it not for these places of evacuation. But I marvel how that should be true though these courtesans were utterly rooted out of the city. For the gentlemen do even coop up their wives always within the walls of their houses for fear of these inconveniencies, as much as if there were no courtesans at all in the city. So that you shall very seldom see a Venetian gentleman's wife but either at the solemnization of a great marriage, or at the christening of a Jew, or late in the evening rowing in a gondola. The second cause is for that the revenues which they pay unto the senate for their toleration, do maintain a dozen of their galleys (as many reported unto me in Venice), and so save them a great charge. The consideration of these two things hath moved them to tolerate for the space of these many hundred years these kind of Laides and Thaides,[16] who may be as fitly termed the stales[17] of Christendom as those were heretofore of Greece. For so infinite are the allurements of the famous Calypsos, that the fame of them hath drawn many to Venice from some of the remotest parts of Christendom, to contemplate their beauties, and enjoy their pleasing dalliances. And indeed such is the variety of the delicious objects they minister to their lovers, that they want[18] nothing tending to delight.

[15] **capricornified:** given horns or cuckolded; a cuckold is a man whose wife is unfaithful.
[16] **Laides and Thaides:** probably Lais, a famous Greek courtesan, and Thais, an Athenian courtesan.   [17] **stales:** prostitutes.   [18] **want:** lack.

→ DUDLEY CARLETON

The English Ambassador's Notes   *1612*

Dudley Carleton (1573–1632) attended Oxford, and between receiving his bachelor's and master's degrees traveled abroad for five years, acquiring a number of European languages. He was elected to Parliament in 1604 and became secretary to the earl of Northumberland. The earl was a conspirator in the Gunpowder Plot, a plan to blow up King James and Parliament; when the plot was discovered, Carleton was implicated but was eventually cleared of wrongdoing. In 1610 Carleton began his career as a diplomat, replacing Henry Wotton as ambassador to Venice. He remained there until 1615, at which point he helped negotiate a peace treaty with Spain and served as ambassador to The Hague. He was later a member of the Privy Council, continued to serve in Parliament, and took part in other foreign negotiations. Carleton left an extensive correspondence

Dudley Carleton, Notes, Public Record Office, London, State Papers 99, file 8, ff. 340–44 reprinted in *Venice: A Documentary History*, ed. David Chambers and Brian Pullan (Oxford: Blackwell, 1992), 27–29.

of letters and dispatches, only a small portion of which have been published. The excerpt that follows on the current state of Venetian trade, which had declined considerably by the early seventeenth century, is taken from notes for a report that Carleton probably compiled in 1612. While ambassadors performed necessary political and diplomatic functions in relations between their ruler and the host country, they also served often as spies, transmitting important information back to their own governments. Carleton writes in a comparative, if not competitive, vein, evaluating the respective strengths and weaknesses of Venetian and English trade. While critical of a number of Venetian practices, he also acknowledges that England's emerging participation in the Levant trade is less profitable than the trade handled by the Venetians. English identification and rivalry with its competitor could have prompted in Shakespeare's audience various responses to Antonio, the Venetian merchant, his friends and colleagues, ranging potentially from sympathy to antagonism.

The English Ambassador's Notes

The opinion of many is that this state can not long continue according to the rule of preserving *iisdem artibus*[1] as of getting, because they here change their manners, they are grown factious, vidicative,[2] loose, and unthrifty. Their former course of life was merchandising: which is now quite left and they look to landward buying house and land, furnishing themselves with coach and horses, and giving themselves the good time with more show and gallantry than was wont. . . . Their wont[3] was to send their sons upon gallies[4] into the Levant to accustom them to navigation and to trade. They now send them to travel and to learn more of the gentlemen than the merchant. . . .

In matter of trade the decay is so manifest that all men conclude within 20 years' space here is not one part left of three. The wonted course of bringing Indian merchandise to Alexandria, from thence hither, and from hence transporting it to other parts was the cause of their greatness of trade. These merchandises being now brought about by sea into our parts hath so changed the case that whereas pepper and other spice was wont to be bought here by our men, it is now brought from our ports and here sold.

The Turkey trade which we are now entered into is likewise a hindrance to these merchants and no great profit to our men: for some are of opinion our merchants thrived better when they bought these wares here of these men, partly by reason of the ability of these people above ours in trading,

---

[1] *iisdem artibus:* by the same skills; refers to the art of raising money, mentioned just before this excerpt begins. [2] **vidicative:** vindictive. [3] **wont:** custom or habit. [4] **gallies:** ships with sails and oars commonly used in the Mediterranean.

partly by the advantage these have of exchanging commodities, by reason they furnish silk and velvets, etc., and when they had the trade to themselves they did manage it better. Our men buying all with ready money make the less gain.[5] . . . In trade there is a manifest decay in Venice, and by consequence of shipping. The decay of trade proceeds of the greatness of the imposts[6] which are 6 in the hundreth of all that comes out of the Levant, from whence likewise none but Venetians may bring merchandise which causeth the trade to run so quick in Florence, Genoa, and Marseilles to which places all men that will may bring, the impost is very small.[7] . . . The decay of shipping is manifest to the eye, there having been 80 great vessels of above 1000 ton apiece in this port and trafficking these seas, whereas now not one is to be seen, some being sold, others cast away, most taken by pirates. In which these Signori[8] are much condemned of incuriousness[9] for not providing some sufficient convoy or setting out some vessels purposely against the Corsari[10] which they never took into consideration, as if therein they had lost their judgments, which commonly fail first when a ruin doth follow, and it is much marveled that seeing the world so much changed in the course of trade yet they will not consent to change their strict laws touching the Levant.

[5] **Our men . . . make the less gain:** the Venetians, who used valuable luxury cloths to barter with merchants in Turkey, had the advantage of trading for foreign goods with the exchange of their own equally or more enticing products. The English are at a disadvantage insofar as they don't have desirable goods to exchange, and must rely on purchase by money, leaving them less leverage in bargaining.   [6] **imposts:** customs duty levied on merchandise.   [7] **the impost is very small:** all merchandise brought from the eastern Mediterranean is taxed and only Venetians are allowed to impost in Venice; the other countries allow any merchant to bring his goods there and they charge a smaller customs tax.   [8] **Signori:** possibly referring to the *Signoria*, the body of lords of the Venetian republic.   [9] **incuriousness:** carelessness.   [10] **Corsari:** pirates.

➔ **WILLIAM BEDELL**

Letter to Adam Newton

1608

William Bedell (1571–1642) began his study at Cambridge at the age of twelve, and went on to receive three degrees there; he was ordained as a priest in the Church of England in 1597. He was a scholar of Greek and Latin, but also knew some Aramaic, Arabic, and Hebrew; he apparently was also talented in modern

William Bedell, From Letter to Adam Newton, Dean of Durham, January 1, 1608, reprinted in *Two Biographies of William Bedell,* ed. E. S. Shuckburgh (Cambridge: Cambridge University Press, 1902), 228–9.

languages, as he produced a book of English grammar for some friends in Venice later in life. After leaving the university, he began his career as a preacher; in 1607 he became chaplain to Henry Wotton, ambassador to Venice, where he lived for more than three years. The Venetian republic was at that time in conflict with the pope, which encouraged Wotton's and Bedell's hopes for a Protestant reformation in Italy. While in Venice, Bedell established contacts with local Jewish scholars and continued to study Hebrew. After returning to England, Bedell resumed his work as a minister and also was elected a member of Parliament. He later became bishop of Kilmore and Ardagh in Ireland, where he fought against the extortion of poor Catholics by the English church. Bedell's lifelong suspicion of ecclesiastical profit at the expense of parishioners voices itself in the text that follows, excerpted from a letter written to a colleague in England. While he adopts a common Protestant criticism of what he sees as idolatrous Catholic practices, he nevertheless admires their inculcation of the young in religious rituals and their care in decorating and maintaining the churches. While *The Merchant of Venice* engages with the theological conflict between Judaism and Christianity, it does not directly identify Venice as a Catholic state, and thus avoids explicitly addressing the volatile topic of religious division brought about by the Reformation. However, Venice *was* a Catholic republic, and while it could be viewed as ripe for religious reform, it could also be criticized for current practices. Hence, from Bedell's perspective, Antonio's condemnation of Shylock's allegedly devilish citing of scripture (1.3.89) could be understood ironically; Protestants consistently charged Catholics with misreading scripture, in contrast to the Jews who were seen as more accurate interpreters of the Bible (see Andrew Willet, *Tetrastylon* p. 270). The Venetian community's condemnation of Shylock's stereotypically "Jewish" avarice might be viewed with similar irony, given another Protestant criticism, presented damningly by Bedell here in the mouths of Jews, of the greed and financial corruption of Christianity by the Catholic Church.

Letter to Adam Newton

Thus is the office of preaching wholly in a manner devolved to the Friars, and when they were here, to the Jesuits.[1] Of which last I can say nothing, since 'twas not my ill-fortune to meet them here at my coming. . . . As for the Friars, which I have heard here, their whole intentions seem to be either to delight or to move: as for teaching, they know not what it means. But to hear their strange wresting[2] of the Holy Scripture, to see the fooleries of their idolatry to the little crucifix that stands at their elbow, the antics of their gesture more than player or fencer-like . . . it is (I assure you, Sir)

[1] **Jesuits:** the Jesuits had been expelled by the Venetian government in retaliation for actions taken against the Republic by the pope.   [2] **wresting:** twisting, forced interpretation.

matter of great patience; and for my part I have found myself better satisfied (at least wise less cloyed) with the sermons of the Jews, than with theirs. And in one thing the Jews condemn them, and not undeservedly, as merchants of God's word. For in the middest of their sermons still the preacher makes some pretty occasion or other to fall into the commonplace of moving them to alms; and three, or four with long canes, and a bag at the end skims over the whole auditory;[3] and the people generally not undevout, being taken in a good mood, while the impression is yet new, are not unliberal. Out of this the preacher hath his share: the rest goes to the collectors, or guardians of the fraternity, or other school that heard him.

And because I am fallen into the mention of the people, that you may at once understand the present face of religion here; if ever there were any city to which the epithets would agree which St. Luke give to Athens, . . . this is it.[4] Such a multitude of idolatrous statues, pictures, relics in every corner, not of their churches only, but houses, chambers, shops, yea the very streets, and in the country, the highways and hedges swarm with them. The sea itself is not free; they are in the ships, boats, and watermarks. And as for their slavery and subjection to them, it is such, as that of paganism came not to the half of it. Whereof to give you such a taste as may be also for some cause of it. No sooner do their children almost creep out of their cradles, but they are taught to be Idolators. They have certain childish processions wherein are carried about certain puppets, made for their Lady,[5] and some boy that is better clerk[6] than his fellows goes before them with the words of the Popish Litany; where the rest of the fry[7] following make up the choir. A great tyrant is custom[8] and a great advantage hath that discipline which is sucked in with the mother's milk. But to convey superstition into the minds of that tender age under the form of sport, and play, which it[9] esteemeth more than meat and drink, is a deeper point of policy, and such as wise men perhaps would profitably suck somewhat out of it for imitation to a right end.

But one thing certainly they go beyond us in: and that is their liberality and cost in the solemn setting forth of their service and adorning their churches, and especially at their feast-days. Wherein if they pass measure also, as possibly they do, yet is that extreme less exceptionable, and always more curable, than our beggary, the scorn of our religion. Not only popular conceits,[10] but the most part of men of whatsoever quality are led much more by shows than substance. And what a disproportion it is, to come from

---

[3] **auditory:** audience.   [4] **this is it:** see Acts 17:16, in which Athens is described as "wholly given to idolatry."   [5] **their Lady:** the Virgin Mary.   [6] **clerk:** scholar or clergyman.   [7] **fry:** children.
[8] **custom:** habit.   [9] **it:** refers to "tender age."   [10] **popular conceits:** public imagination, minds of the common people.

ours so unhonestly kept[11] (for in buildings we generally go beyond them) to the glittering churches and monasteries of Italy, you may easily discern. Truly, sir, I have heard some wise men account this as no small cause of the perversion[12] of so many of our young gentlemen that come into these parts.

[11] so unhonestly kept: in an unpleasant, filthy manner.   [12] perversion: turning to Catholicism.

✦ A Discovery of the Great Subtlety and Wonderful Wisdom of the Italians   *1591*

This anonymous tract combines a powerful anti-Catholicism with anti-Italian sentiment, demonstrating that not all things Italian garnered English admiration. The author posits a "subtle" (i.e., deceitful) and greedy national character, a result, he claims, of the Italian climate and propensity for trade, which aims to dominate and profit from surrounding countries. While he presents these qualities in national or racial terms, he ultimately imagines these flaws as finding their source in religion; here Italy, the home of the pope, embodies Catholicism and all the author dislikes about it. Common Protestant characterizations of Catholics as hypocritical and materialistic emerge here along with allegations of manipulation of doctrine for political purposes, especially to the end of dominating and profiting from other countries. *A Discovery* makes the radical claim that the Catholic Church seeks a kind of global economic monopoly, a view that many in early modern England probably did not share. Nevertheless, given that such opinion did exist, would it influence members of the play's audience to see the Catholics of Shakespeare's Venice as desiring dominion and the acquisition of wealth?

CHAPTER 1

A Description of Italy and the Causes of the Subtlety of That Nation

That which produceth such effects in Italy is the moderate temperature of the climate, situated in a subtle air near unto the sea everywhere, without any excess heat or cold. And beside, another cause is the trading and great dealing that the Italians have with the people of Asia, of Africa, and Europe, as also with Flanders, or a great part of them with whom they haunt and live. By reason whereof, besides that they are of themselves very witty and subtle headed, all cunning flights, crafty conveyances, and deceitful coz-

---

A Discovery of the Great Subtlety and Wonderful Wisdom of the Italians, Whereby They Bear Sway over the Most Part of Christendom, and Cunningly Behave Themselves to Fetch the Quintessence out of the People's Purses (London, 1591), 1–2, 19, 23, 31, 35.

enages[1] are so proper and common to them, whereby they can fetch under[2] other people, and are so cunning to finger from them their money, and can, moreover, so closely cover their actions that of a thousand hardly one could ever come within them to perceive their juggling.[3] For as any deceit or cozenage finely handled is not permitted but of those which know it, and look very near unto it, deceiving those which have their eye but on the natural and external show. So there are none but those which curiously seek out the beginning, the progress and advancements of the Roman and Italian government, and the means whereby they have drawn money from other nations of the earth since the time of Romulus to this present day, who can find out their fetches and shifts,[4] or discover the masks where with they are disguised to advance and enrich themselves by the overthrow and pillage of others.

CHAPTER 15

That They Care Not at Rome for Any Diversity of Religions, So They Tend Only to Maintain Their Domination

This is most manifest, for if any man invent any new form of religion never known before in the world, appareling himself after a strange fashion never seen before, using gestures altogether ridiculous and foolish, living after a most austere, too cruel and brutish manner (as do the Capuchians, Sicilians, and such like foolish orders of friars), all shall be approved and received by the Italians with great plaudities,[5] so that such religions will serve them for a wall and defense for their kingdom and gain. But if anyone appear or come near them that dare speak against such abuses, and touch them to the quick[6] a little, leaning upon the pillar of that doctrine which has been given from the terrestrial paradise,[7] they will shake a heaven and earth, and remove all a world, to stop his mouth and kill him with great exclamations, that he went about to sow new doctrines, and begin some new sect of religion. Whereby we may closely see, that under this cloak of religion, they do but aspire to be rulers, and to finger money from other peoples and nations: and that all the wars and civil broils which have so long troubled Germany, Switzerland, all Flanders and France have been broached and begun by them for such matters, albeit, those who had the conduct thereof, enterprised them for the zeal they bore to their own religion.[8]

---

[1] cozenages: frauds. [2] fetch under: cheat. [3] juggling: practice of deception. [4] fetches and shifts: tricks and stratagems. [5] plaudities: applause or approval. [6] quick: the central or most important part. [7] terrestrial paradise: Eden, i.e., since the time of Adam and Eve. [8] those who had the conduct thereof . . . religion: although the Italians instigated the wars for profit, the contenders fought for the sake of the cause.

CHAPTER 18

The Counsel of Rome Setteth Kings and Christian Princes Together by the Ears, and the Way How They Discover All Their Counsels and Enterprises

The Roman counsel hath got some more advantage by this their policy and device, since the creation of their Pope: for pretending to bring all kings, peoples and princes of Christianity, Catholics, as well spiritual as temporal unto their subjection, under cloak of their bishops, they have spun the thread of all the wars which have been between the emperors, and other Christian princes, so to bring them on their knees, and to undermine them that way, that they must still of force have recourse to them, as to a place of refuge and protection, the only mean to pull down their haughty stomachs, and to set the Italian far above them. To pass over this plank well, they have given the name of cardinals to the priests of Rome, granting the sons and brethren of kings and Christian princes to be honored with the same title and estate, suffering them also to remain with their train[9] amongst them. To the end that by these their creatures they might handle and turn the other at their devotion, make them arm their people and march against their neighbors at their pleasure, and disarm again and retire their forces when they should serve God. Considering also that by such persons, the Counsel's wills and determinations of kings and Christian princes to the which they are called, should thereby be revealed and manifested to them.[10] The quick spirited Venetians, having long time since discovered this policy, being a people of Italy very subtle, and well advised, would never permit that any ecclesiastical person should be admitted into their counsel, because they had all taken the oath of the Pope of Rome.

CHAPTER 22

The Causes Why the Money That Is Transported to Rome Is Called by the Name of Quintessence

Now we have sufficiently spoken of the dominations of the Romans, we will pass to the next point touching the extraction of the *quintessence* of purses, where ye must note that I take purses for the gold and silver that is put in

---

[9] **train:** entourage.   [10] **the Counsel's wills . . . manifested to them:** the sense seems to be that the cardinals, after being called to consult with their rulers, will then reveal all to Rome.

them, speaking by a figurative speech, for I borrow these terms by a similitude. For even as in all liquors, be it of wines, oils, plants, sulphurs, alums,[11] antimony,[12] and other minerals, they who are skilled to draw out the tinctures or properties, and can separate them from the terrestrial and elemental body . . . [t]hese have the sweet fruition of the finest and most precious part of the bodies and natural substances, whereof the more celestial and spiritual part is called *quintessence*. In like manner the Romans having learned by incomparable skill and artistry to draw unto themselves the most noble portable, the most desirable and fairest coin of all Christendom (leaving the grosser and more terrestrial sort of baser moneys to the kings, princes, and people of Christendom for their usage), appropriating to themselves the more spiritual and celestial part, which they can tell how to separate from the temporal and earthly, are very well said to draw the *quintessence* out of their purses. . . .

He that would be employed for Rome, she had to command[13] all over the world, and had the bestowing of bishops, abbeys, priories, and parsonages, to advance and make happy all those that embraced their part. Now of a great number, few or none stand out with them, but are overcome with such temptations, and if there be any that will not yield to them, we see by and by that the bishops and their officials have them in chase, and do their uttermost to inflame the magistrate against them to put them to death, as did the priests of the old law, against Elijah,[14] Jeremiah,[15] and the rest of the prophets, and their successors, against Christ. But if the magistrates chance to wink at[16] them, the Counsel of Rome has caused ecclesiastical counselors to be admitted into all courts of parliament to advise, stir up and inflame all . . . other counselors to make open war with fire and sword against such people, not without sore charges to the judges, to look unto them, menacing their negligence herein with double punishment.[17] By reason whereof, it shall be hard for them to escape their hands. Whereby appeareth how great the carnal wisdom of this nation is, to maintain and hold fast their revenues and domination.

---

[11] **alums:** a double sulphate of aluminum and potassium.   [12] **antimony:** a brittle metallic element.
[13] **she had to command:** Rome was able to control the world.   [14] **as did the priests of the old law, against Elijah:** see 1 Kings 17 and 18.   [15] **Jeremiah:** see Jeremiah 20:1–18.   [16] **wink at:** ignore.   [17] **charges to the judges . . . double punishment:** giving instructions to attend to those refusing to assist Rome and threatening the judges with punishment if they neglect to do so.

→ ROBERT WILSON

From The Three Ladies of London   *1584*

Robert Wilson (d. 1600), an actor and playwright, was a member of Queen Elizabeth's acting company and, later, the Lord Chamberlain's Men, the group to which Shakespeare also belonged. He had a reputation as an actor for improvisation and comic performance. His plays are modeled on the form of drama popular in the medieval period: the morality play, which represents allegorical figures of virtue and vice, rather than interaction between human characters. The earliest of Wilson's extant plays is *A Right Excellent and Famous Comedy Called the Three Ladies of London,* which represents the corrupting effects of money, personified by Lady Lucar. The play, which Shakespeare almost certainly knew, stages a conflict between an Italian debtor and a Jewish creditor. However, in striking contrast to *The Merchant of Venice,* the Jewish moneylender shows himself merciful in forgiving a loan rather than forcing the debtor into apostasy, while the Italian borrower is a fraudulent, greedy opportunist. This unflattering picture of the Italian merchant, representative of attitudes in the earlier part of the sixteenth century, draws on negative associations with merchants in general, and resident Italian merchants who were seen as competing with local businessmen. (See the proclamation on aliens that follows this excerpt and the discussion of merchants in the introduction to Chapter 2, Finance.)

Enter Mercadorus[1] the Merchant and Gerontus[2] the Jew

GERONTUS: But senior Mercadorus tell me, did ye serve me well or no?
That having gotten my money would seem the country to forgo:[3]
You know I lent you two thousand ducats for three months space,
And ere[4] the time came you got another thousand by flattery and
        your smooth face.
So when the time came that I should have rescued my money,
You were not to be found but was fled out of the country:
Surely if we that be Jews should deal so one with another,
We should not be trusted again of our own brother:
But many of you Christians make no conscience to falsify your faith
        and break your day.

---

[1] *Mercadorus:* the Latin *mercator* means "merchant."   [2] *Gerontus:* could either signify "old age" or may refer to Gernutus, a Jewish character in a ballad, "Gernutus, the Jew of Venice," who pledges a pound of his own flesh against the loss of property he has insured.   [3] **forgo:** leave. [4] **ere:** before.

Robert Wilson, *A Right Excellent and Famous Comedy Called the Three Ladies of London* (London, 1584), D3v–D4r, E3r, E4v–F1v.

I should have been paid at three months' end, and now it is two year
    you have been away.

Well I am glad you be come again to Turkey, now I trust I shall
    receive the interest of you so well as the principal.

MERCADORUS: A good a master Geronto pra hartly bear a me[5] a little
    while,

And me shall pay ye all without any deceit or guile:

Me have a much business for by pretty knacks to send to England,

Good a sir, bear a me four or five days, me'll dispatch your money out
    of hand.

GERONTUS: Senior Mercadore, I know no reason why, because you
    have dealt with me so ill,

Sure you did it not for need, but of set purpose and will:

And I tell ye to bear with ye four or five days goes sore against my mind,

Lest you should steal away and forget to leave my money behind.

MERCADORUS: Pra hartly do tink a no such ting[6] my good friend a me,

Be me trot and fayt[7] me'll pay you all every penny.

GERONTUS: Well I'll take your faith and troth once more, I'll trust to
    your honesty

In hope that for my long tarrying you will deal well with me:

Tell me what ware you would buy for England, such necessaries as
    they lack?

MERCADORUS: O no lack, some pretty fine toy or some fantastic new
    knack,

For da gentelwomans in England buy much tings for fantasy:

Your pleasure a me sir, what me mean a dare buy.[8]

GERONTUS: I understand you sir, but keep touch with me, and I'll bring
    you to great store,

Such as I perceive you came to this country for:

Wherein I perceive consisteth that country gentlewoman's joys.

Besides I have diamonds, rubies, emeralds, sapphires, smaradines,[9]
    opals, onacles,[10] jacinths,[11] agates, turquoise, and almost all kind of
    precious stones:

And many more fit things to suck away money from such green-
    headed wantons.[12]

---

[5] **pra hartly bear a me:** Mercadorus speaks in an "Italian" dialect; "pray heartily, bear with me . . ."; I have relied in part on the notes from Mithal's edition of the play in my glosses.
[6] **tink no such ting:** think no such thing.   [7] **Be me trot and fayt:** by my troth and faith, truly.
[8] **Your pleasure me a sir, what me mean a dare buy:** please tell me what I should buy to sell in England.   [9] smaradines: smaragdite, a light green stone.   [10] **onacles:** onyx.   [11] **jacinths:** reddish-orange gem.   [12] **green-headed wantons:** immature, spoiled persons.

MERCADORUS: Fatta[13] my good friend me tank you most hartly alway,
me shall a content your debt within dis two or three day.

GERONTUS: Well look you do keep your promise, and another time you
shall command me:

Come, go we home where our commodities you may at pleasure
see. . . .   [*Exeunt*]

*Enter Mercadorus reading a letter to himself, and let Gerontus the Jew
follow him, and speak as followeth:*

GERONTUS: Senior Mercadore, why do you not pay me? Think you I
will be mocked in this sort?

This is three times you have flouted[14] me, it seems you make thereat a
sport.

Truly pay me my money, and that even now presently,

Or by mighty Mohammad I swear, I will forthwith arrest ye.

MERCADORUS: Ha pray a bear wit me three or four days, me have
much business in hand:

Me be troubled with letters you see here, dat[15] comes from England.

GERONTUS: Tush this is not my matter, I have nothing therewith to do,

Pay me my money or I'll make you, before to your lodging you go.

I have officers stand watching for you, so that you cannot pass by,

Therefore you were best to pay me, or else in prison you shall lie.

MERCADORUS: Arrest me dou skal knave,[16] marry do and if thou dare,

Me will not pay de one penny, arrest me, do, me do not care.

Me will be a Turk, me came hedar[17] for dat cause,[18]

Darefore me care not for de so much as two straws.

GERONTUS: This is but your words, because you would defeat me,

I cannot think you will forsake your faith so lightly.

But seeing you drive me to doubt, I'll try your honesty:

Therefore be sure of this, I'll go about it presently.   *Exit.*

MERCADORUS: Marry farewell and be hanged, sitten,[19] scald[20] drunken
Jew.

I warrant you me shall be able very well to pay you.

My Lady Lucar have sent me here this letter,

Praying me to cozen[21] de Jew for love a her.

Darefore me'll go to get a some Turk's apparel,

Dat me may cozen da Jew, and end dis quarrel.   *Exit.*

---

[13] **Fatta:** faith, as in "by my faith."   [14] **flouted:** expressed contempt for.   [15] **dat:** that, "d" often
stands for "th."   [16] **dou skal knave:** you contemptible rascal.   [17] **hedar:** hither, here.   [18] **Me
will be a Turk, . . . cause:** converting to Islam would release him from his debts; see p. 157.
[19] **sitten:** covered with excrement.   [20] **scald:** scabby or scaly.   [21] **cozen:** cheat.

Enter the Judge of Turkey, with Gerontus and Mercadorus

JUDGE: Sir Gerontus, because you are the plaintiff, you first your mind shall say,
Declare the cause you did arrest this merchant yesterday.
GERONTUS: Then learned Judge attend. This Mercadorus whom you see in place,
Did borrow two thousand ducats of me, but for a five weeks space.
Then Sir, before the day came, by his flattery he obtained one thousand more,
And promised me at two months end I should receive my store:
But before the time expired, he was closely fled away,
So that I never heard of him at least this two years' day:
Till at last I met with him, and my money did demand,
Who swore to me at five days' end, he would pay me out of hand.
The five days came, and three days more, then one day he requested,
I perceiving that he flouted me, have got him thus arrested:
And now he comes in Turkish weeds[22] to defeat me of my money,
But I trow[23] he will not forsake his faith, I deem he hath more honesty.
JUDGE: Sir Gerontus you know, if any man forsake his faith, king, country, and become a Mohammed,
All debts are paid, 'tis the law of our realm, and you may not gainsay it.
GERONTUS: Most true (reverent Judge) we may not, nor I will not against our laws grudge.
JUDGE: Senior Mercadorus is this true that Gerontus doth tell?
MERCADORUS: My lord Judge, de matter, and de circumstance be true me know well.
But me will be a Turk, and for dat cause me came here.
JUDGE: Then it is but a folly to make many words. Senior Mercadorus draw near.
Lay your hand upon this book, and say after me.
MERCADORUS: With a good will my lord Judge, me be all ready.
GERONTUS: Not for any devotion, but for Lucar's[24] sake of money.
JUDGE AND MERCADORUS: Say, "I, Mercadorus, do utterly renounce before all the world, my duty to my Prince, my honor to my parents, and my good will to my country: Furthermore I protest and swear to be true to this country during life, and thereupon I forsake my Christian faith."

---

[22] **weeds:** clothes.   [23] **trow:** trust.   [24] **Lucar:** wealth, also a character in this play.

GERONTUS: Stay there[25] most puissant[26] Judge. Senior Mercadorus,
    consider what you do,
  Pay me the principal, as for the interest, I forgive it you:
  And yet the interest is allowed amongst you Christians, as well as in
    Turkey,
  Therefore respect your faith, and do not seem to deceive me.

MERCADORUS: No point da interest, no point da principal.

GERONTUS: Then pay me the one half, if you will not pay me all.

MERCADORUS: No point da half, no point denere,[27] me will be a Turk I say,
  Me be weary of my Christian religion, and for dat me come away.

GERONTUS: Well, seeing it is so, I would be loath to hear the people
    say, it was long of me
  Thou forsakest thy faith, wherefore I forgive thee frank and free:
  Protesting before the Judge, and all the world, never to demand
    penny nor halfpenny.

MERCADORUS: O Sir Gerontus, me take a your proffer, and tank you
    most heartily.

JUDGE: But senior Mercadorus, I trow ye will be a Turk for all this.

MERCADORUS: Senior no, not for all da good in the world, me forsake
    my Christ.

JUDGE: Why then it is as Sir Gerontus said, you did more for the
    greediness of money,
  Than for any zeal or good will you bear Turkey.

MERCADORUS: Oh Sir, you make a great offense,
  You must not judge a my conscience.

JUDGE: One may judge and speak truth, as appears by this,
  Jews seek to excel in Christianity, and the Christians in Jewishness.

                                              Exit.

MERCADORUS: Vell vell, but me tank you Sir Gerontus with all my very
    heart.

GERONTUS: Much good it may do you sir, I repent it not for my part.
  But yet I would not have this bolden you to serve another so,
  Seek to pay, and keep day with men, so a good name you will go. *Exit.*

MERCADORUS: You say vel Sir: it does me good, dat me have cozened
    de Jew,
  Faith I would my Lady Lucar de whole matter now knew.
  What is dat me will not do for her sweet sake.
  But now me will provide my journey toward England to take.

---

[25] **Stay there:** wait a moment.   [26] **puissant:** powerful.   [27] **denere:** denier, a small copper coin.

Me be a Turk, no, it will make my Lady Lucar to smile,
When she knows how me did da scall[28] Jew beguile.   *Exit.*

[28] **scall:** scabby or scaly.

→ **QUEEN ELIZABETH I**

Proclamation Ordering Peace Kept in London   *1559*

Tensions between strangers and the native population in sixteenth-century
England grew out of economic competition over merchandise and labor. While
aliens lived and worked under extremely restrictive laws, certain groups (such as
the Italian Lombards in the early part of the sixteenth century) were granted
special rights from time to time, which angered local craftsmen and merchants.
Riots against aliens broke out regularly in London, particularly toward the end
of Elizabeth's reign. She sought to quell them by a number of means, including
the following proclamation, a decree issued by the by the ruler which attempted
to legislate without consulting Parliament. This proclamation seeks to discourage individuals from settling disputes privately by means of "reproaches of
words or . . . quarrels" or worse, rather than allowing the courts to handle such
quarrels. In order to maintain the peace, the state had to insure that official
mechanisms for redressing wrongs would be available to natives and strangers
alike. Although early modern England had a reputation for xenophobia (a reputation that Fynes Moryson tries to dispel in his description of England, p. 162),
the state was clearly concerned that its officers hear and judge cases fairly and
impartially to ensure justice and safety for resident aliens. Does Portia's handling of the dispute between Antonio and Shylock seem to reflect the desire for
equity under the law or does her treatment reflect resistance to this desire?

The Queen's majesty commandeth all manner her subjects, of what degree
soever they be, to keep the peace as they be bound, and specially towards all
manner of persons of strange nations within her majesty's city of London or
elsewhere, without reproaches of words or like quarrels, and to remit the avenge
of all quarrels past of late[1] in the same city to the ordinary justicers. And the like
also her majesty commandeth to all strangers born, to be observed on their part.

    And for the satisfaction of all sorts, both English and strangers, her
majesty most straightly[2] commandeth her Mayor of London, and all other

---

[1] **past of late:** occurring recently.   [2] **straightly:** strictly.

Queen Elizabeth I, Proclamation 462, "Ordering Peace Kept in London," Hampton Court,
August 13, 1559, reprinted in *Tudor Royal Proclamations,* ed. Paul L. Hughes and James F. Larkin
(New Haven: Yale University Press, 1964–69), 2:134.

justicers, as they will answer at their perils, duly to see, with indifference and without any partiality, the first occasioners next hereafter of quarrels and frays to be severely punished, to the example of others.

Like as it is presently ordered by her majesty that the whole circumstance of certain frays in London betwixt her subjects and certain strangers shall be duly examined and tried, and according to the laws of the realm judged and determined. For this is her highness' determination, that no partial favor be showed to English or stranger, but that every of them shall live in the safety and protection of her laws.

→ SIR EDWARD COKE

From The Reports   *1680*

Edward Coke (1552–1634) studied at Cambridge and pursued his education as a lawyer at the Inner Temple (one of the oldest inns of court, a collection of law schools situated outside of London). He began practicing law in 1578. He is the author of collection of legal reports that have the distinction of being called *The Reports;* while they were being published, no other such collection appeared. Law reports served to record the details of common law cases for both practitioners and students of the law in early modern England. While official court documents contained basic information on the outcome of cases, reports were written by individuals who attended trials in order to record the legal strategies and arguments deployed by the lawyers, as well as indicate the judge's reasoning in the final decision. Coke covers more material than previous reporters, not only recording the arguments and judgment of a particular case, but also including earlier cases, presenting their arguments and explaining their judgments. His practice of offering a short treatise on the issue in question served to educate students on the structure and functioning of the law; however, he often blurs the line between his own opinion and the facts of the case. Calvin's case deals with the status of people born in Scotland after its union with England at the accession of James I (formerly James VI of Scotland); were they to be considered aliens in England or natives? The judges of this case, over which Coke presided as chief justice of the court of common pleas, ruled that such persons should not be considered aliens (*Dictionary of National Biography* [*DNB*], s.v. Coke). The excerpt that follows provides an explanation of the legal status and restrictions on aliens in England in this period. Aliens on good terms with the state were accorded certain legal protections with regard to property rights and access to the courts; sometimes they could become denizens (something like a resident alien) or even naturalized citizens and acquire greater rights. Friendly

Sir Edward Coke, *The Reports* (London, 1680); 2nd edition in English, Part 7, 443.

aliens, however, could become enemies; Coke also describes a category of perpetual enemies, which includes Jews, to whom no legal protection would be accorded. Here we see Coke's opinion distorting the law, insofar as Jews *were* given legal rights in medieval England; as the author himself noted in *The Fourth Part of the Institutes,* the Jews were provided with their own court to expedite financial litigation. In a dispute litigated in 1596, a witness describes, in the course of his testimony, a Passover service in his employer's home; not only are the Jewish litigants not persecuted for heresy, the court, "being moved with the losses and troubles which the poor Strangers indured, persuaded [the opposing party] . . . to deal charitably" with them (Baron 15: 127). Nevertheless, Coke's view of Jews as enemies could have carried a certain measure of cultural acceptance. In *The Merchant of Venice,* although Shylock appears to be considered an alien friend in Venice, since he is ostensibly protected by the law, certain characters and comments do reflect Coke's view. At a tense moment in their negotiations over the terms of the loan, Antonio suggests that he does view Shylock as an enemy when the merchant challenges the Jew to make the loan interest bearing (1.3.123–28); the medieval church permitted interest to be charged on loans to infidels because they were considered perpetual enemies (Nerlich 240, n. 116; Nelson 4–8). Shylock's status an alien in Venice is somewhat unclear: on what grounds is he considered an alien — because he is Jewish? an immigrant? And while the play's representation of Venetian law suggests that it offers protection for aliens, the law in the play also contains a special proviso that allows Portia to get the better of Shylock precisely because he is an alien (4.1.342 ff.)

From *The Reports*

CALVIN'S CASE (1609)

Every alien is either a friend that is in league, etc. or an enemy that is in open war, etc. Every alien enemy is either *pro tempore,* temporary for a time, or *perpetuus,* perpetual, or *specialiter permissus,* permitted especially. Every subject is either *natus,* born, or *datus,* given or made: and of these briefly in their order. An alien friend, as at this time, a German, a Frenchman, a Spaniard, etc. (all the Kings and Princes in Christendom being now in league with our Sovereign . . .) may by the common law have, acquire, and get within this realm, by gift, trade, or other, lawful means, any treasure, or goods personal whatsoever, as well as an Englishman, and may maintain any action[1] for the same. But lands within this realm, or houses (but for their necessary habitation only) alien friends cannot acquire, or get, nor maintain any action real or personal, for any land or house unless the house be for

---

[1] **maintain an action:** conduct a lawsuit.

their necessary habitation. For if they should be disabled to acquire and maintain these things, it were in effect to deny unto them trade and traffic, which is the life of every island. But if this alien become an enemy (as alien friends may) then he is utterly disabled to maintain any action, or get anything within this realm. And this is to be understood of a temporary alien, that being an enemy may be a friend, or being a friend may be an enemy. But a perpetual enemy (though there be no wars by fire and sword between them) cannot maintain any action, or get any thing within this Realm. All infidels are in law . . . perpetual enemies (for the law presumes not that they will be converted, that being . . . a remote possibility) for between them as with the devils, whose subjects they be, and Christian, there is perpetual hostility, and can be no peace; for as the Apostle sayeth, 2 Corinthians 6:15 *Que autem conventio Christi ad Belial, aut que pars fideli cum infideli,*[2] and the Law sayeth, *Judeo Christianum nullum serviat mancipium, nefas enim est quem Christus redemit blasphemum Christi in servitutis vinculis detinere.*[3] . . . *Infideles sunt Christi et Christianorum inimici.*[4] And herewith agreeth the Book in 12 *Hen.* 8 fol. 4. where it is holden that a pagan cannot have or maintain any action at all.

[2] *Que . . . infideli:* "What accord hath Christ with Belial [the Devil, or generally, a devil]? Or what part hath he that believeth with an infidel?" [3] *Judeo . . . detinere:* "No Christian slave should serve a Jew, it is certainly a sin for a blasphemer of Jesus to detain in chains of servitude one whom Jesus redeemed." [4] *Infideles . . . inimici:* "Infidels are the enemies of Jesus and of Christians."

→ FYNES MORYSON

From An Itinerary   1617

Fynes Moryson (1566–1630) received a bachelor's and a master's degree from Cambridge University; although he studied civil law, travel was his true passion. In 1591 he embarked on a four-year trip through Europe, meeting along the way diverse notables, including the Catholic theologians William Allen and Robert Bellarmine in Rome and a leader of the Protestant Reformation, Theodore Beza, in Geneva. He returned to England briefly in 1595 and departed again that year for a two-year sojourn through the Middle East. Then Moryson went to Ireland, where his brother was a colonial governor, and served as chief secretary of the Lord Deputy, the head of the English colonial government. In 1606 Moryson began writing an account of his travels, which he planned to arrange

Fynes Moryson, *An Itinerary* (1617), Book 4, part 3 (London, 1617), 177–78, 232, 235. Unpublished chapters reprinted in *Shakespeare's Europe,* ed. Charles Hughes, 2nd ed. (New York: Benjamin Blom, 1967), Book 3, Ch. I, 290, 408–11, 415–16, 423–24, 463, 473, 487.

in five parts. He published the first three parts as *An Itinerary* in 1617: the first section deals with his journeys through Europe, Scotland, and Ireland; the second with Tyrone's rebellion in Ireland; and the third with the customs and geography of various countries. The fourth part covers similar material, but was not published in Moryson's lifetime. The excerpts below, taken from both the *Itinerary* and the unpublished fourth section, provide an array of national stereotypes, including one for England, as formulated by an Englishman. When we look at *The Merchant of Venice* through the lens of Moryson's work, it is interesting to compare his accounts of national "characters" with representations of different nationalities in the play. When we see his views diverging from those expressed in the play, how does this cause us to reevaluate the play's representation of national others?

Caesar reports that the old Britans were appareled in skins, and wore long hair, with the beard all shaven, but the upper lip. Now the English in their apparel are become more light[1] than the lightest French, and more sumptuous than the proudest Persians. More light I say than the French, because with singular inconstancy they have in this one age worn out all the fashions of France and all the nations of Europe, and tired their own inventions, which are no less busy in finding out new and ridiculous fashions, than in scraping up money for such idle expenses. Yea, the tailors and shopkeepers daily invent fantastical fashions for hats, and like new fashions and names for stuffs.[2] Some may think that I play the poet, in relating wonderful but incredible things, but men of experience know that I write with historical truth. That the English by God's goodness abounding at home with great variety of things to be worn, are not only not content therewith, and not only seek new garments from the furthest East, but are besides so light and vain, as they suffer themselves to be abused by the English merchants, who nourishing this general folly of their countrymen, to their own gain, daily in foreign parts cause such new colors and stuff to be made, as their masters send painted out[3] of England to them, teaching strangers to serve our lightness with such inventions as themselves never knew before.

PALATINE

The Count Palatine of the Rhine, by old institution is chief among the temporal Electors,[4] and is of the same family, of which the Dukes of Bavaria descend. The pedigree of them both, is derived from the Emperor Charles the Great. . . .

---

[1] **light:** frivolous. [2] **stuffs:** materials. [3] **send painted out:** send out designs or patterns.
[4] **Elector:** a prince of Germany, entitled to take part in the election of the Emperor.

The xxvii. Chapter treateth of Fraunce and of our pro
uinces the whyche be vnder Fraunce and of
the naturall dysposicyon of the peo=
ple, and of ther money and
of theyr
speche.

I am a frenche man lusty and stout
My rayment is iagged & kut round about
I am ful of new inuencyons
And dayly I do make new toyes and fashions
Al nacions of me example do take
Whan any garment they go about to make.
R.i.

FIGURE 8 *A Frenchman, from Andrew Boorde,* The First Book of the Introduction
of Knowledge *(1542). The accompanying verse describes the Frenchman's clothes as
"jagged and cut round about [as decoration] / I am full of new inventions / And daily
I do make new toys and fashions. / All nations of me example do take / When any garment they go about to make."*

⟨The fyꝛst chapter treateth of the naturall dyſpoſicion of an Engliſh man, and of the noble realme of England, ⁊ of the money that there is vſed.

⟨I am an Engliſh man, and naked I ſtand here
Muſyng in my mynde, what rayment I ſhal were
For now I wyll were thys and now I wyl were that
Now I wyl were I cannot tel what
All new faſhyons be pleſaunt to me
I wyl haue them, whether I thꝛyue or thee
Now I am a fryſker, all men doth on me looke
What ſhould I do, but ſet cocks on the hoope
what do I care, yf all the woꝛlde me fayle
I wyll get a garment, ſhal reche to my tayle
Than I am a minion, foꝛ I were the new gyſe

⟨The

FIGURE 9 *An Englishman, from Andrew Boorde,* The First Book of the Introduction of Knowledge *(1542). The accompanying verse reads, in part, "I am an English man, and naked I stand here / Musing in my mind what raiment [clothing] I shall wear / For now I will wear this and now I will wear that / Now I will wear I cannot tell what. / All new fashions be pleasant to me."*

Frederick the Fourth, Palatine and Elector, being under age, had Duke John Casimire, his father's brother, for his tutor. . . . [H]e brought up his nephew wisely and religiously, appointing him his diet[5] apart with his teachers and the Steward of his court, to whose table one professor of the university was daily invited, who had charge to propound a question to the Prince, out of the histories, and the controversies of religion. And the Prince did not presently make answer, except it were in a common subject, but asked time to consider of it, and consulting apart with his teachers, after some half hour returned to give his answer. Thus by daily practice the chief accidents of histories and controversies of religion were made familiar to him. . . . The said Elector Frederick the Fourth, being a pupil, was after the foresaid manner brought up in the reformed religion, according to the doctrine of Calvin. . . . His court was not great, nor any way comparable to that of the Elector of Saxony. For he had scarce thirty gentlemen to attend him, and to them he gave no more than some twenty-five guldens for stipend, which they spent upon their servants that attended them and kept their horses. And he had no more than eight yeomen for the guard of his body. Wine was sparingly drawn, and all expenses made with great frugality. But the fame of this Elector's wisdom and affability made him much esteemed of strangers. And while he conversed with his citizens, often coming to the public place for exercise of the piece[6] and crossbow, and being easy of access, yet carried himself like a grave and noble Prince, he became dear to his subjects.

NATURE AND MANNERS [OF THE GERMANS]

All the Germans have one national vice of drunkenness in such excess (especially the Saxons), as it stains all their national virtues, and makes them often offensive to friends and much more to strangers. But it is a great reproach for any woman to be drunken or to drink in any the least excess. . . . They are by nature placable,[7] and far from malice or treason to their enemies. When they dispute they neither have nor need any moderator, but coldly[8] urge their arguments, and are soon satisfied with the answer. And when they fight they neither have nor need any to part them, but themselves will gently take up the quarrel. They chide rudely more than they fight, for general all, but especially the Saxons, and above all the coachmen and common people are rude in behavior and words; they will not stay a minute in the inn nor by the way, upon any occasion for a companion in

---

[5] diet: meals.   [6] piece: another term for crossbow.   [7] placable: agreeable.   [8] coldly: calmly.

the coach, and when they are heated with drink, they are apt to give rude, yea, reproachful words, especially to strangers (whose best course is to pass them over, as not understood). But even among themselves this rude speech and drunkenness, and especially the small danger in fighting (where it is a villainy to thrust, and a small cut or slash is the worst can befall them) causes many quarrels. . . .

NATURE AND MANNERS [OF THE ITALIANS]

Thus the Italians being by nature false dissemblers in their own actions, are also most distrustful of others with whom they deal or converse, thinking that no man is so foolish to deal plainly, and to mean as he speaks, for which cause the Pope and the Princes of Italy never take Italians for the guard of their bodies, but only Swiss or Germans which nations they repute faithfully minded, [and] free from treasons. . . .

For fleshy lusts, the very Turks (whose carnal religion alloweth them) are not so much transported therewith, as the Italians are (in their restraint of civil laws and the dreadful law of God). A man of these Northerly parts can hardly believe without the testimony of his own eyes and ears, how chastity is laughed at among them, and hissed out of all good company, or how desperate adventures they will make to achieve disordinate desire in these kinds. . . . In Italy marriage is indeed a yoke, and that not easy, but so grievous, as brethren no where better agreeing, yet contend among themselves to be free from marriage, and he that of free will or by persuasion will take a wife to continue their posterity, shall be sure to have his wife and her honor as much respected by the rest, as if she were their own wife or sister, beside their liberal contribution to maintain her,[9] so as themselves may be free to take the pleasure of women at large. By which liberty (if men only respect this world) they live more happily than other nations. For in those frugal commonwealths the unmarried live at a small rate of expenses, and they make small conscience of fornication, esteemed a small sin and easily remitted by confessors. Whereas other nations will live at any charge to be married, and will labor and suffer wants, yea, beg with a wife, rather than have the sting of conscience and infamy by whoring. The women of honor in Italy, I mean wives and virgins, are much sooner inflamed with love, be it lawful or unlawful, than the women of other nations. For being locked up at home, and covered with veils when they go abroad, and kept from any conversation with men . . . they are more stirred up with the sight and much

---

[9] **maintain her:** marriage is so acutely avoided that brothers will help support the wife of their sibling who does agree to marry.

more with the flattering and dissembling speeches of men, and more credulous in flattering their own desires, by thinking the said poor actions of wooing to be signs of true love, than the women of other nations having free conversation with men. In general the men of all sorts are carried with fierce affections to forbidden lusts, and to those most which are most forbidden, most kept from them, and with greatest cost and danger to be obtained. And because they are barred not only the speech and conversation but the least sight of their love (all which are allowed men of other nations) they are carried rather with a blind rage of passion and a strong imagination of their own brain, than with the true contemplation of virtues, or the power of beauty, to adore them as images, rather than love them as women. And as now they spare no cost, and will run great dangers to obtain their lustful desires, so would they pursue them to very madness, had they not the most natural remedy of this passion ready at hand to allay their desires, namely harlots, whom they call courtesans, having beauty and youth and whatsoever they can imagine in their mistress, besides the pleasure of change more to delight them, so driving out love with love, as one nail with another. This makes them little regard their wives' beauty or manners, and to marry for dowry, parentage, and procreation women unknown and almost unseen, resolving . . . to satisfy the humors of love (be they of conversation, of beauty, or of disordinate lusts in the diverse and some bestial kinds of enjoying that pleasure) by the freedom of the Stewes.[10] While courtesans walk and ride in coaches at liberty, and freely saluted and honored by all men passing by them, their wives and virgins are locked up at home, watched by their women attending them abroad, have their faces covered with a veil not to be seen, and it is death by private revenge for any man to salute them or make the least show of love to them; if it be perceived by any of the kindred, who will not fail to kill him (for their revenge is never less than death). In regard of this jealousy, that the young women may not be defiled, nor the old women their keepers hired to be bawds[11] to them, no women go to market, but only men, and the most rich disdain not to buy all necessities for their own families, in which few have any men or at least they come not near the women. Yet for all this care, the Italians many times wear the fatal horns[12] they so much detest, because women thus kept from men, think it simplicity[13] to lose any opportunity offered, though it be with the meanest servant, and because there want not[14] men as watchful to betray their chastity, as their husbands are to keep it, but especially because snares are laid for them

---

[10] **Stewes:** brothels.   [11] **bawds:** pimps.   [12] **fatal horns:** a husband with an unfaithful wife was said to have grown horns.   [13] **simplicity:** foolishness.   [14] **there want not:** there is no lack of.

in the very churches, and more specially in the nunneries, whether they cannot deny their wives and daughters to repair upon festival days of devotion. . . .

Touching the manners of the Italians. They are for the outside by nature's gift excellently composed. By sweetness of language, and singular art in seasoning their talk and behavior with great ostentation of courtesy, they make their conversation sweet and pleasing to all men, easily gaining the good will of those with whom they live. But no trust is to be reposed in their words, the flattering tongue having small acquaintance with a sincere heart, especially among the Italians, who will offer courtesies freely, and press the acceptance vehemently, only to squeeze out compliment on both sides, they neither meaning to perform them, nor yet daring to accept them, because in that case they would repute the acceptor ignorant and uncivil, forever avoiding his conversation as burdensome to them. And indeed in these fair speeches which we call courting, they so transcend all golden mediocrity,[15] as they are reputed the authors of all flattery spread through all our transalpine nations, especially in salutations by word of mouth, and epistles, forced with hyperbolical protestations and more than due titles to all degrees. . . . Yea the gentlemen of Venice, proud above all others, will be called in ordinary salutations *Clarissimi*, that is, most bright or famous, and challenge this title peculiar to themselves, not communicable to any other gentlemen whatsoever, so as if a man say that a *Clarissimo* without name did or said this or that, he is understood to say that a gentleman of Venice did or said it. . . .

Music

The Italians, and especially the Venetians, have in all times excelled [in the art of music], and most at this day, not in light tunes and hard striking of the strings (which they dislike), nor in companies of wandering fiddlers (whereof they have none or very few single men of small skill), but in consorts of grave solemn music, sometimes running so sweetly with soft touching of the strings as may seem to ravish the hearer's spirit from his body. Which music they use at many private and public meetings, but especially in their churches, where they join with it wind instruments, and most pleasant voices of boys and men, being indeed such excellent music as cannot but stir up devotion in the hearers. For the nature of music being not to provoke

---

[15]**mediocrity:** moderation.

new but to elevate present affections, and the greatest or best sort coming to church for devotion, such music cannot but increase the same. Only the Pope's Chapel hath no instruments of music, but only most excellent voices of men and boys.

CLOTHING

The harlots called courtesans commonly wear doublets[16] and breeches[17] under their women's gowns. Yea, I have seen some of them (as at Padua) go in the company of young men to the tennis court in men's apparel and rackets in their hands, most commonly wearing doublets and hose[18] of carnation satin, with gold buttons from the chin round to the waist behind, and silk stockings, and great garters with gold lace both of the same color. And I met a Duchess carried in a horse litter to Rome, whose gentlewomen and ladies of honor rode astride upon ordinary hackney horses, in doublets and breeches of the said stuff,[19] fashion and color. But all these had their heads attired like women. And I observed them to be thus appareled at ordinary times of the year, besides the foresaid liberty of carnival,[20] when men and women masked walk the streets at pleasure in men's or women's apparel.

ENGLAND

This discourse . . . I will refer to the intended treatise of England and Scotland upon these subjects, more exactly to be written, to avoid the imputation of ignorance in affairs at home, while I assert knowledge of foreign states.

Yet in the meantime til that treatise be compiled I desire leave for strangers' sake, briefly to note some singularities of England. . . . And first for the satisfaction of strangers, who say that old writers have taxed[21] the old Britains to have been cruel and inhospitable by nature towards strangers, and that to this day they find by experience the English to be insolent and rude towards them. For the old Britain's cruelty and inhospitality, generally the most barbarous people are most cruel, but in the time of Caesar, himself witnesseth that the inhabitants of Kent were most courteous and full of humanity, and as shortly after they embraced Christian religion in the primitive Church, and had then famous universities, before France had any (that of Paris being founded in the model and imitation of them), so no doubt they were far from barbarism.

---

[16] **doublets:** close-fitting body garment worn by men.   [17] **breeches:** pants.   [18] **hose:** leggings or stockings worn by men.   [19] **stuff:** material.   [20] **carnival:** Mardi Gras.   [21] **taxed:** accused.

FIGURE 10 *Jew Selling Clothes, from Pietro Bertelli,* Diversarum Nationem *(1594). Although he is in Italy, he is represented as speaking Yiddish, a combination of Hebrew and German. He is asking, "What will you give?" presumably a typical call of a salesman. In the course of the sixteenth century, Jews increasingly abandoned moneylending as a profession and turned instead to trading.*

JEWS

The Jews are a nation incredibly despised among all Christians, and of the Turks also. And were dispersed throughout the face of the world, save[22] that they have been long banished out of some Christian kingdoms, as England, France, and Netherland, where notwithstanding they lurk disguised, though they be not allowed any habitation by the state. And where they are allowed to dwell, they live upon usury and selling of frippery wares,[23] as brokers,

---

[22] **save:** except.   [23] **frippery wares:** old clothes.

therein permitted by Christian Princes for private gain to use horrible extortions upon their subjects, but are not allowed to buy any lands, houses, or stable inheritances, neither have they any coin of their own, but use the coins of Princes where they live. The ten tribes of the kingdom of Israel were long since carried captive and dispersed in the furthest east, and are not known where they live, having no commerce with the Jews known to us.

Touching those of the kingdom they had at Jerusalem, they are thought to be mingled in their tribes and families, but the general opinion is, that those of the Tribe of Judah live in Turkey, and those of the Tribe of Benjamin live in Italy, Germany, and Poland. They are a miserable nation and most miserable in that they cannot see the cause thereof, being the curse of the blood of their Messiah, which they took upon themselves and their children, whose coming they still expect, saying it is thus long deferred for their sins, but they look for his coming from the East before and towards the end of the world.

→ JOHN LEO

From A Geographical History of Africa   *1600*
Translated by John Pory

John Leo (1495?–1552?) or Al-Hassan Ibn Mohammed Al-Wezaz, Al-Fasi, was born in Granada shortly after it was regained by the Spanish from the Muslims. His family emigrated to Fez, in the empire of Morocco, sometime thereafter. He studied at the university in Fez, and subsequently served as a notary, writing up legal documents and even serving as a judge in some of his travels to outlying areas. At the age of fifteen, he was entitled to wear the white robes denoting a learned man. He also appears to have traveled extensively as a merchant and a soldier; in 1512 he accompanied the King of Fez on an expedition. On his return from a trip to Constantinople in the early 1520s, he was captured by Christian pirates, who brought him to Pope Leo X in Rome. The pope gave Al-Hassan his freedom and a pension to induce him to stay, and ultimately convinced him to convert, serving as godfather and giving his own name to the convert. It is difficult to determine whether or not John Leo retained his faith in Islam in making this conversion (Robert Brown i–v, xxii–xxx, xli–xlii).

While in Rome, Leo learned Latin and taught Arabic; in 1526 he wrote his *Geographical History of Africa*. The text was originally drafted in Arabic, then translated into Italian; both originals are lost, though we have a version redacted

---

John Leo, *A Geographical History of Africa* (1526), trans. John Pory (London, 1600), 6–7, 68, 368, 371–72.

by the travel writer Giovanni Battista Ramusio (1485–1557) (Robert Brown xlv–xlvi, i). John Pory's translation into English was published in 1600; the portion excerpted below gives several accounts of the origins of the "moors" of northern Africa. Leo recalls the former splendor of Morocco — an empire and a city — and describes its decay in the sixteenth century. However, he also presents the Xeriffo, or current ruler of the kingdoms of Morocco and Fez, as an extremely powerful leader. Does Shakespeare's representation of the Prince of Morocco focus on the weakness of the contemporary city of Morocco, emphasize the power associated with the Xeriffo, or both? Why does Portia comment on the Prince of Morocco's complexion, but not on that of any of the other suitors? What is the color of his skin, and what difference would this make to an early modern audience?

About the original[1] of the Africans, our historiographers do much disagree. For some will have them to be derived from the inhabitants of Palestine; because (as they say) being expelled out of their own country by the Assyrians, they came at length into Africa, and seeing the fruitfulness of the soil, chose it to be their place of habitation. . . . Some others report that the Africans descended from certain people of Asia, who being chased thence by reason of wars which were waged against them, fled into Greece, which at the same time had no inhabitants at all. Howbeit the enemy still pursuing them, they were forced to cross the sea of Morea,[2] and being arrived in Africa, to settle themselves there: but their enemies abode still in Greece. All which opinions and reports are to be understood only of the original of the tawny people, that is to say, of the Numidians and Barbarians.[3] For all the Negroes or black Moors[4] take their descent from Chus, the son of Cham, who was the son of Noa.[5] But whatsoever difference there be between the Negroes and the tawny Moors, certain it is that they had all one beginning. For the Negroes are descended of the Philistims, and the Philistims of Mesraim[6] the son of Chus: but the tawny Moors fetch their pedigree from the Sabeans, and it is evident that Saba[7] was begotten of Rama, which was the eldest son of Chus. Diverse other opinions there be as touching this matter: which because they seem not so necessary we have purposely omitted.

[1] **original:** origin.   [2] **sea of Morea:** the Ionian Sea, west of Greece.   [3] **Numidians and Barbarians:** Numidia and Barbary refer to the area of northwestern Africa.   [4] **Moor:** term used to refer to Muslims, usually inhabiting northwest Africa, and implying either a "tawny" light-brown complexion or a darker one.   [5] **Chus . . . Noa:** these names refer to Cush, Ham, and Noah; the genealogy traced here is taken from Genesis 10.   [6] **Philistims . . . Mesraim:** referring to the Philistines and Mitzraim, or Egypt, respectively; see Genesis 10:13–14.   [7] **Saba:** Sheba is listed as the son of Raamah in Genesis 10:7.

A division of the tawny Moors into sundry tribes or nations.

The tawny Moors are divided into five separate people or tribes: the Zanhagi, Musmudi, Zeneti, Hacari, and Gumeri. The tribe of Musmudi inhabit the western part of mount Atlas, from the province of Hea to the river of Seruan. Likewise they dwell upon the south part of the said mountain, and upon all the inward plains of that region. These Musmudae have four provinces under them: namely Hea, Sus, Guzala, and the territory of Morocco.

A most exact description of the great and famous city of Morocco.[8]

This noble city of Morocco in Africa is accounted to be one of the greatest cities in the whole world. It is built upon a most large field, being about fourteen miles distant from Atlas. One Joseph the son of Tesfin, and king of the tribe or people called Luntuna, is reported to have been the founder of this city, at that very time when he conducted his troups into the region of Morocco, and settled himself not far from the common highway, which stretcheth from Agmet over the mountains of Atlas, to those deserts where the foresaid tribe or people do usually inhabit. Here may you behold most stately and wonderful workmanship: for all their buildings are so cunningly and artificially contrived, that a man cannot easily describe the same. This huge and mighty city, at such time as it was governed by Hali the son of king Joseph, contained more than 100,000 families. It had four and twenty gates belonging thereto, and a wall of great strength and thickness, which was built of white stone and lime. . . . Yea a great part of this city, especially about the foresaid temple[9] [at present] lieth so desolate and void of inhabitants, that a man cannot without great difficulty pass, by reason of the ruins of many houses lying in the way. Under the porch of this temple it is reported that in old time there were almost [a hundred] shops of sale books, and as many on the other side over against them: but at this time I think there is not one book-seller in all the whole city to be found. And scarcely is the third part of this city inhabited. Within the walls of Morocco are vines, palm trees, great gardens, and most fruitful cornfields: for without their walls they can till no ground, by reason of the Arabian's often inroads.[10] Know ye this for a certainty, that the said city is grown to untimely decay and old age: for scarcely five hundreth and six years are past, since the first building thereof, forasmuch as the foundations thereof were laid in the time

---

[8] **Morocco:** both the city, Marrakesh, and the region.   [9] **foresaid temple:** Robert Brown identifies this as El Kutubia (352).   [10] **inroads:** raids.

of Joseph the son of Tesfin, that is to say, in the 424 year of the Hegeira.[11] Which decay I can impute to none other cause, but to the injury of continual wars, and to the often alterations of magistrates and of the commonwealth.

The Xeriffo, commonly called the king of Morocco, Sus, and Fez.

Among all the princes of Africa, I suppose that there is not anyone who, in richness of state, or greatness of power, may be preferred before the Xeriffo, in that his dominion, which comprehendeth all that part of Mauritania, called by the Romans Tingitana, extendeth itself north and south from Capo Boiador,[12] even to Tanger,[13] and east and west from the Atlantic Ocean, as far as the river Muluia,[14] and somewhat further also, in which space is comprehended the fairest, fruitfullest, best inhabited, and most civil part of all Africa, and among the other states, the famous kingdoms of Morocco and Fez. . . .

Let thus much suffice to have been spoken of the Xeriffo, whose proceedings appear much like to those of Ismael the Sophy[15] of Persia. Both of them procured followers by blood and the cloak of religion; both of them subdued in short time many countries; both of them grew great by the ruin of their neighbors; both of them received grievous checks by the Turks, and lost a part of their states. . . .

The Xeriffo his revenues, or comings in.

The Xeriffo is absolute lord of all his subjects' goods, yea and of their persons also. For though he charge them with never so burdensome tributes and impositions, yet dare they not so much as open their mouths. He receiveth from his tributary vassals, the tenths and first fruits of their corn and cattle.[16] . . . Besides these revenues, the king hath the tolls and customs of Fez, and of other cities; for at the entering of their goods, the natural citizen payeth two in the hundred, and the stranger ten.[17]

---

[11] **Hegeira:** the Hejeira, which means "flight," refers to the year 622 C.E., when Mohammed was forced to flee from Mecca to Medina. The Muslim calendar counts years from this date, so 424 would correspond to 1046 C.E.   [12] **Capo Boiador:** midway down the coast of western Sahara, just south of Morocco.   [13] **Tanger:** Tangier, city on the northern coast of Morocco. [14] **Muluia:** the river Moulouya, which flows to the eastern border of Morocco.   [15] **Sophy:** the supreme ruler of Persia.   [16] **the tenths . . . cattle:** a tenth of their income and the first produce or offspring of their crops or animals.   [17] **natural citizen . . . ten:** natives pay 2 percent customs on bringing their goods into the city of Fez while strangers pay 10 percent.

→ GEORGE BEST

From A True Discourse of the Late Voyages of Discovery, for the Finding of a Passage to Cathaya   *1578*

George Best (d. 1584) was a navigator who accompanied English navigator and explorer Martin Frobisher (1535–1594) on three voyages across the Atlantic (1576, 1577, and 1578) to discover a northwest passage to China. In 1578 he published an account of his travels, *A True Discourse of the Late Voyages of Discovery, for the Finding of a Passage to Cathaya*. The three parts of the book correspond to the events of the three voyages, although the first part also sets out a geography of and a commentary on the countries of the world. Best was apparently killed in a duel in 1584. The excerpt below considers theories of race current in the period; Best supports the idea that complexion is linked to morality, and, relying on biblical texts, contends that racial hierarchy is divinely determined. What effect would using biblical authority to justify these theories have on an early modern reader? How do Best's theories compare with Portia's comments about the Prince of Morocco's "complexion"? Best associates race not only with skin color and nationality but religion as well: "of this black and cursed Chus came all these black Moors which are in Africa" (p. 177); this curse is attributed not only to Africans, but explicitly to African Muslims. Shakespeare similarly, albeit implicitly, conflates nation, religion, and race both in his representation of Morocco and in Jessica's identification of Chus as one of Shylock's "countrymen" (3.2.283). How does relating him to Chus shape our understanding of Shylock and Jessica's moral, national, religious, and racial status in the play?

There is some other cause than the climate or the sun's perpendicular reflection, that should cause the Ethiopians great blackness. And the most probable cause to my judgment is, that this blackness proceedeth of some natural infection of the first inhabitants of the country, and so all the whole progeny of them descended, are still polluted with the same blot of infection. Therefore it shall not be far from our purpose to examine the first original of these blackmen, and how by a lineal descent they have hitherto continued thus black.

It manifestly and plainly appeareth by holy scripture, that after the general inundation and overflowing of the earth, there remained no more men alive, but Noah and his three sons, Shem, Cham, and Japhet, who only were

George Best, *A True Discourse of the Late Voyages of Discovery, for the Finding of a Passage to Cathaya* . . . , in *The Principal Navigations, Voyages, Traffics, and Discoveries of the English Nation*, ed. Richard Hakluyt (1598–1600), 3:52–53.

left to possess and inhabit the whole face of the earth. Therefore all the sundry descents[1] that until this present day have inhabited the whole earth, must needs come of the off-spring either of Shem, Cham, or Japhet, as the only sons of Noah, who all three being white, and their wives also, by course of nature should have begotten and brought forth white children. But the envy of our great and continual enemy the wicked spirit is such, that as he could not suffer our old father Adam to live in the felicity and angelic state wherein he was first created, but tempting him, sought and procured his ruin and fall, so again, finding at this flood none but a father and three sons living, he so caused one of them to transgress and disobey his father's commandment, that after him all his posterity should be accursed. The fact of disobedience was this: when Noah at the commandment of God had made the ark and entered therein, and the flood-gates of heaven were opened, so that the whole face of the earth, every tree and mountain was covered with abundance of water, he straightly[2] commanded his sons and wives, that they should with reverence and fear behold the justice and mighty power of God, and that during the time of the flood while they remained in the ark, they should use continency, and abstain from carnal copulation with their wives. And many other precepts he gave unto them, and admonitions touching the justice of God, in revenging sin, and his mercy in delivering them, who nothing deserved it. Which good instructions and exhortations notwithstanding, his wicked son Cham disobeyed, and being persuaded that the first child born after the flood (by right and law of nature) should inherit and possess all the dominions of the earth, he contrary to his father's commandment while they were yet in the ark, used company with his wife, and craftily went about thereby to disinherit the offspring of his other two brethren. For the which wicked and detestable fact, as an example for contempt of Almighty God, and disobedience of parents, God would a son should be born whose name was Chus, who not only itself, but all his posterity after him should be so black and loathsome, that it might remain a spectacle of disobedience to all the world. And of this black and cursed Chus came all these black Moors which are in Africa, for after the water was banished from the face of the earth, and that the land was dry, Shem chose that part of the land to inhabit in, which now is called Asia, and Japhet had that which now is called Europa, wherein we dwell, and Africa remained for Cham and his black son Chus, and was called Chamesis after the father's name, being perhaps a cursed, dry, sandy, and unfruitful ground, fit for such a generation to inhabit in.

---

[1] **descents:** descendants.   [2] **straightly:** strictly.

Thus you see, that the cause of the Ethiopians' blackness is the curse and natural infection of blood, and not the distemperature of the climate. Which also may be proved by this example: that these black men are found in all parts of Africa, as well without the Tropics, as within, even unto *Capo de buona Speranza*[3] southward, where, by reason of the Sphere, should be the same temperature that is in Sicilia, Morea, and Candie,[4] where all be of very good complexions. Wherefore I conclude, that the blackness proceeds not of the hotness of the clime, but as I said, of the infection of the blood, and therefore this their argument gathered of the Africans blackness is not able to destroy the temperature of the middle zone.

[3] *Capo de buona Speranza:* the Cape of Good Hope, the southern tip of Africa. [4] Sicilia . . . Candie: Sicily, Northwest Africa, and Crete are in approximately the same zone as Africa.

→ JOHN LEO

From A Geographical History of Africa *1600*

Translated by John Pory

John Leo (1495?–1552?) gives a very cynical and hostile account of Mohammed in his *Geographical History of Africa.* Judging from the tone of this description, Leo would seem to have completely absorbed a Christian viewpoint, although he avoids the stock accusations of idolatry that contemporary Christian descriptions of Islam often include. He also refrains from associating race and religion in his account, in contrast to Shakespeare's representation of the Prince of Morocco in *The Merchant of Venice.* What information do we get about Morocco's religious beliefs? Does race play a role in the depiction of his religion? Leo reportedly returned to Tunis later in life and reverted to Islam, but it is difficult to know if this information is accurate or reflects the popular sixteenth-century fantasy of the Muslim as a facile changer of faiths (Robert Brown xlvii–xlviii).

Of Mahumet, and of his accursed religion in general.

Mahumet his father,[1] was a certain profane idolater called Abdala, of the stock of Ismael,[2] and his mother was one Hennina a Jew, both of them being of very humble and poor condition. He was born in the year of our Lord 562 and was endowed with a grave countenance and a quick wit. Being

[1] **Mahumet his father:** Mohammed's father. [2] **Ismael:** son of Abraham, Genesis 16.

John Leo, *A Geographical History of Africa* (1526), trans. John Pory (London, 1600), 380–81.

grown to man's estate,[3] the Scenite Arabians, accustomed to rob, and run all over the country, took him prisoner, and sold him to a Persian merchant, who discerning him to be apt, and subtle about business, affected and held him in such account, that after his death his mistress remaining a widow, scorned not to take him for her husband. Being therefore enriched by this means with goods and credit, he raised up his mind to greater matters. The times then answered very fitly for one that would disturb or work any innovation. For the Arabians upon some evil entreaty[4] were malcontented with the Emperor Heraclius.[5] The heresies of Arius and Nestorius,[6] had in a miserable sort shaken and annoyed the church of God. The Jews, though they wanted[7] power, yet amounted they to a great number. The Saracens prevailed mightily, both in number and force. And the Roman Empire was full of slaves. Mahumet therefore taking hold on this opportunity, framed a law, wherein all of them should have some part, or prerogative. In this, two Apostata Jews,[8] and two heretics, assisted him: of which one was John, being a scholar of Nestorius' school; and the other Sergius, of the sect of Arius. Whereupon the principal intention of this cursed law was wholly aimed against the divinity of our Savior Jesus Christ, wickedly oppugned[9] by the Jews and Arians. He persuaded[10] this law, first by giving his wife to understand, and his neighbors by her means, and by little and little others also, that he conversed with the angel Gabriel, unto whose brightness he ascribed the falling sickness,[11] which many times prostrated him upon the earth; dilating and amplifying the same in like sort, by permitting all that which was plausible to sense and the flesh; as also by offering liberty to all slaves that would come to him and receive his law. Wherefore being prosecuted hard by the masters of those fugitive slaves led away by him, he fled to Medina Talnabi, and there remained some time. From this flight the Mahumetans fetch the original of their Hegeira.[12] But questionless there was nothing that furthered more the enlargement of the Mahumetan sect than prosperity in arms and the multitude of victories; whereby Mahumet overthrew the Persians, became lord of Arabia, and drove the Romans out of

---

[3] **man's estate:** adulthood.   [4] **entreaty:** treatment.   [5] **Heraclius:** Eastern Roman emperor (610–641).   [6] **Arius and Nestorius:** Arius, a presbyter of Alexandria in the fourth century, denied that Jesus Christ was consubstantial, or of the same essence or substance with God *(OED)*. Nestorius, fourth-century bishop of Constantinople, believed that Jesus was divided into two persons, one human and the other divine. Christian doctrine teaches that Jesus's divine and human natures are united in one person (*Encyclopedia Britannica Online* [*EBO*], s.v. Nestorius).   [7] **wanted:** lacked.   [8] **Apostata Jews:** Jews who have converted to another religion.   [9] **oppugned:** attacked.   [10] **persuaded:** commended to acceptance.   [11] **falling sickness:** he represented his epileptic fits as a response to visions of the angel Gabriel.   [12] **Hegeira:** the Hejeira was Mohammed's flight from Mecca to Medina in 622 C.E., the date at which the Muslim calendar begins.

Syria. And his successors afterwards extended their empire from Euphrates to the Atlantic Ocean, and from the river Niger to the Pirenei[13] mountains, and beyond. They occupied Sicily, assailed Italy, and with continual prosperity, as it were, for three hundred years, either subdued, or encumbered, both the east and west. But to return to Mahumet his law, it embraceth circumcision and maketh a difference between meats pure and unpure, partly to allure the Jews. It denieth[14] the divinity of Christ, to reconcile the Arians, who were then most mighty; it foisteth in[15] many frivolous fables, that it might fit the Gentiles; and looseth the bridle to the flesh, which is a thing acceptable to the greatest part of men.

[13] **Pirenei:** probably the Pyrenees, which lie on the border of France and Spain. [14] **denieth:** denies. [15] **foisteth in:** introduce surreptitiously.

→ SEBASTIAN MUNSTER

From The Messiah of the Christians and the Jews   *1655*
Translated by Paul Isaiah

Sebastian Munster (1488–1552) was born in Ingelheim, Germany. He was a scholar of Hebrew, which he taught at the University of Basel in Switzerland. In 1530 he published the *Dictionarium Trilingue,* a Latin, Greek, and Hebrew dictionary; in 1534–35 he produced a two-volume edition of the Hebrew Bible, which included a Latin translation and annotations. He was also a geographer, publishing a Latin edition of Ptolemy's *Geographia* in 1540 and his own *Cosmographia* in 1544, the earliest German description of the world (*EBO,* s.v. Sebastian Munster).

   Munster was interested in converting Jews to Christianity and translated several books of the New Testament into Hebrew to aid that endeavor. His missionary tract, *Messias Christianorum et Judaeorum* (1539) was first published in Hebrew and Latin (Baron 13: 233). In the excerpt below, taken from the seventeenth-century English translation, Munster represents Jews as belonging to a distinct race, which manifests its religious affiliation by the darkness of its complexion. Does *The Merchant of Venice* suggest Jews are racially different from Christians? In addition to the mention of Chus (see 3.2.283 and Best's discussion, p. 176), does the play employ other ways of marking racial difference? Are Shylock and Jessica seen as racially similar or different? Why? How does gender influence ideas about race, religion, beauty, and goodness both in Munster's discussion and in the play?

Sebastian Munster, *The Messiah of the Christians and the Jews* (1539), trans. Paul Isaiah (London, 1655), 1–3, 8–9.

A Disputation of a Christian with an Obstinate Jew

CHRISTIAN: Is that man who comes to meet me a Jew? Truly his face and form show him to be a man. I know what to do, I will salute him in Hebrew, and I shall easily know whether he be a Jew or not. If he be a Jew, he will answer in Hebrew, but if he be not a Jew, he will hold his peace, not knowing what I say. God save you, O Jew.

JEW: And God save you; how know you me to be a Jew, that you speak so in Hebrew with me? Art thou a Jew and one of our people?

CHRISTIAN: I am not a Jew, neither of thy people, neither am I acquainted with you; but from the form of your face, I knew you to be a Jew: For you Jews have a peculiar color of face, different from the form and figure of other men; which thing hath often filled me with admiration, for you are black and uncomely, and not white as other men.

JEW: It is a wonder, if we be uncomely, why you Christians do so love our women, and they seem to you more beautiful than your own.

CHRISTIAN: Your women indeed are more comely than your men, but you seduce them most corruptly.

JEW: Nay, feeling we are the elect people of God, and his inheritance, we are more comely than all the Nations of the earth; as it is written in Daniel 1.15, that the countenances of the Jewish children appeared fairer and fatter in flesh than all the other children.

CHRISTIAN: You have not yet told me, why the women amongst you are more beautiful than the men, and are not so easily known as the men.

JEW: Because you Christians do not so much reproach them, as the men; moreover we adorn them with excellent apparel, that they may find favor in your eyes, and we by them obtain what we desire, and it may be well with us by reason of them, and our souls may live for their sakes among the gentiles; as it is written, and they entreated[1] Abraham well for Sarah's sake, because she was a woman of a comely countenance, Genesis 12.16.

---

[1]**entreated:** treated.

A maiden Iewe of Andrinople.

FIGURE 11 *A Maiden Jew Elaborately Dressed, from Nicolas de Nicolay,* The Navigations, Peregrinations, and Voyages Made into Turkey *(1585).*

→ ANDREW WILLET

From Concerning the Universal and Final Vocation of the Jews
1590

Andrew Willet (1562–1621) received his bachelor and master of arts degrees at Cambridge and became a minister in the Church of England in 1587; he received his doctorate of divinity in 1601. He was a well-regarded preacher and eventually took over his father's parish in 1598. While attending to the needs of his parishioners and raising a family (eighteen children, of whom thirteen survived) he wrote forty-two books and treatises. His contemporaries considered him a "walking library" because of his voracious and wide-ranging reading habits. Willet also preached before the court, and served as tutor to Prince Henry, eldest son of King James I. Among his writings is the *De Universali et Novissima Iudeaorum Vocatione, Secundum Apertissimam Divi Pauli Prophetiam, in Ultimis Hisce Diebus Praestanda* (Concerning the Universal and Final Vocation of the Jews That Is to Be Apparent in the Last Days, According to the Extremely Lucid Prophecy of St. Paul), an excerpt of which is presented in translation below. The vocation of the Jews in Christian thought represents the event in the Messianic era when they will be called to return to God and convert to Christianity. In this text, composed in dense Latin prose, Willet considers, poses, and answers a number of questions on the subject, frequently citing biblical evidence. In the course of his discussion, he takes up the question of the continued existence of the Jews despite their dispersion, noting that they maintain a racial and ethnic distinction rather than assimilate, unlike transplanted European Christians. Like Best, he proposes divine motives, some positive and some negative, behind maintaining a distinct Jewish race. Reflecting on Willet, we might ask if Jewish racial difference is represented positively or negatively in *The Merchant of Venice*? Does conversion erase this difference for Jessica? Does this idea of racial difference affect our understanding of Shylock's conversion?

Now we are at the point where we must consider the following question, namely, to what end it is directed, since the inward parts of the Levitical stock[1] are not altogether eradicated, but still thrive and even flourish, if not to this end, that a rebellious and degenerate youth of the family, a shoot producing offspring, having been snatched from the lamentable darkness, might take delight and revive under the most pleasant glance of God. For

---

[1] Levitical stock: the Jews.

Andrew Willet, *Concerning the Universal and Final Vocation of the Jews That Is to Be Apparent in the Last Days, According to the Extremely Lucid Prophecy of St. Paul* (Cambridge, 1590), 25–26. Translated from the Latin for this volume by Andrew Dinan (2000).

how are we to explain the fact that all the remaining nations who have abandoned their ancestral residences and have landed at one time on foreign shores, after the span of a few years, have forgotten their own country and race and have degenerated by their connection with other nations; nevertheless the entire Judaic race alone, passing through the regions of the earth, is joined together as a nation with no other family, nor are they infected with the blood and line of barbarians, as they themselves say. . . .

Our England was first settled by the British. Then the Saxons overran it, but they themselves later on, in truth, became British (only the name was changed). Finally the Gauls, having invaded this country of ours, the most famous Anglia, with force and arms, they themselves even passed over into [becoming] English. Historians say that the same thing happened rather frequently and quite openly even in other kingdoms, whose inhabitants very often were altered and uprooted, but nevertheless the nature and condition of the race remained the same. If an English person should go to Spain, his heirs would be reckoned among the Spanish, although he would not lose his native affinity. If a Scot should carry off his family and migrate into Gaul[2] his posterity would take on Gallic characteristics, inasmuch as it would no longer be accustomed to Scottish ways. Nevertheless a Jew, whether he should go on a journey to Spain or Gaul or whether he should set out into any other region, would declare that is he neither Spanish or Gallic, but Jewish.

Now this induces perplexity and gives rise to the question as to the source of this great distinction. Perhaps the difference is thought to lie in this fact: although all the other nations indiscriminately join themselves to each other in marriage, and thus they grow together into one stock, by the blending and commingling of their people, the Jews nevertheless shudder at this outrage (for that is what they consider it) and marry only their own women, seeking no others in marriage. I answer that this reason indeed, which is now set forth, is not without merit if it relies on a certain probability. Now, however, the matter is quite otherwise. For although Moses gave them a command not to defile themselves by marriages with Gentiles, nevertheless they quickly discarded their concern to uphold his commands. For although the form of the state remained rather pure and intact up until that point, in the time of Ezra,[3] of course, when the captives returned as if to their restoration, many were found to have furnished foreign wives for themselves, and this resulted even in children; nor were the priests them-

---

[2] **Gaul:** France. [3] **Ezra:** Ezra lived in the fifth century B.C.E., during the Babylonian exile and subsequent return to Israel. Willet cites Ezra 10:44 in the margin.

selves immune from this crime. How much more likely it is then, given that they succumbed to this [practice] in that Golden Age, that this corruption was predominant among the Jewish people during the coming of the Iron Age.[4] I do not say this on the basis of my own conjecture, but in fact, the history of the facts from earlier times attests to this. A decree survives from the Council of Averno[5] that states that any woman who had become engaged to a Jewish man and later had married him was not only to be driven from the gatherings and assemblies of the Christians, but she was also to be barred from the fellowship of Holy Communion. It was also ratified in another council[6] that a Jew would not choose for himself a wife from the Christians. And in another session of the same senate it was decreed that if a Christian wife should have associated herself with a Jewish man, and if the man did not convert to the same faith, the marriage was broken up by the separation of either party. It is clearly demonstrated from these sanctions and edicts that the Jews were not held absolutely bound by these laws, and that the force of that most ancient Mosaic law was diminished over time. Therefore, that reason which was introduced to explain the continuity and preservation of the Jewish stock seems to be wrong. Other explanations must be sought.

I see three effective explanations that can be advanced; and no more can properly be presented. First, Divine Providence most clearly stands out and shines through in the preservation of that race, because the sacred volume of the Old Books,[7] which the Ethnics[8] and heretics very often tried to obliterate and contaminate with their own calumnies[9] and fictions, was transmitted to us by the industry of the Jews. They were able to testify with us against the Gentiles that those divine words, which are held in our hands, have come down from God, so that the accusation of novelty which the profane men contrive against the Christians may be attacked with such a clear proof that it loses its force and grows weak At this point there arises yet another conjecture with regard to the great old age of the Jews, that God did not allow their name to be covered up in silence and to perish so that the unwelcome distinction of everlasting disgrace might cling to it, with nothing

---

[4] **Golden Age . . . Iron Age:** presumably, the Golden Age for the Jews was the era in which they still enjoyed God's favor. For Willet, and many Christians in the early modern period, the Jews were seen as punished by God for their rejection of Jesus and hence entered into an Iron Age after the crucifixion. [5] **Council of Averno:** a council that passed legislation on ecclesiastical matters. [6] **another council:** Willet's marginal note refers to a specific law passed by the Council of Toledo, 3. decreto. 14. The council of Toledo, another ecclesiastical council, passed this law in 589 C.E. (Linder 484–85). [7] **Old Books:** the Hebrew Bible. [8] **Ethnics:** pagans. [9] **calumnies:** slanders.

to be destroyed by the decay of old age, which more easily perhaps had been able to be washed off if it[10] should perish together with the limit of some time, however long.[11] But whoever predicts better things for this race and asserts that it has been defended with such careful protection so that it might at last ascend more quickly the most eminent citadel of the Church will divine much more truly and properly concerning this hidden matter.

[10] it: the Jewish people.   [11] At this point . . . long: the sense of this complex sentence seems to be that the preservation of the Jewish people prevented their disgrace from fading. A tradition in Christian teaching argued that the Jews were responsible for the death of Jesus and, as a result, suffered from a subsequent curse and its attendant disgrace.

CHAPTER 2

Finance

><

The financial concerns of *The Merchant of Venice*, like those of early modern England, are shaped by a tradition of religious teachings on poverty, lending, and trade. The sixteenth century brought substantial religious and economic change, which required renegotiations of the relationship between religion and the economic order. The Reformation altered institutions for and attitudes toward the poor: the dissolution of the monasteries resulted in a reduction of charitable institutions, while new beliefs that linked morality to success suggested that poverty was the result of sin and hence deserved (Clay 1: 219; Thomas 102). Poverty increased during this same period, with shifts in the agrarian economy forcing more people to urban centers in search of employment (Clay 1; 216 ff). While England was still primarily an agricultural society, industry and trade in urban centers began to rise, especially in London (Clay 1: 165 ff, 197 ff). Whereas in the past they had primarily exported raw materials and relied on German and Italian merchants, the English began to export finished cloth and establish native merchant companies (Clay 2: 103, 105–06, 108 ff). Attitudes toward lending and interest also changed during this period, as Protestant doctrines reinterpreted and extenuated the concept of charging interest (Nelson 73 ff). The negative association of Jews with usury derives from Mosaic (or Hebrew) laws regulating the charging of interest as well as the fact that Jews were permitted periodically to charge interest throughout the medieval and early modern periods. Charging interest was generally condemned up to the early modern period. However, as economic development made it an increasingly necessary transaction, public attitudes toward the charging of interest became more accepting. Hence, merchants and moneylenders, whose moneylending was regularly condemned in the Middle Ages, were viewed in an increasingly positive light. Public opinion changed, however, only through the process of vigorous debate; the older perspectives did not disappear overnight.

Usury

The biblical prohibitions against usury posed significant theological, economic, social, and political challenges for both Jews and Christians for centuries before reaching a kind of crisis and initial resolution in the early modern era. The late sixteenth and early seventeenth centuries witnessed significant shifts in views among European Christian and Jewish thinkers, as economic pressures made interest-bearing loans more necessary, and hence acceptable. Medieval and early modern debates on usury, particularly in the Christian tradition, focused on how the prohibition is worded in the book of Deuteronomy (Nelson). While Exodus and Leviticus prohibit lending to the poor on interest, Deuteronomy forbids charging interest of one's brother, regardless of economic status, but permits it with a stranger: "Thou shalt not lend upon usury to thy brother; usury of money, usury of victual, usury of any thing that is lent upon usury. Unto a stranger thou mayest lend upon usury; but unto thy brother thou shalt not lend upon usury" (23:19–20). The debate over charging interest hinges on the significance of the Hebrew term for usury, נשך (neshech), which means "biting," and on the identity of the "stranger." In early modern England, brotherhood in theory applied to everyone, which suggested a universal prohibition against usury; however, questions about what constituted usury were raised in this period, paving the way toward licensing the charging of interest. Usury was usually distinguished from interest as being a higher and harsher rate, but this difference was not consistently maintained. In his *Discourse upon Usury* (1572), Thomas Wilson presents a popular and widespread argument against the charging of any interest, though this perspective is already losing ground by the second half of the sixteenth century (see p. 195). The *Discourse* takes the form of a dialogue in which different characters voice arguments for and against charging interest; Gromel Gayner (whose name means miser), a merchant who is also a moneylender, speaks in favor of taking interest (which he distinguishes from usury), arguing: "if you forbid gain, you destroy intercourse of merchandize, you overthrow bargaining and you bring all trading betwixt man and man to . . . confusion" (56v). While the preacher, Okerfoe (whose name means enemy to usury), ultimately convinces his opponents of the viciousness of lending at interest, Wilson himself is not able to convince his fellow members of Parliament when he advances his arguments during the debate over the 1571 bill permitting usury under 10 percent. While no one, not even Gromel Gayner, admits to being in favor of usury, most people in the period concluded that charging a limited amount of interest was probably a necessary evil. By the early seventeenth century, Francis Bacon could examine the pros and cons of usury without referring to its biblical or religious significance (see p. 207); it had begun to emerge from its association with sin. In *The Merchant of Venice*, Antonio clearly argues against usury while Shylock advocates its use. Whose view is presented as more valid in the play? How would an audience familiar with the current debates over charging interest judge the "winning" view?

Jews

Sir Edward Coke, writing around the time Bacon published his defense of charging interest, linked the sinful nature of usury with its practice by the Jews. His account of usury presents it as unequivocally evil (see p. 210). Coke blames Jewish usury for substantial material and spiritual ills in medieval England: the forfeiture of estates and the encouragement of sin in opposition to God's will. He also suggests that no compromise on charging interest was possible, alleging that Henry III and Edward I attempted "to use some mean and moderation herein, but in the end it was found there was no mean in mischief" (p. 211). The statute of 1290, which prohibited Jews from charging interest, resulted, in Coke's opinion, in the expulsion of the Jews that took place that year; in prohibiting their main means of livelihood, he reasons, the law in effect made it impossible for them to live there. While this explanation of the statute represents the Jews as cruel, odious, and infidel usurers, Coke also acknowledges the profit that accrues to non-Jewish rulers who appropriate the proceeds of the Jewish moneylenders. How does his view of Jewish usury compare with Antonio's assessment of Shylock's practice of charging interest? Given his views, why does Antonio seek to borrow money from Shylock?

Coke would have been surprised to discover that rabbinic law[1] agreed with him on the subject of the evil of charging interest. While biblical law allowed Jews to charge interest of "others," which could be interpreted as permitting interest-bearing loans to Christians, rabbinic law sought to forbid charging any direct interest, either to strangers or brothers. While the Mishnah states that "One may borrow from and lend money to them [non-Jews] on interest" the Gemara significantly limits the biblical permission, asking "[What!] Cannot one do without?" (Baba Metzia 70b). The reasons follow: "R. Hiyya, the son of R. Huna, said: This [permission] is granted only [up to] the [minimum] requirements of a livelihood. Rabina said: Here [in the Mishnah] the reference is to scholars. For why did the Rabbis enact this precautionary measure? Lest he learn of his ways. But being a scholar, he will [certainly] not learn of his ways" (Baba Metzia, 70b–71a). The rabbis argue that the permission to charge interest of strangers applies only to support one's basic needs; they also seek to restrict this activity to scholars, for fear that business transactions between unlearned Jews and non-Jews might lead the former to idolatry and apostasy, or conversion away from the faith. In the sixteenth century, Rabbi Yehiel Nissim da Pisa writes *Chayei Olam,* a treatise explaining the Jewish laws regulating lending; he harshly condemns charging any direct interest in loans between Jews, and, following the Talmud and later legal decisions, he suggests that while lending on interest to non-Jews might be permitted in a limited way, the practice should be avoided: "Blessed be He who . . . foresaw what would result from lending on interest to gentiles. It is quite clear that as a result of [this] habit . . . [Jewish lenders] fall into the error of lending on interest and usury to Jews" (p. 217). This issue is considered by other Jews in the sixteenth and seventeenth centuries. David de Pomis, writing in defense of Jewish physicians in 1587, digresses from his main topic to explore the questions of the identity of the "stranger" in the biblical text, and the definition of usury. Drawing on the arguments of the Dominican theologian Sixtus Medices, de Pomis offers numerous proofs that Jews and Christians are brothers and that Jews should not charge them interest. However, he acknowledges that Jews sometimes do "'bite' the Christians with usury," though only out of necessity, which might excuse it (p. 219). While he admits that this action may be improper, he goes on to suggest that, if structured as a contract, a non-biting interest-bearing loan between Jew and Christian might be permissible. In

---

[1] Also referred to as the oral law, rabbinic law was codified between 200 and 500 C.E. in the Talmud. It consists of two parts: the Mishnah, a commentary on biblical (or written) law, and the Gemara, a commentary on the Mishnah.

fact, he states that "modern rabbis" allow this for the purpose of raising income for the poor, though he cautions that "this must be understood intelligently" (p. 220). He concludes that usury is licit in a Christian state, but locates this permission in the arguments of Medices rather than in his own. Leon Modena, a prominent seventeenth-century rabbi, in his explanation of Jewish law to Christians, seeks to respond to misrepresentations of the Jewish laws governing usury. He challenges the notion that Jewish law permits the malicious practice of fraud against Christians, though he concedes that living in the diaspora has damaged "the ancient Israelitish uprightness" (p. 222). However, he implicitly attributes this degradation to the legal restrictions placed by Christian rulers on the ability of Jews to hold property and practice their vocations. Although Modena also believes that Christians are brothers who should not be charged interest, he points out that because Jews "have no other way of livelihood left them, but only this of usury, [the Rabbis] allege it to be lawful for them to do this, as well as for the rest of their brothers by nature" (pp. 222–23). Does the play's representation of Shylock's view of usury represent it as *the* Jewish view? What difference does it make for our understanding of the play to realize that in the early modern period there was a wide range of views on the subject of charging interest? How does the fact that Jews and Christians agree in some respects about usury affect our interpretation of the conflict between Shylock and Antonio? What do we make of the fact that Shylock lends money to Antonio on the latter's terms, that is, without interest?

While Modena is anxious to deny that Jews aspire to extort money from Christians, Sir Thomas Sherley is eager to demonstrate the "profit that may be raised . . . out of the Jews" in his proposal offered to King James in 1607 (p. 224). Christians' extorting money from Jews has a long history in Europe, and Sherley does not hesitate to advance a plan to exploit Jews in return for allowing them to settle in Ireland or trade in England. He implies the vocation of these Jews is that of merchant, though, and not usurer; profits to England would accrue by the payment of tributes, customs, and fines by the Jews. In addition, the English would be able to borrow large sums from them. By the end of the sixteenth century, Christians were increasingly encountering Jewish merchants rather than moneylenders. In his description of the Jewish community of Constantinople, Nicolas de Nicolay acknowledges the presence of Jewish lenders there, but concentrates on their increasing involvement in "trade and traffic of merchandise" to the point of dominating the Levant market (p. 226). English trading in the Mediterranean market during the same period would have brought them in increasing contact with merchant Jews. In what ways does the Christian community in *The Merchant of Venice* profit from Shylock's wealth? Does it change our perceptions of Shylock to realize that an early modern audience might have seen his representation as a usurer as old-fashioned?

Merchants

While English merchant companies existed in some form in the late fif-teenth century, they increased substantially in number and scope by the end of the sixteenth century. In the early sixteenth century, England relied heav-ily on German and Italian merchants, not only to import goods but also to export raw materials and cloth; foreign traders handled half of all English imports *and* exports (Clay 2: 106). Anti-alien sentiment and associations of merchants with greed combined to produce the disparaging representation of the Italian merchant, as seen in Robert Wilson's *The Three Ladies of Lon-don* (p. 154). Still, local merchants did not always fare better, as Thomas Wil-son's Gromel Gaynor (p. 194) indicates. However, merchants gained respect as their prominence in the economy grew. While national and religious con-flicts depressed trade in the second half of the sixteenth century, England increased both imports and exports of goods by developing direct trade rela-tions in the Baltic and the Mediterranean, expanding the geographical scope of commerce (Clay 2: 126–27, 131). Although most English overseas trade was conducted by individual merchants or small, temporary partner-ships, the development of trade with distant countries and voyages of dis-covery required more substantial investments and led to the development of joint stock companies "in which the necessary capital was raised by the pooling of a relatively large number of relatively small contributions" (Clay 2: 191, 193). If the company sought and received a charter from the crown, it usually acquired a monopoly on the commerce of a particular item or route that promised greater profits (Clay 2: 193). The Turkey Company, which later became the Levant Company, obtained in their charter such exclusive rights to trade in the territories of the Grand Signoir of Turkey without competition from the ruler or any of her subjects. In her letter granting the charter, Elizabeth I emphasizes the desert of the merchants requesting the monopoly, who have "travailed, and caused travail to be taken, as well by secret and good means, as by dangerous ways and passages both by land and sea, to find out and set upon a trade of merchandise and traffic into the lands, islands, dominions, and territories of the great Turk" (p. 228). The risk and effort they invested served to justify the grant they were receiving. The value of such an enterprise was viewed not only in financial terms, but in political and religious ones as well: "Whereby there is good and apparent hope and likelihood both that many good offices may be done for the peace of Christendom, and relief of many Christians that be or may happen to be in thralldom or necessity under the said Grand Signoir, his vassals or subjects, and also good and profitable vent and utterance may be had of the commodities of our realm, and sundry other great benefits to the advancement of our honor, and dignity Royal, the increase of revenues of our crown, and general wealth of our realm" (p. 228). The merchant hence becomes an agent of relief to his fellow Christians, a promoter of national honor, and a benefactor to his fellow countrymen by bringing back goods and profit. The Merchant Adventurers' Company, which held a monopoly on wool export, was already in operation by the beginning of the sixteenth century. When its secretary, John Wheeler, writes in the beginning of the seventeenth century, he draws on positive views of merchants currently circulating in order to counter increasing disaffection with monopolies (see p. 231). He too emphasizes the effort expended by the merchant to accommodate the needs of the commonwealth, and suggests that trade is so natural an activity as to be practiced in some way by almost everyone. Daniel Price sought to emphasize the spiritual aspects of trade in his sermon on Matthew addressed to the London Company of Merchants. Again, the labor of the merchant is associated with risk and diligence, which makes him an allegory for "the true Christian, [who] seeketh, [and] taketh pains, . . . to follow hard to the mark, . . . no peril, no danger, no cost, no temptation, no opposition can confront him" (p. 237). With regard to *The Merchant of Venice,* does it change our view of Antonio to know that merchants also functioned as moneylenders and that negative stereotypes of trade circulated at the time the play was performed? How do the more positive representations of merchants, as articulated by Wheeler and Price, resonate in the character of Antonio?

➔ Biblical Laws *1611*

The Hebrew Bible regulated the making of loans, and in most instances forbade the taking of interest, described both as נשך (*neshech*) or biting, usually translated as usury, and תרבית (*tarbit*) or increase. Exodus and Leviticus prohibit charging interest on loans to the poor, whether a stranger or a brother, whereas Deuteronomy permits interest to be charged of a stranger, though presumably not a poor one. All subsequent discussion and regulation of interest on loans in

King James Bible (London, 1611).

the Jewish and Christian traditions, examples of which are cited in subsequent excerpts, draw on the principles established in these verses. While Shakespeare does not directly cite any of these biblical texts, questions arising from these laws — who is a brother? who is a stranger? in what circumstances can interest be charged? — not only appear in Shylock and Antonio's heated exchange in act 1, scene 3, but also circulate throughout the play.

Exodus 22

25. If thou lend money to any of my people that is poor by thee, thou shalt not be to him as an usurer, neither shalt thou lay upon him usury.

Leviticus 25

35. And if thy brother be waxen poor, and fallen in decay with thee; then thou shalt relieve him: yea, though he be a stranger, or a sojourner; that he may live with thee. 36. Take thou no usury of him, or increase: but fear thy God; that thy brother may live with thee. 37. Thou shalt not give him thy money upon usury, nor lend him thy victuals for increase.

Deuteronomy 23

19. Thou shalt not lend upon usury to thy brother; usury of money, usury of victuals, usury of any thing that is lent upon usury. 20. Unto a stranger thou mayest lend upon usury; but unto thy brother thou shalt not lend upon usury: that the Lord thy God may bless thee in all that thou settest thine hand to in the land wither thou goest to possess it.

→ THOMAS WILSON

From A Discourse upon Usury by Way of Dialogue and Orations   *1572*

Thomas Wilson (1525?–1581) was educated at Cambridge and wrote *The Art of Rhetoric* (1553), which rejected Continental idioms in favor of English terms. He dedicated the book to John Dudley, whose father, the duke of Northumberland, was de facto ruler for Edward VI, whose youth prevented him from ruling in his

---

Thomas Wilson, *A Discourse upon Usury by Way of Dialogue and Orations, for the Better Variety, and More Delight of All Those, That Shall Read This Treatise* (1572), 32r–33r, 35r–38r, 93v, 97r–v, 178v–180r, 184v–185r, 192v–193v.

own right. Northumberland was the most powerful man in England at that time and a committed Protestant who opposed Mary Tudor's succession to the throne. When Northumberland was executed by Mary, Wilson fled to Italy, where he was eventually arrested and charged before the Inquisition with heresy for his books on rhetoric and logic. He escaped during a riot and eventually returned to England in 1560 after Elizabeth I's ascension to the throne, where he was named Master of Requests, a leading officer in the court which dealt with the suits of the poor. He was also a member of Parliament and was sent to Portugal to negotiate commercial matters. Wilson wrote *A Discourse upon Usury by Way of Dialogue and Orations* (1572) in the late 1560s, as he was translating Demosthenes' orations. He continued in various important diplomatic missions and eventually was named privy councillor and secretary of state in 1577. In *A Discourse upon Usury* he presents the views of a "rich worldly merchant," a "godly and zealous preacher," and two experts in the civil and common law, respectively. The preacher, who rejects any type of interest, wins over his fellows in the debate. It is possible that Wilson's experience in the Court of Requests made him particularly sensitive to the hardships interest posed to poor borrowers. Antonio's opinion on moneylending essentially echoes Wilson's; while this view still circulated widely at the end of the sixteenth century, it was increasingly ignored in practice.

THE PREACHER'S ORATION

I know a gentleman born to five hundred pound land,[1] and entering into usury upon pawn of his land did never receive above a thousand pound of net money, and within certain years running still upon usury, and double usury, the merchants terming it usance and double usance by a more cleanly name, he did owe to master usurer five thousand pound at the last, borrowing but one thousand pound at first; so that his land was clean gone, being five hundredth pounds inheritance, for one thousand pound in money, and the usury of the same money for so few years, and the man now beggeth. I will not say but this gentleman was an unthrift[2] diverse ways in good cheer, nay in evil cheer I may call it, in wearing gay and costly apparel, in roistering[3] with many servants more than needed, and with mustering in monstrous great house,[4] in haunting evil company, and lashing out fondly[5] and wastefully at cards and dice, as time served. And yet I do say, he lost more by the usurer than he did by all those unthrifty means: for his vain expenses was not much more than a thousand pound, because he had no more: whereas the usurer had not only his thousand pound again, but four times more,

---

[1] **five hundred pound land:** possibly refers to income of the land per year.   [2] **unthrift:** not thrifty.   [3] **roistering:** given to noisy reveling.   [4] **mustering in monstrous great house:** entertaining liberally.   [5] **lashing out fondly:** squandering.

FIGURE 12 *An English Usurer, from John Blaxton,* The English Usurer *(1634). The demon poised behind the usurer suggests his association with sin.*

which is five thousand pound in the whole, and for want of this payment the five hundred pound land was wholly his. And this gain only he had for time. They say time is precious. He may well say, time was precious to him that payed so dearly for it; or rather the usurer may say that time was very precious to him that took[6] so much unto him. . . .

And therefore I say and maintain it constantly that all lending in respect of time for any gain, be it never so little, is usury, and so wickedness before god and man, and a damnable deed in itself, because we are commanded to lend freely, and to look for nothing over and above that we lend. But, as I said before, this usury is the daughter of covetousness, a monstrous daugh-

[6] **took:** brought.

ter, I say, of an horrible foul foster dame.[7] Seneca said wisely: . . . the covetous man doth nothing well, but when he dies. Which sentence may most aptly be said of the wretched usurer, who in his lifetime is the cause of all hurt, for by leaving his goods behind, there may some good come by his death when his wealth is dispersed abroad amongst others, and his unmerciful dealing brought to an end. . . .

The scripture commandeth: thou shalt not steal, thou shalt not kill, thou shalt not commit adultery, thou shall not bear false witness, thou shalt not lend out thy money for gain, to take anything for the loan of it, and yet we do all these things, as though there were neither scripture that forbade us, nor heaven for us to desire, nor hell to eschew, nor god to honor, nor devil to dread. And this last horrible offense, which I count greater, or as great, as any of the rest, is so common amongst us, that we have no sense to take it for sin, but count it lawful bargaining, and judge them goodly wise men that having great masses of money by them will never adventure[8] any jot thereof in lawful occupying,[9] either to carry out our plenty, or to bring in our want,[10] as good merchants use and ought to do, but living idle at home will set out their money for profit, and so enrich themselves with the labor and travail of others, and being themselves none other than drones that suck the honey which other painful[11] bees gather with their continual travail[12] of diverse flowers in every field. And whether these men be profitable or tolerable to a commonwealth, or no, I report me to you.[13] Besides that, god doth utterly forbid them, whose commandment ought to be obeyed if we be Christians and of god, as we profess to be.

And therefore for my part, I will wish some penal law of death to be made against those usurers, as well as against thieves and murderers, for that they deserve death much more than such men do, for these usurers destroy and devour up, not only whole families, but also whole countries, and bring all folk to beggary that have to do with them, and therefore are much worse than thieves or murderers, because their offence hurteth more universally and toucheth a greater number, the one offending for need, and the other upon willfulness. And that which is worst, under the color of friendship, men's throats are cut, and the doers counted for honest and wise men amongst others that have so ungodly gathered goods together. What is the matter that Jews are so universally hated wheresoever they come? Forsooth, usury is one of the chief causes, for they rob all men that deal with them, and

---

[7] **foster dame:** foster mother. [8] **adventure:** risk. [9] **occupying:** occupation, business. [10] **either . . . to bring in our want:** export surplus local goods and/or import goods not available locally. [11] **painful:** diligent. [12] **travail:** journeying or toil. [13] **I report me to you:** I appeal to you (to decide).

undo them in the end. And for this cause they were hated in England, and so banished worthily, with whom I would wish all these Englishmen were sent that lend their money or their goods whatsoever for gain, for I take them to be no better than Jews. Nay, shall I say: they are worse than Jews. For go whither you will throughout Christendom, and deal with them,[14] and you shall have under ten in the hundredth, yea sometimes for six at their hands,[15] whereas English usurers exceed all god's mercy, and will take they care not how much, without respect had to the party that borrowed, what loss, danger, or hindrance soever the borrower sustaineth. And how can these men be of god that are so far from charity, that care not how they get goods so they may have them? . . .

So that by these two idle occupations, great usury and many flocks of sheep and herds of beasts, this noble country is made in a manner a forest, and brought to great ruin and decay, through dispeopling of men,[16] over-throwing of towns, and oppressing of the poor, with intolerable usury. And, I pray you, what is more against nature, than that money should beget or bring forth money, which was ordained to be a pledge or right betwixt man and man, in contracts and bargaining, as a just measure and proportion in bargaining, and not to increase itself, as a woman does, that brings forth a child, clean contrary to the first institution of money?[17]. . .

Yea, anything above the principal is excess, because it is an injury or wrong done to another man, by usurping his goods, and so a deadly sin. . . .

Those other learned men . . . Calvin and Bucer,[18] did somewhat enlarge this law by a charitable exposition, for the hardness of men's hearts void from mercy, and also for very necessity sake, to help the needy banished men then dwelling amongst them, not but that they would have all men lend freely as god hath commanded.[19] And yet what warranty have they for so expounding the scriptures of god? But go we on. The lord god saith in his Ten Commandments, Thou shalt not covet thy neighbor's goods, his ox, his ass, or anything else that is thy neighbor's. And what other thing, I pray you, do these usurers that take overplus, but covet their neighbors' goods? For theirs it is not, because they did lend it, and more than they did lend they should not have in right,[20] so that they break the law . . . thou shalt not

---

[14] **them:** Jewish usurers. [15] **ten . . . hands:** ten or six percent interest. [16] **dispeopling of men:** depopulation. [17] **the first institution of money:** the original purpose of money. [18] **Calvin and Bucer:** John Calvin was a French Protestant theologian of the Reformation; Martin Bucer was a German Protestant reformer who taught in England at Cambridge University. Both Calvin and Bucer allow for taking of interest in appropriate circumstances (*EBO*, s.v. John Calvin; Nelson). [19] **for the hardness . . . commanded:** if people are too hard-hearted to lend without interest, then usury could be permitted; but they prefer free lending.
[20] Since Wilson considers the loan the property of the borrower, the lender has no real claim to it, much less to additional interest.

covet. Now, lord in heaven! Who, hearing these speeches, and having the fear of god before his eyes, will of set purpose and willful madness become a usurer? For, as I can perceive, the greatest harm in this offense is that men have not the sense or feeling of this heinous fault within their hearts, but think they may lend their money and goods for gain, without committing sin or offending god at all, whereas of other sins they are ashamed, although through frailty they do offend. For no man durst ever stand in the defense of theft, adultery, or murder, and yet usury is as great and as horrible before god as any of them all, according to the opinion of all good and learned men in all ages. Yea, so horrible is this sin, that amongst all other sins it makes men to forget god, or rather to think there is no god.

Marcilius Ficinus[21] in his book *De Christiana Religione* [On Christian Religion], the eleventh chapter, sayeth thus: ... Only usurers amongst all others (says he) being bond slaves to covetousness cannot think well of heavenly things of god; and the reason is for that they are worldly, and makes the world their god. And truly as a Christian is no ethnic,[22] Paynim,[23] nor Jew, so is he no usurer; and this will I boldly say, that Turks, Tartarians,[24] and ethnics indeed are pure angels in comparison of these worldly usurers, and less harm do they to Christians than usurers do, and are less cruel or unmerciful. For the Paynim sometimes showeth mercy, but the usurer never shows mercy at all, but would be lord of the whole world himself alone if he might. ...

Some learned fathers,[25] yea, and some great philosophers also, think it a thing almost impossible for a great rich merchant that is a mighty occupier[26] to be a good Christian. But I am not altogether of the mind, especially if the merchants that are great occupiers do live in any fear of god. For I think there be of merchants, as there be of other sorts, and surely all such as deal lawfully in their allowed trades, as I have said, are honest, just, upright and worshipful, and not inferiors to other men of any calling. True it is, we are all sinners more and less, and therefore everyone of us had need to call to god for mercy. And assured we are that god, being the searcher of our hearts, will take an account of all our doings. ...

For, if the merchant may be allowed to make gain of his money, he will rather use the certain and assured way than[27] dangerously adventure the seas, and so the Queen shall lose her gain and right of inheritance, and the State shall be undone. The plough man will no more turn up the ground for

---

[21] **Marcilius Ficinus:** Marcilio Ficino was an Italian philosopher, theologian, and linguist (*EBO*, s.v. Marcilio Ficino).   [22] **ethnic:** heathen.   [23] **Paynim:** pagan.   [24] **Tartarians:** Central Asian people, associated with savagery.   [25] **learned fathers:** Church Fathers, early eminent leaders and scholars of Christianity.   [26] **occupier:** businessman.   [27] **than:** text reads "that."

uncertain gain, when he may make an assured profit of his money that lies by him. The artificer[28] will leave his working. The clothier will cease his making of clothes, because these trades are painful and chargeable.[29] Yea, all men will give themselves wholly to live an idle life by their money, if they have any. . . .

Whereof, considering god, nature, reason, all scripture, all law, all authors, all doctors, yea all counsels besides, are utterly against usury, if you love god and his kingdom, my masters, if your natural country be dear unto you, if you think to have merry days in this world, and to live in joy, you and your children after you, both now and ever, yea, if you have care of your own souls, for Christ's sake abhor this ugly usury and loathe with all your hearts this cursed lime[30] of the devil, and lend indeed freely, as god has commanded you, . . . having always a charitable intention with you to help your poor neighbors with part of your plenty, and make no merchandises hereafter by the sun shining and moon shining, by years, by months, by days and by hours,[31] lest god take his light from you, and shorten all your days and all your hours. And as you lend freely, as occasion serveth, to all men, of what estate or condition soever they be, so give to the miserable folk willingly, and help the poor householder frankly. And then shall god bless you and all yours, and give you the peace of conscience which passes all treasure, and make you inheritors of his kingdom, after this transitory life. Now god for his mercy's sake open all your eyes, and give you the right understanding of this his will, for his dear son's sake, Jesus Christ, our lord and savior. Amen.

[28] **artificer:** craftsman.   [29] **chargeable:** burdensome.   [30] **lime:** sticky substance used to trap birds.   [31] **by years . . . by hours:** lending on interest for a specified time.

→ Debate on the Usury Bill   *1571*

The debate on the Usury Bill reflects the shift of early modern attitudes as charging interest becomes morally acceptable and economically necessary. Thomas Wilson, member of Parliament and author of *A Discourse upon Usury,* expresses in the discussion below the older view that charging any kind of interest is a sin. While others agree with him regarding the sinfulness of usury, nevertheless several see usury as inevitable and thus are willing to allow its practice within a limited scope. In spite of the numerous arguments published against

Debate on the Usury Bill in the House of Commons, 1571, in *The Journals of All the Parliaments during the Reign of Queen Elizabeth, Both of the House of Lords and House of Commons,* collected by Sir Simonds D'Ewes (London: Paul Bowes, 1682), 171–74.

usury in this period, it was, in effect, permitted insofar as those charging rates below 10 percent per year were not punished. The text of the statute that emerged from this debate follows this excerpt.

The Bill against Usury was read the second time, whereupon ensued diverse arguments and speeches. . . .

Mr. Molley[1] first learnedly and artificially[2] making an introduction to the matter, showed, what it might be thought on[3] for any man to endeavor the defense of that which every preacher at all times, following the letter of the Book, did speak against; yet saith he, it is convenient, and being in some sort used, it is not repugnant to the word of God. Experience hath proved the great mischief which doth grow by reason of excessive taking, to the destruction of young gentlemen and otherwise infinitely: but the mischief is of the excess, not otherwise. Since to take reasonably, or so that both parties might do good, was not hurtful; for to have any man lend his money without any commodity, hardly should you bring that to pass.[4] And since every man is not an occupier[5] who hath money, and some which have not money may yet have skill to use money, except[6] you should take away or hinder good trades, bargaining and contracting cannot be.[7] God did not so hate it, that he did utterly forbid it, but to the Jews amongst themselves only, for that he willed they should lend as brethren together; for unto all others they were at large,[8] and therefore to this day they are the greatest usurers in the world. But be it, as indeed it is, evil, and that men are men, no saints, to do all these things perfectly, uprightly and brotherly; yet [from two evils, the lesser one is to be preferred] and better may it be born[9] to permit a little, than utterly to take and prohibit traffic, which hardly may be maintained generally without this.

But it may be said, it[10] is contrary to the direct word of God, and therefore an ill law. If it were to appoint[11] men to take usury, it were to be disliked; but the difference is great between that and permitting . . . a matter to be unpunished. . . . We are not, quoth he, so straightened[12] to the word of God, that every transgression should be surely punished here. Every vain word is here forbidden by God, yet the temporal law doth not so utterly condemn it. As for the words of the Scripture, he saith, the Hebrew soundeth thus in answer of this question: [that he should not give his money

---

[1] **Mr. Molley:** presumably a member of Parliament; I have not been able to identify him further. [2] **artificially:** skillfully. [3] **on:** of. [4] **hardly should you bring that to pass:** it would be difficult to bring this about. [5] **occupier:** businessman. [6] **except:** unless. [7] **except . . . cannot be:** meaning unclear, but the sense seems to be that trade will be hindered if interest is not allowed. [8] **unto all others they were at large:** free to lend upon interest. [9] **born:** endured. [10] **it:** the proposed bill. [11] **appoint:** decree. [12] **straightened:** strictly bound.

with biting]:[13] so it is the biting and over-sharp dealing which is disliked and nothing else. And this, he said, was the opinion and interpretation of the most famous learned man Beza, and in these days, of Bellarmine[14] and diverse others; who say, that the true interpretation of the Hebrew word is not *usura* [usury], but *morsus* [biting].

Doctor Wilson,[15] Master of the Requests, said that in a matter of so great weight he could not shortly speak, and acknowledging that he had thoroughly studied the matter, desired the patience of the House. And first he endeavored to prove that the common state may be without usury; then he showed how even men that have been ignorant of God or his laws, finding the evils thereof by their laws, redressed it, and utterly prohibited the use thereof. As the Athenians caused all the writings taken for interest money to be burnt. . . . He then made a definition of usury, showing it was taking of any reward, or price or sum, over and above the due debt. To make anything of that which is not mine, it is robbery. Forthwith upon the delivery[16] of the loan money, it is not mine. . . . He remembered, out of Ezekiel and other the Prophets, sundry places of Scripture, and vouched St. Augustine's saying, that to take but a cup of wine, is usury and damnable. This he seemed to say in answer to that which had been before pronounced, that it was not usury except it were *morsus* [biting].

He showed that loss may grow by usury; first, to the Queen then to the Commonwealth. To the Queen in this, that men not using their own money, but finding great gain in usury, do employ the same that way; so that her customs[17] must decrease. To the Commonwealth, for that who so shall give hire for money, is to raise the same in the sale of his commodity.[18] All trades shall be taken away, all occupations lost; for most men seeking most ease, and greatest gain, without hazard or venture, will forthwith employ their money to such use. He showed it to be so hateful in the judgment of the common law, that a usurer was not admitted to be a witness, nor after his death to the common sepulcher of Christians.[19] And for that his discourse had been long, he inserted (as he said) this tale for recreation of the hearers.

In Italy, quoth he, a great known usurer being dead, the curate denied him the common place of burial. His friends made suit,[20] the priest would not hear; in fine,[21] the suitors bethought them of a policy to bring it to pass

---

[13] that . . . biting: paraphrase of Leviticus 25:37 (see p. 194).   [14] Beza . . . Bellarmine: Theodore Beza was a Protestant Reformation theologian; Robert Bellarmine was an Italian Catholic theologian and a specialist in Hebrew studies.   [15] Doctor Wilson: Thomas Wilson.   [16] delivery: to the borrower.   [17] customs: duties levied on goods.   [18] who . . . commodity: i.e., charging interest forces up the price of goods.   [19] admitted to be a witness . . . common sepulcher of Christians: allowed neither to serve as a witness in a court of law nor to be buried in a Christian cemetery.   [20] made suit: entreated.   [21] in fine: in conclusion.

that he might be buried in the church, which was this. The parson of the church did accustomably use to carry his books daily from his house to the church on his ass; and the ass by often going needed not to be driven, but knowing his journey, as soon as he was laden, would of himself go to the church door. They desired the parson, his ass might carry the dead body,[22] and where it should stay,[23] the body to be buried. To so fond[24] a request the priest agreed, the body was laid on the ass, who feeling a greater burden than he was used to bear, did run towards the town, never staying until he came to the common place of execution.

This tale merrily told, he again entered to his matter, and proved the condemnation of usury and usurers, by the authority of the Nicene[25] and diverse other councils: he showed that the Divines[26] do call usury a spider, a canker, an asp, a serpent, and a devil; he showed how in nature the offenses of homicide and usury are to be compared, and by examples proved the ruins of diverse commonwealths, when such practices for gain are suffered, as that of the commonwealth of Rome, etc. The manner of exchange now used in London, and how much abuse he showed, a thing in old time not practiced, but by the King, as in Edward III's time, when thereby the King obtained such treasure, and such excessive wealth, that it was first wondered at, then guessed that it grew by the science of alchemy. . . . He concluded that the offense in his conscience should be judged felony. . . .

After which one (whose name is not expressed . . .) endeavored the answer of Mr. Wilson, but with a protestation of his insufficiency, and then he showed, how the divines have not agreed what is usury, but for his own part, he was to incline to the opinion of the learned of these days . . . which took that for no usury which is without grievance.[27] . . . To kill is prohibited, yet sometimes not to kill is evil. Phineas killed, and was therefore commended.[28] And thefts at times have been in scriptures approved. So likewise usury is allowed of in the scriptures, but that it might be used to strangers only; albeit the chosen children of God amongst themselves might not use it. But let be, whether it be utterly unlawful, or in some sort to be tolerated, it is a question; and until it be determined for the common commodity and maintenance, let it be as hitherto it hath been used. And for the common sort of bargains of corn for cloth, silk for land, etc. what they be, whether

---

[22] **They desired . . . dead body:** the friends of the deceased asked the parson to let his ass carry the body. [23] **stay:** come to a stop. [24] **fond:** foolish. [25] **Nicene:** the Council of Nicea (325) was the first ecumenical council of the Christian Church. It dealt with a number of matters and issued many decrees, including one condemning the lending of money at interest by the clergy (*EBO*, s.v. Council of Nicea). [26] **Divines:** clergymen. [27] **grievance:** complaint. [28] **Phineas . . . commended:** Numbers 25:6–13 relates Phineas's killing of a man and women engaged in public promiscuity associated with idol worship.

usury or no, we know not. That all should be well, it is to be wished; that all may be done well among men, it is beyond hope, for we are no saints, we are not of perfection to follow the letter of the gospel. . . . To do that therefore to another which we would to ourselves (the state, circumstance, and case to ourselves considered) is commendable, or not to be reproved; if we ourselves be to borrow, who is it that would not in extremity give a little to save much money? It is said, the usurer doth or may grow rich: Who hath disliked in a commonwealth that there should be *homines boni frugi*?[29] They may be considered, and may be good, more than for one purpose. He further stood on this, that God did not absolutely forbid usury, which surely if it had been utterly ill, he would have done. And he added, that the common laws were cruel in their censures, and wished that they should be no more remembered than they are followed.

Sergeant Lovelace[30] argued to this effect, that usury was of money only, protesting that he hated all kind of usury, but yet the greater the ill was, the more and more greatly did he hate the same. But to prohibit it with so sharp and extreme a law as to lose all, he thought it would be the ground of greater covetousness; withal he added, to prohibit the ill of covetousness in generality were rash, void and frivolous; since that the speech and the act itself is indefinite, comprehending all our actions and doings; and therefore, as utterly vain to prohibit it, in vain words of generality. To prohibit drunkenness, pride, envy, surfeiting,[31] etc., were somewhat in some particular sort;[32] to do it in generality . . . were but folly. Of these great evils, . . . when we may not reach to the best, furthest and uttermost, we must do, as we may say, by degrees. . . . Whereupon he concluded, that there should be degrees in punishing of usury; as he that should take so much, to lose, or be punished thus; he that shall take more, more deeply. . . .

Mr. Norton[33] showed that all usury is biting: as in the word *steal* is contained all kind of injurious taking away of a man's goods: and as slanderizing is said to be murdering or homicide; so is usury justly ever to be said biting, they being both so correlated or knit together, that the one may not be without the other. He concluded, that since it is doubtful what is good, we should be mindful of the old saying, [Do not do that which is dubious], and for that [that which is not from faith is sin]; therefore he wished that no allowance should be of it.

---

[29] *homines boni frugi:* successful or wealthy men.   [30] **Sergeant Lovelace:** probably William Lovelace, d. 1577; sergeants were an elite class of lawyers, comparable in status to knights, who had a monopoly over civil litigation (Baker 180, Milsom 40, Baker and Milsom, 668).   [31] **surfeiting:** gluttony.   [32] **somewhat in some particular sort:** was worthwhile in particular instances.   [33] **Mr. Norton:** Thomas Norton (1532–1584) poet, lawyer, and outspoken member of Parliament.

→ Usury Bill

Whereas in the parliament holden the seven and thirty year of the reign of our late sovereign lord, King Henry the Eighth of famous memory, there was then made and established one good act for the reformation of usury; by which act the vice of usury was well repressed, and specially the corrupt chevisance[1] and bargaining by the way of sale of wares and shifts of interest; and where since that time by one other act made in the fifth and sixth years of the reign of our late sovereign lord, King Edward the Sixth, the said former act was repelled,[2] and new provisos for repressing of usury devised and enacted; which said latter act had not done so much good as was hoped it should, but rather the said vice of usury, and especially by way of sale of wares and shifts of interest, hath much more exceedingly abounded, to the utter undoing of many gentleman, merchants, occupiers, and other, and to the importable[3] hurt of the commonwealth, as well for that in the said latter act there is no provision against such corrupt shifts and sales of wares, as also for that there is no difference of pain, forfeiture, or punishment upon the greater or lesser exactions and oppressions by reason of loans upon usury: be it therefore enacted, that the said latter statute made in the fifth and sixth years of the Reign of King Edward the Sixth, and every branch and article of the same, from and after the five and twenty day of June next coming shall be utterly abrogated, repelled, and made void; and that the said act made in the said seven and thirty year of King Henry the Eighth, from and after the said five and twenty day of June next coming shall be revived and stand in full force, strength, and effect.

II. And be it further enacted, that all bonds, contracts and assurances, collateral or other, to be made for payment of any principal or money to be lent, or covenant to be performed, upon or for any usury in lending or doing of any thing against the said act now revived, upon or by which loan or doing there shall be reserved or taken above the rate of ten pounds for the hundred[4] for one year, shall be utterly void.

III. And be it further enacted, that all brokers, solicitors, and drivers of bargains for contracts or other doings against the said statute now revived, whereupon shall be reserved or taken more than after the rate of ten pounds for the loan of a hundred pounds for a year, shall be to all intents and purposes

---

[1] **chevisance:** dealing for profit.   [2] **repelled:** repealed.   [3] **importable:** unbearable, unendurable.
[4] **ten pounds for the hundred:** 10 percent per year.

Usury Not Exceeding Ten Per Cent Permitted (13 Elizabeth, c. 8), 1571, *Statutes of the Realm* (London: G. Eyre and A. Strahan, 1810–22), vol. 4, pt. 1, 542–43.

judged, punished and used as counselors, attorneys or advocates in any case of premunire.[5]

IV. And forasmuch as all usury being forbidden by the law of God is sin and detestable; be it enacted, that all usury, loan and forbearing[6] of money, or giving days for forbearing of money, by way of loan, chevisance, shifts, sale of wares, contract or other doings whatsoever for gain mentioned in the said statute which is now revived, whereupon is not reserved or taken or covenanted to be reserved, paid or given to the lender, contractor, shifter, forbearer, or deliverer above the sum of ten pounds for the loan or forbearing of a hundred pounds for one year, or after that rate for a more or lesser sum or time,[7] shall be, from the five and twenty day of June next coming, punished in form following, that is to say: that every such offender against this branch of this present statute shall forfeit so much as shall be reserved by way of usury above the principal for any money so to be lent or forborne; all such forfeitures to be recovered and employed as is limited for forfeitures by the said former statute now revived.

V. And be it further enacted that justices of oyer and terminer and justices of assise[8] in their circuits, justices of peace in their sessions, mayors, sheriffs, and bailiffs[9] of cities, shall also have full power and authority to inquire, hear, and determine of all and singular offenses committed against the said statute now revived.

VI. And be it further enacted, that the said statute now revived shall be most largely and strongly construed for the repressing of usury, and against all persons that shall offend against the true meaning of the said statute by any way or device directly or indirectly.

VII. Provided always, that this statute doth not extend nor shall be expounded to extend unto any allowances or payments for the finding of orphans according to the ancient rates or customs of the city of London, or any other city where like order is for the custody of orphans and their goods, as is in the said city of London.[10]

---

[5] **premunire:** an action taken against a party, and his or her legal representative, for bringing suit in the wrong court; in this instance, not only the lender, but his or her agents would also be liable to punishment for transgressing the usury statute.   [6] **forbearing:** delaying the payment.   [7] **above the sum . . . time:** i.e., interest above ten percent per year.   [8] **justices of oyer and terminer . . . assise:** judges from the royal courts sent into the country to "hear and determine," that is, inquire into the alleged offense, and to try cases (Baker 20, 24–25).   [9] **mayors, sheriffs, and bailiffs:** various local law enforcement officials.   [10] **Provided always . . . London:** the capital belonging to orphans was loaned out at interest until they came of age and could receive their inheritances (Jones 29).

→ FRANCIS BACON

Of Usury 1625

Francis Bacon (1561–1626) attended Cambridge University at the age of twelve and studied there for two years before being admitted to Gray's Inn, where he studied law. He was a great parliamentarian and statesman, serving as advisor to both Elizabeth I and James I, and held the posts of Attorney-General, Lord Keeper, and Lord Chancellor. His career was a stormy one, despite (or perhaps because of) the high positions he held, and he was ultimately convicted of accepting bribes and barred from further public service. He wrote prolifically on numerous subjects, including science, philosophy, and law. His *Essays or Counsels, Civil and Moral* present brief, personal considerations of different topics; the first edition appeared in 1597, to which emendations and additional essays were added in subsequent editions. In his discussion of usury, presented below, Bacon dispenses with conventional views and rationally analyzes the practice in terms of its advantages and disadvantages. It is striking to compare his discussion of usury with that of Coke in the *Institutes* (p. 210); although they were contemporaries, they take diametrically opposed views here as well as in their debates over matters of state. While Bacon echoes some of the charges against usury made by Wilson and others, he does not situate his critique within the context of sin and Biblical law. He concludes that charging interest is unavoidable and proceeds pragmatically to structure it in such a way as to minimize its "incommodities" (106).

Many have made witty invectives against usury . . . but few have spoken of usury usefully. It is good to set before us the incommodities and commodities of usury, that the good may be either weighed out or culled out; and warily to provide, that while we make forth to that which is better, we meet not with that which is worse.

The discommodities of usury are, first, that it makes fewer merchants. For were it not for this lazy trade of usury, money would not lie still, but would in great part be employed upon merchandising, which is the *vena porta*[1] of wealth in a state. The second, that it makes poor merchants. For as a farmer cannot husband his ground so well if he sit at a great rent;[2] so the merchant cannot drive his trade so well, if he sit at a great usury. The third is incident to the other two; and that is the decay of customs[3] of kings or

---

[1] *vena porta:* a major vein, here used figuratively for significant source or conduit.   [2] **For . . . rent:** a farmer who pays a large rent reduces the profit he can make from his produce.   [3] **customs:** duties paid on merchandise.

Francis Bacon, "Of Usury," *The Essays or Counsels, Civil and Moral, of Francis Bacon* (London, 1625), 239–46.

states, which ebb or flow with merchandising. The fourth that it bringeth the treasure of a realm or state into a few hands. For the usurer being at certainties, and others at uncertainties, at the end of the game most of the money will be in the box; and ever a state flourisheth when wealth is more equally spread. The fifth, that it beats down the price of land, for the employment of money is chiefly either merchandising or purchasing, and usury waylays both. The sixth, that it doth dull and damp all industries, improvements, and new inventions, wherein money would be stirring, if it were not for this slug. The last, that it is the canker and ruin of men's estates, which in process of time breeds a public poverty.

On the other side, the commodities of usury are, first, that howsoever usury in some respect hinders merchandising, yet in some other it advances it; for it is certain that the greatest part of trade is driven by young merchants upon borrowing at interest; so as if the usurer either call in or keep back his money, there will ensue presently a great stand[4] of trade. The second is, that were it not for this easy borrowing upon interest, men's necessities would draw upon them a most sudden undoing; in that they would be forced to sell their means (be it lands or goods) far under foot;[5] and so, whereas usury doth but gnaw upon them, bad markets would swallow them quite up. As for mortgaging or pawning, it will little mend the matter: for either men will not take pawns without use; or if they do, they will look precisely for the forfeiture.[6] I remember a cruel, monied man in the country, that would say, "The devil take this usury, it keeps us from forfeitures of mortgages and bonds." The third and last is, that it is a vanity to conceive that there would be ordinary borrowing without profit; and it is impossible to conceive the number of inconveniences that will ensue, if borrowing be cramped. Therefore to speak of the abolishing of usury is idle. All states have ever had it, in one kind or rate, or other. So as that opinion must be sent to Utopia.

To speak now of the reformation and reiglement[7] of usury; how the discommodities of it may be best avoided, and the commodities retained. It appears by the balance of commodities and discommodities of usury, two things are to be reconciled. The one, that the tooth of usury be grinded, that it bite not too much; the other, that there be left open a means to invite monied men to lend to the merchants for the continuing and quickening of trade. This cannot be done, except[8] you introduce two several sorts of usury,

---

[4] **stand:** halt.   [5] **far under foot:** below true value.   [6] **pawning . . . forfeiture:** items were pawned in exchange for money, but often served as collateral while interest was also charged.   [7] **reiglement:** regulation.   [8] **except:** unless.

a less and a greater. For if you reduce usury to one low rate, it will ease the common borrower, but the merchant will be to seek for money.[9] And it is to be noted, that the trade of merchandise, being the most lucrative, may bear usury at a good rate: other contracts not so.

To serve both intentions, the way would be briefly thus. That there be two rates of usury: the one free, and general for all; the other under license only, to certain persons and in certain places of merchandising. First therefore, let usury in general be reduced to five in the hundred;[10] and let that rate be proclaimed to be free and current; and let the state shut itself out to take[11] any penalty for the same. This will preserve borrowing from any general stop or dryness. This will ease infinite borrowers in the country. This will, in good part, raise the price of land, because land purchased at sixteen years' purchase will yield six in the hundred,[12] and somewhat more; whereas this rate of interest yields but five. This by like reason will encourage and edge industrious and profitable improvements; because many will rather venture in that kind than take five in the hundred, especially having been used to greater profit. Secondly, let there be certain persons licensed to lend to known merchants upon usury at a higher rate; and let it be with the cautions following. Let the rate be, even with the merchant himself, somewhat more easy than that he used formerly to pay; for by that means all borrowers shall have some ease by this reformation, be he merchant, or whosoever. Let it be no bank or common stock, but every man be master of his own money. Not that I altogether dislike banks, but they will hardly be brooked,[13] in regard of certain suspicions. Let the state be answered some small matter[14] for the license, and the rest left to the lender; for if the abatement[15] be but small, it will no whit discourage the lender. For he, for example, that took before ten or nine in the hundred, will sooner descend to eight in the hundred, than give over his trade of usury, and go from certain gains to gains of hazard. Let these licensed lenders be in number indefinite, but restrained to certain principal cities and towns of merchandising. . . .

If it be objected that this doth in a sort authorize usury, which before was in some places but permissive, the answer is, that it is better to mitigate usury by declaration, than to suffer it to rage by connivance.

---

[9] **be to seek for money:** have to seek.   [10] **five in the hundred:** 5 percent.   [11] **shut itself out to take:** prohibit itself from taking.   [12] **six in the hundred:** 6 percent.   [13] **brooked:** tolerated.   [14] **be answered some small matter:** receive a small fee.   [15] **abatement:** reduction in interest rate.

→ SIR EDWARD COKE

From The Institutes of the Laws of England   *1642*

Edward Coke (1552–1634), after being trained in the law, rose very quickly in public life, eventually holding such important positions as speaker of the House of Commons, Attorney-General, and Chief Justice of Common Pleas and King's Bench, two central courts of law. He had a tumultuous career, often clashing with the king, James I; a strong rivalry existed between Coke and Francis Bacon. Coke is perhaps better known as an author of legal texts than as a politician. In addition to his thirteen volume *Reports* (eleven of which were published in his lifetime), Coke's other major work is *The Institutes of the Laws of England*, which is divided into four parts: (1) commentary on a treatise on land tenures; (2) statutes; (3) criminal law; and (4) jurisdiction of different courts of law. Although his writing often reflects his strong biases and contains errors, it made older legal knowledge accessible; he firmly believed in the importance of studying legal origins. Hence it is characteristic of him to consider medieval laws pertaining to Jews even though they were not officially allowed to live in England during his lifetime. In his discussion of the *Statutum de Judaismo* (the statute regarding Judaism), which forbids usury, he argues that this prohibition was the catalyst that forced the Jews to leave England in the expulsion of 1290. Rather than presenting the eviction as resulting from a state order to that effect, Coke suggests it was a voluntary exodus that Jews undertook in reaction to this law, which prohibited their only means of livelihood. His tone is quite hostile toward both usury and the Jews, though he also mentions the financial interest that kings had in Jewish prosperity. *The Merchant of Venice* enacts a similar ambiguity in representing this negative association with Jews and usury concurrent with the notion of Shylock's being a source of wealth to be tapped by Christians.

Statutum de Judaismo, 18 Edward I (1290)

Two great mischiefs did follow before the making of this statute upon Jewish usury; now the difficulty was how the same should be remedied. The mischiefs were these:

1. The evils and disherisons[1] of the good men of the land.
2. That many of the sins or offences of the realm had risen and been committed by reason thereof, to the great dishonor of Almighty God.

The difficulty how to apply a remedy was, considering what great yearly revenue the King had by the usury of the Jews and how necessary it was that

---

[1] **disherisons:** the act of disinheritance.

Sir Edward Coke, *The Institutes of the Laws of England*, Part 2 (London, 1642), 506–07.

the King should be supplied with treasure. What benefit the Crown had before the making of this act appeareth by former records, as take one for many: from the 17th of December in the 50 year of King Henry III until the Tuesday in Shrovetide[2] the second year of King Edward I, which was about seven years, the Crown had four hundred and twenty thousand pounds fifteen shillings and four pence . . . at what time the ounce of silver was but 20 shillings and now it is more than treble so much. . . .

Many prohibitions were made both by this King and others, some time they [the Jews] were banished, but their cruel usury continued, and soon after they returned. And for respect of lucre[3] and gain, King John in the second year of his reign granted unto them large liberties and privileges, whereby the mischiefs rehearsed in this act without measure multiplied.

Our noble King Edward I and his father King Henry III before him, fought by diverse acts and ordinances to use some mean and moderation herein, but in the end it was found that there was no mean in mischief.[4] . . . And therefore King Edward I as this act said, in the honor of God, and for the common profit of his people, without all respect (in the respect of these) of the filling of his own coffers, did ordain, that no Jew from thenceforth would make any bargain, or contract for usury, nor upon any former contract should take any usury; . . . so in effect all Jewish usury was forbidden. . . .

This law struck at the root of this pestilent weed, for hereby usury itself was forbidden; and thereupon the cruel Jews thirsting after wicked gain, to the number of 15,060, departed out of this realm into foreign parts, where they might use their Jewish trade of usury, and from that time that nation never returned again into this realm.

Some are of opinion (and so it is said in some of our histories) that it was decreed by the authority of Parliament that the usurious Jews should be banished out of the realm; but the truth is that their usury was banished by this act of Parliament, and that was the cause that they banished themselves into foreign countries, where they might live by their usury. And for that they were odious both to God and man, that they might pass out of the realm in safety, they made petition to the King, that a certain day might be preferred[5] to them to depart the realm, to the end that they might have the King's writ to his sheriffs for their safe conduct, and that no injury, molestation, damage, or grievance be offered to them in the meantime. . . .

And thus this noble King by this means banished forever these infidel usurious Jews; the number of which Jews thus banished was fifteen thousand and threescore.

---

[2] **Shrovetide:** the three-day period before Lent, beginning on Sunday and running through Tuesday, or Mardi Gras.   [3] **lucre:** money.   [4] **no mean in mischief:** no accommodating evil.   [5] **preferred:** appointed.

→ YEHIEL NISSIM DA PISA

From The Eternal Life   *1559*

Yehiel Nissim da Pisa (1507–1573?) was born in Pisa into a family of bankers and scholars. He was educated in religious and secular studies and trained as a banker. He was also a skillful calligrapher and copied, among other things, Aristotle's *Physics*. His own writings treated a range of subjects, including philosophy, Jewish law, and Kabbalah (mysticism). His text on usury, חיי עולם (*Chayei Olam*, or *The Eternal Life*) was written in Hebrew in 1559 but was not published until the twentieth century (Rosenthal 23–30). Da Pisa situates his discussion of the laws of moneylending in the context of Jewish history and observance of the Torah (the Hebrew Bible or the first five books thereof; also the laws contained therein). Like Leon Modena (p. 221) and Samuel Usque (p. 288), he acknowledges the effects of persecution on the Jewish people, which he attributes to their refusal to follow God's law. In writing *The Eternal Life*, da Pisa hopes, by clearly explaining the law of taking interest, to encourage Jews in their observance of the commandments. Unlike David de Pomis (p. 217), he prohibits Jews from charging interest of each other, though the treatise does delineate other types of transactions in which a lender might earn a profit by sharing in the risk of the borrower. He considers Christians strangers who could therefore be charged interest according to biblical law. Nevertheless, he follows the opinion of the Talmud (the oral law, comprising two parts: the Mishnah, a commentary on biblical law, and the Gemara, a commentary on the Mishnah), which prohibits charging interest of strangers, contrary to the biblical ruling. Such transactions were forbidden by the oral law in order to keep Jews from mingling with pagans, and thus running the risk of falling into idolatry, in the course of business deals. Da Pisa reinterprets the rationale for the Talmudic law to respond to present conditions, arguing that, although permitted, interest should not be charged of non-Jews because it leads to carelessness in observing the prohibition of lending to Jews.

Thus spoke the author:

From the entire human race He chose the Hebrew People and lifted them on the wings of eagles and took them from the hands of the Egyptians to freedom. Israel is the Lord's hallowed portion, His first fruits, and Jacob is the lot of His inheritance. He gave them the Torah of truth and righteous laws which they should observe in order to gain eternal life. He revealed to them His secrets, yea, a multitude of glad tidings, something He did not do

---

Yehiel Nissim da Pisa, *The Eternal Life*, in *Banking and Finance among Jews in Renaissance Italy*, ed. and trans. Gilbert S. Rosenthal (New York: Bloch Publishing Company, 1962), 33–36, 40–45, 47–48, 89–91.

for the other nations who are like dying coals. He gave the Jewish people a living land and a Holy City in its midst, a peaceful habitation, a tent that shall not be moved. Therein is the Ark and Veil, and its people are blessed in the gates of the mighty. Their foundation is on the Holy Mountain where each one performs his worship.

After their exile from the Holy Land, they intermingled among the nations and went from calamity to calamity among the isles, and they[1] became a curse in their[2] midst. For their generation was carried away and exiled and they learned the ways of the gentiles and did not incline their ears to the voice of the righteous Torah. This have we seen and we shall recount it, for they have not recognized His ways and they know not His paths. They have violated several light and grave precepts and have rebelled against God. They take delight in the broods[3] of aliens and have not set their hearts on His pure commandments, nor on the words of the wise men or sages.

And now, I have indeed seen the poor of my people in great difficulties, walking as blind men in the dark, neglecting His righteous laws, with all their intricacies. They find license to violate the most grave transgression in the entire Torah, namely, they lend money to their brother on usury and interest and they violate the laws formulated by our sages. . . .

They have found great loopholes in order to permit practices without any rhyme or reason, for the lust for money has overpowered them. I declare that the staff of the Torah is broken with its glory as well as all of the teachings and law that follow in its light. The people of Israel who used to draw inspiration from its source have devised new devices and have multiplied vanities. They have turned unto it their backs and not their faces and they have removed the speech of men of trust and taken away the sense of elders. Doth not the ear test words in order to elevate the glory of Torah among the masses even as the palate tastes its food? For the heights of the mountains are also His.

Therefore I have girded my loins in order to compose this brief treatise with all of the logical arguments. It will contain all of the laws of interest — both prohibited and permitted — and may they be planted in our hearts like nails. May nothing be lacking and may we be faithful, wise, and understanding. . . .

The purpose of this treatise is to explain the matter of interest and usury prohibited to us by our holy Torah and to clarify all of the details connected with it, both from the Torah as well as from our sages,[4] with all of the subdivisions, for they are as numerous as locusts and have many branches. . . .

---

[1] they: the Jews.   [2] their: the other nations.   [3] broods: children.   [4] sages: refers to the compilers of the Talmud, and thus the Talmud itself.

Indeed its purpose will be to deal with the problem that in these lands, the practice of lending money to gentiles is more widespread than in the rest of Israel's diaspora. Since it seems that there are infringements on the prohibitions of the Torah, we have felt constrained to compose this special treatise, so as to define the limitations and enable everyone to study it without the task of consulting other words or codifiers.[5] Since, in their days, this illegal practice was not so widespread as to merit special treatment, they[6] merely included it among the other precepts. But in these days and in these lands, since the matter of loans has become the common vocation of the people and the chief source of their livelihood, it became essential to compose a short treatise that would, as best as possible, include in it all the laws of interest, so that he who soweth righteousness will have a surer reward and by the wisdom of the righteous will escape from this serious iniquity.

I am not boasting or showing off when I add a word of my own opinion, because everything has been clearly explained by the scholars of the Talmud and by the sages and codifiers. . . .

Many will become wise through this treatise and wise men who understand the laws of the Torah and its preeminence will realize that it is merely a brief summary, so as to enable everyone to take it to heart always as a remembrance between his eyes and as a chain on his neck. And let him know the path which he should follow and how he should conduct his business truthfully and faithfully, lest he stray to the right or left from the teachings of the sages of Israel.

I have called this treatise, *Eternal Life,* the reason being that our prophets cautioned our people against this great sin and listed amongst the most serious transgressions, the matter of interest.

They interpreted the verse, "And he shall surely not live" [Exodus 18:13], to mean that the usurer shall live neither in this world nor the next and they added that whoever lends on interest will not be resurrected. The converse is equally true that whoever is careful not to lend on interest to an Israelite will inherit both worlds and will "dwell in His tents forever" [Psalm 61:5], both in this world and the next and he will arise at the end of days.

CHAPTER I

Therefore, we shall, by way of introduction, offer two introductory statements that will serve as the fundamentals and bases on which our treatise will be built, with the help of the Almighty.

---

[5] **codifiers:** post-Talmudic rabbinic authorities. [6] **they:** other codifiers.

The first principle is that whosoever goes by the name of an Israelite and enters the religion of Moses, of blessed memory, is consequently obliged to observe the positive and the negative commandments of the Torah. Such a person may not evade his responsibility to observe the precepts of the Torah. Indeed, all Israel accepted the [yoke] of the commandments of God when they accepted the Torah and declared, "We will do that which we hear" [Exodus 24:7]. They further received the Torah with a solemn oath and warning that was binding upon their seed until the end of time. All Jews were at that glorious revelation, and in fact, the future generations were there, as Moses himself declared when he said, "I make this covenant and this oath with him that standeth here this day . . . and also with him that is not here" [Deuteronomy 29:14]. Now since no one was left in Egypt, the verse must refer to the coming generations, for the Torah was given to all generations and it is eternally binding, whether we reside in our land or whether we are exiled among the nations and we are required to observe all of its precepts and warnings. In truth, there are certain commandments which are only observed in the land of Israel, . . . which we can only observe when Israel is in control of its land and when the Temple is standing. But the rest of the commandments, such as the Sabbath, the Festivals, the prohibited foods, sexual morality, civil law, and the other prohibitions are incumbent upon us even in exile.

As to what was said about Israel's obligation to observe the Torah and commandments, I do not merely mean the Written Torah[7] alone, but the Oral Torah[8] as well since it is the fundamental interpretation of the commandments and all their details and minutiae which were given to Moses our teacher, and which he handed over to Joshua, and so on through the ages until the days of Rabbi Judah the Prince, who is called Our Holy Teacher, the author of the Mishnah — the interpretation of the commandments.[9] After that came the rest of the scholars who interpreted the Mishnah and called their work the Talmud. After them came the Geonim[10] and the rest of the rabbis, of blessed memory. And were it not for the Oral Torah, we would have walked in the way of the commandments as blind men in the dark. . . .

Whoever willfully forsakes the yoke of a single positive or negative commandment has left the fold of Israel on that matter and is called an "apostate because of one transgression." This applies only if he habitually violates this

---

[7] **Written Torah:** the Hebrew Bible.   [8] **Oral Torah:** the Talmud.   [9] **Moses . . . commandments:** Traditional Judaism teaches that the oral law was given to Moses along with the written law on Mount Sinai, and he passed it along to his disciple Joshua. It is held to have been handed down orally until R. Judah the Prince wrote it down around the third century C.E.
[10] **Geonim:** Babylonian and Palestinian rabbis of the seventh to thirteenth centuries C.E.

law and is firm in his transgression and has completely abrogated the commandment. But if he transgresses occasionally because his evil urge leads him astray, he is not called an apostate in regard to that transgression. . . .

The second principle is that the matter of lending on interest between Jew and . . . Jew is prohibited from the Torah. It is impossible to refute this and it has been explained explicitly in the section of Mishpatim, and in Behar Sinai, and in Ki Tetse.[11] Consequently, the prohibition against lending on interest from one Jew to another Jew is a negative precept from the Torah.

After having prefaced our remarks with these two fundamental principles, we can come to the correct and truthful conclusion that by virtue of these two irrefutable premises, there emerges a truthful conclusion by analogy since every Jew is obliged to observe the commandments of the Torah. Since the prohibition of interest between Jew and Jew is Biblically prohibited, it follows that every Jew is obliged to refrain from taking interest from his fellow Jew. The major and minor premises are obvious, and that which follows from them is equally necessary and true.

CHAPTER 10

In explanation of those to whom it is prohibited to lend on interest.

Interest which is proscribed by the Torah is that which is lent from one Jew to another Jew, as it is written, "Thou shalt not lend on interest to thy brother." This means that anyone who enters into the realm of the Torah of Moses and takes upon himself the yoke of God's commandments, even if he be originally a gentile and subsequently became a righteous proselyte,[12] is in the category of "thy brother" and one may not lend him upon interest. However, one may lend on interest to a gentile for the Torah has specifically permitted such a practice by stating, "Unto a foreigner thou may lend upon interest." Though we say that one may not defraud a gentile, one may lend him on interest since he has voluntarily and of his own accord chosen to pay the sum. The codifiers are divided on this point: Maimonides[13] counted this amongst the positive commandments and indeed, he listed it in his *Book of Precepts* as precept 198. We quote from his book:

---

[11] **Mishpatim . . . Ki Tetse:** a section of the first five books of the Torah is read every Sabbath; these are given names from a word or words in the first verses of the sections. The three sections mentioned here correspond to Exodus 21:1–24:18, Leviticus 25:1–26:2, and Deuteronomy 21:10–25:19.   [12] **proselyte:** convert.   [13] **Maimonides:** (1135–1204) preeminent medieval codifier of Jewish law; da Pisa here refers to his *Sefer haMitzvot,* a listing of the commandments.

Precept 198 teaches us that God commanded us to require interest from a gentile and then we may lend to him . . . for God has commanded us, "Unto a foreigner thou may lend upon interest," and this is a positive commandment.

To be sure, Rabbi Moses of Coucy[14] and others did not count this among the positive precepts, but have declared it to be merely a permissive act, that is, when one lends to a gentile, one may receive interest from him. Since our Talmud is the basic law, it refutes the opinion that the lending of money on interest to a gentile is a positive commandment. In the Gemara Metsia[15] it is written: "Cannot one do without [lending on interest to gentiles]?" That is to say, must one be compelled to take interest from non-Jews? Indeed, the lender has the option to take it, for no one can force him to do so. . . . My opinion is that this is merely a permissive act and even though the Torah permitted it, our sages prohibited us from lending to gentiles on interest as a precautionary measure. They later eased the law so that we may lend to gentiles in order to make a living. In our times, it is the practice to permit lending on interest to gentiles for more than a mere living. In all of the nations of Christendom, the practice of lending money is as widespread as other business enterprises.

Blessed be He who chose them and their teachings and who foresaw what would result from lending on interest to gentiles. It is quite clear that as a result of the habit of lending to gentiles, these people[16] fall into the error of lending on interest and usury to Jews. They can no longer distinguish between truth and falsehood or between the permitted and prohibited practices because habit is a powerful force that makes a great impression on the beliefs of people.

[14] **Moses of Coucy:** Moses ben Jacob of Coucy was a thirteenth-century French scholar who authored the *Sefer Mitzvot Gadol* (The Great Book of Commandments), which lists the negative and positive precepts; while he relies on the work of Maimonides, Moses does not always agree with him (*Encyclopedia Judaica* 12: 418–20).   [15] **Gemara Metsia:** tractate or section of the Talmud dealing with damages; the reference is to 70b–71a.   [16] **these people:** Jewish lenders.

→ **DAVID DE POMIS**

From De Medico Hebraeo   *1587*

David de Pomis (b. 1525) was born in Spoleto to a family who traced its lineage to a prisoner brought back by the Emperor Titus after the destruction of Jerusalem in the first century C.E. He received a degree in medicine at the University of Perugia in 1551 and worked as a rabbi and doctor until prevented by

---

David de Pomis, *De Medico Hebraeo* (Venice, 1587), in H. Friedenwald, *The Jews and Medicine* (Baltimore: Johns Hopkins University Press, 1944), 40–44.

Pope Paul IV's bull (papal edict) that prohibited Christians from hiring Jewish physicians. Shortly after Gregory XIII reissued this ban, de Pomis wrote *De Medico Hebraeo Ennarratio Apologica* (1587), a defense of Jewish physicians. After serving several powerful noblemen, de Pomis eventually obtained permission to practice medicine generally among Christians. Like Leon Modena's text (p. 221), *De Medico Hebraeo* asserts the honor and integrity of Jews and Judaism, and challenges Christian antipathy and suspicion toward them (Friedenwald 1: 31–33). In this excerpt, provided in a modern translation of the Latin, de Pomis disputes the notion that Jews and Christians are enemies. He uses as support a text on moneylending written by Sixtus Medices, a Dominican theologian, who argues (in *De Foenore Iudaeorum Libri Tres,* published in Venice in 1555 [Nelson 22n]) that Jews and Christians are brothers and therefore should not lend on interest to each other. De Pomis, and apparently Sixtus Medices, suggest that Jews are allowed to charge interest out of necessity, not enmity toward Christians, and argues that through a contract, the sin of usury is avoided such that Jews could lend on interest to both their Christian and Jewish brethren.

And all the more should we show respect to the Christians (who, as we have said, differ so little from the Jews), since they profess to observe not only the seven precepts,[1] but indeed diligently follow (or should follow) the ten that were handed down by Moses. Thus, there is no reason for interpreting whatever is said in the book against the "foreign cult"[2] as being directed against the Christians; rather it is directed against those seven tribes, worshipers of idols, dwelling at Judea at the time — in order that the Jews might not be corrupted by contact with them. As, in truth, it befell since these tribes were not completely removed from the Holy Land, as God had ordered. Thus it is evident that Israel sinned.

Again it is said, and made a matter of discussion, that there are some things to be found in the Gemara[3] written against Jesus and against those following his teachings. But to this we counter that the real meaning should be understood as referring, not to the Nazarene,[4] but to another man named Jesus who dwelt in the condemned city of Lud. He was the son of Papos who lived much earlier, at the time of Judas Maccabee;[5] this was obviously many years before Jesus of Nazareth. . . .

---

[1] **the seven precepts:** the seven commandments of the sons of Noah; a rabbinical concept of natural law, which they understood to be incumbent upon all peoples. [2] **book against the "foreign cult":** Avoda Zarah, a section of the Talmud that deals with idol worship. [3] **Gemara:** a commentary on the Mishnah, which is a commentary on the Hebrew Bible. Together the Gemara and Mishnah make up the Talmud or the oral law. [4] **Nazarene:** Jesus Christ. [5] **Judas Maccabee:** Jewish military leader who died 161 or 160 B.C.E. (*EBO*, s.v. Judas Maccabeus).

But they say: do not the Jews believe that the Christians are descendants of the stock of Esau?[6] And in regard to this stock the Prophet has said (Obadaiah 1:18): "nothing will be left of the House of Esau." To this we answer: This saying of the prophet, however, must be interpreted intelligently; it refers to those who follow the practices of Esau and follow in his footsteps (as is explained by the Rabbis). . . . For they say that Esau committed fornication and murder, he denied the resurrection[7] and even thought little of the birthright. Now, whoever engages in such things is condemned equally by the Christians: the Christian, as Christian, abhors such evil things.

And if the Christian be of the lineage of Esau, he is thereby a brother of the Jew, and may not be harmed by us. For it is written: "when you cross the boundaries of Esau, take care that you do no evil." What is more, we are not obliged to hate them, as we have stated clearly in these sections above. Likewise it has been said by the Reverend Sixtus [Medices], the Dominican theologian, who proves, in his treatise on the money lending of Jews, that the Jews could not think of the Christians as stranger[s] . . . indeed, according to him, in such maxims as "you shall not lend to your brother for interest, shall not 'bite'[8] him," one must understand that Christians as well as Jews are involved. . . . From all these things, it may at least be concluded, as he says (i.e. the Dominican Sixtus), that it is in no way allowed to the Jews that they should practice usury, either among themselves or among the Christians who are their brothers; it may, however, be permitted to practice this among foreigners (or for foreigners to practice it among themselves). Thus, when it is written (Deuteronomy 23:20–21): "Thou shalt not lend upon interest to thy brother . . . unto a foreigner thou mayest lend upon interest," it is to be understood that the Jew must by no means "bite" with usury either the Jew or the Christian. But those who are sprung from the stock of Ishmael[9] or others (so long as they be no Christians) may be "bitten" by the Jew (and he by them), for they are not brothers but foreigners. And if indeed the Jews do "bite" the Christians with usury, this is not by permission of the law, but by a cogent necessity which, as it is thought, may make this excusable. . . .

---

[6] **Esau:** one of the sons of Isaac in the book of Genesis; his relationship with his brother Jacob, who is considered the founder of the house of Israel, was an antagonistic one. [7] **resurrection:** the resurrection of the dead in the messianic era. [8] **bite:** the Hebrew term for usury is derived from the word נשׁך, *neshech,* meaning "to bite." [9] **Ishmael:** the half-brother of Isaac and son of Abraham in the book of Genesis. These brothers also had a troubled relationship. Ishmael is sometimes understood as representing the forefather of Islam.

All these details we have dealt with in order that what we have already explained should be reinforced by this Christian witness [Sixtus], worthy of all good faith; that is, that the Christians are the brothers of the Jews, at least according to the flesh, nor should that affinity be scorned which they possess in spiritual matters, as we have said above. And . . . in the second book, at the end of the fourth chapter, [Sixtus] says: And those Jews who fear God have always, even up to our own times, rejected and abhorred the charge of usury. And all those who are sprung from Abraham are joined together in the greatest affinity, and each one should help the other as much as possible. Concerning Abraham it is written: "for the father of a multitude of nations have I made thee" [Genesis 17:5; Romans 4:17]. . . .

Again it is said: If the Jews and Christians are brothers, wherefore does the Jew practice usury against him? For it is written: Thou shalt not practice usury against thy brother. To this objection (which indeed is not to be scorned lightly), we might give answer by a double method: first that the Jew practices usury upon the Christian improperly (as David declares in the Psalm [15:5]: "He that putteth not out his money on interest"); secondly, that the money-lending of the Jews does not come under the name of usury, but rather of a contract agreed upon between an assembly of Christians and the Jew — publicly, and after mature consideration; moreover the permission of the ruling lord was received. Nor does the Jew take under the name of usury the money he receives; but only under that of contract drawn up between them publicly after mature consideration. Moreover, one can only think that things which are allowed to spread by public consent must have been instituted to suit the convenience of all. Thus the Jew does not practice usury upon the Christian as upon a stranger, but he does this under the cloak of some contract. And according to modern rabbis, he could do the same thing with another Jew (this must be understood intelligently), as, for example, for the purpose of helping the poor. Thus the theologian mentioned above[10] concludes that in the Christian state usury may be permitted.

[10] theologian mentioned above: Sixtus.

→ LEON MODENA

From The History of Rites, Customs, and Manners of Life, of the Present Jews, throughout the World

1650

Translated by Edmund Chilmead

Leon Modena (1571–1648) was born to a family renowned for scholarship and wealth; they lived for a while in Modena, hence the family name, but Leon was born in Venice, where he lived much of his adult life. He was an extremely precocious student and preacher, accomplished in Jewish learning, poetry, music, and Latin; he made a living through teaching and writing poems, letters, and texts on Jewish and secular subjects. Modena was ordained a rabbi in Venice where he served as cantor (one who leads services) and as a very popular preacher; both Jews and Christians flocked to his sermons. He was acquainted with Henry Wotton, the English ambassador to Venice, and his chaplain, William Bedell; he may also have been the rabbi Thomas Coryate argues with in his account of a visit to the Venetian ghetto (p. 137). In 1614, Modena was commissioned, probably by Wotton, to write a description of Jewish practices for James I, entitled *Vita, Riti, e Costumi de gl' Hebrei*. However, the text was not published until 1637 as *Historia de gli Riti Hebraici;* the excerpt below is drawn from the English translation published in 1650. Modena had many dealings with Christians and was interested in religious polemic; in addition to the *Historia*, he authored numerous poems, rabbinic decisions, popular books, texts on rabbinic authority and Christianity, and an autobiography. *The History of Rites* is the first account of Jewish life written in the vernacular by a Jew for a Christian audience (Adelman 19–35 passim). Among the topics he considers, Modena includes usury, challenging Christian assumptions that Jewish law encourages fraud. He also argues that Christians do not fall into the category of those to whom it is lawful to charge interest in Jewish law; however, owing to the restrictions Christians place on professions Jews may undertake, usury has been permitted to allow Jews to make a living.

CHAPTER V

Of their dealing in worldly affairs, and of their usury.

They are commanded, both by the Law of Moses, as also by that of the Rabbis, or the Traditional Law,[1] to carry themselves most uprightly in their

---

[1] **Law of Moses . . . Traditional Law:** refers to the first five books of the Hebrew Bible; the Traditional or rabbinic law is the oral tradition recorded in the Talmud.

Leon Modena, *The History of Rites, Customs, and Manners of Life, of the Present Jews, throughout the World* (1637), trans. Edmund Chilmead (London: J. L., 1650), 73–77.

dealings, and to defraud, or cozen[2] no man, neither Jew, nor other, observing always, and with all sorts of men, those just ways of dealing, which are commanded them in many places of the Scripture, and particularly in Leviticus 19:11, 13, 15, 33–37. . . .

2. And those men that, have given out of them,[3] some in speeches and others in writing, that they swear every day (and account it a godly work) to endeavor to defraud, and cheat the Christians, [they foster] a most gross untruth, and scatter [it] abroad . . . only to render [Jews] more odious[4] among the nations than they are. Whereas, in truth, many of the Rabbis have commanded them the clean contrary[5] in their writings: out of which, Rabbi Bachya hath made a full collection of the passages that concern this particular in his book entitled, *Kad HaKemach*,[6] under the letter *gimel*,[7] *gezelah*:[8] where he says, that it is a far greater sin, to defraud one that is not a Jew, than to defraud a Jew; in respect of the scandal which by this means is given, besides the wickedness of the act itself. And this they call חלל השם, *chillul Hashem, to profane the name (of God)*, which is one of the greatest sins that can be. So that if there chance to be found any among them who is a fraudulent, cheating person, it must be imputed to the dishonesty and baseness of that particular man's disposition, and not that he is any way prompted thereto by any encouragements, either found in the law, or any way given him by the Rabbis.

3. True it is, that by reason of the distress unto which their so long captivity has brought them and their not being suffered[9] to enjoy any lands or possessions at all, or to exercise many other kinds of merchandisings or ways of traffic that are of reputation and profit, they are at length become much abased in spirit, and have degenerated from the ancient Israelitish uprightness, and sincerity.

4. In like manner as, for the same reason, they have made it lawful to take use-money,[10] notwithstanding that which is said, Deuteronomy 23:19–20: *Thou shalt not lend upon usury to thy brother, etc. Unto a stranger thou mayst lend upon usury, etc.* Where, by the word *stranger*, it is clear, that no other could be meant, but only those seven nations, of the Hittites, Amorites, Jebusites, etc. from whom God commanded them to take even their lives also.[11] But for as much as now they have no other way of livelihood left

---

[2] **cozen:** deceive. [3] **have given out of them:** have claimed about the Jews. [4] **odious:** hated. [5] **clean contrary:** the exact opposite. [6] *Kad HaKemach:* כד הקמח, "The Jar of Flour," written by Bachya ben Asher, a fourteenth-century Spanish kabbalist and Bible-commentator (Marcus 439). [7] *gimel:* the third letter of the Hebrew alphabet. [8] *gezelah:* transliterated Hebrew word for theft. [9] **suffered:** permitted. [10] **use-money:** interest. [11] **to take even their lives also:** see Deuteronomy 7:1–5.

them, but only this of usury, they[12] allege it to be lawful for them to do this, as well as for the rest of their brothers by nature.

5. And only these seven nations are meant in all those places wherever the Rabbis permit any usury, or any extortion to be used (seeing that the same is so often permitted unto them in the Holy Scripture) and, without all doubt, none of those nations, among whom they are at this present dispersed and suffered to inhabit, and do likewise receive all courteous usage, from the princes of the several nations, especially among the Christians: seeing that, this would be a crime, not only against the written law, but against that of nature also.

[12] they: the rabbis.

→ **SIR THOMAS SHERLEY**

The Profit That May Be Raised to Your Majesty out of the Jews
c. 1607

Thomas Sherley or Shirley (1564–1630) attended Oxford but left without taking a degree. He saw military service in Ireland and the Low Countries (the area comprising what is today Holland, Belgium, and Luxembourg) and in the 1590s undertook a career of privateering, plundering vessels carrying Spanish merchandise and pillaging towns on the Portuguese coast. In 1602 he directed his efforts against the Turks, but was captured and imprisoned in Constantinople; King James and others pleaded for his release, which was granted in 1605 with the payment of a large ransom. In the fall of 1607 he was imprisoned for illegally interfering with the operations of the Levant Company and again in 1611 for debt, which plagued him and his family throughout his life. Sherley's financial misadventures and choice of career probably influenced his thinking in the proposal below, which outlines the "profit that may be raised to your Majesty out of the Jews"; the date of this manuscript follows shortly upon his release from the Turks and just prior to his conflict over the Levant Company. He may have dreamed up this idea while in Constantinople, where he had the opportunity to come in contact with the local Jewish community. His naked desire to exploit the Jews coexists unpleasantly with his reluctance to offer religious freedom; this attitude bears a striking resemblance to that of medieval English kings who both vilified and profited from the Jews residing in their realm.

Sir Thomas Sherley, "The Profit That May Be Raised to Your Majesty out of the Jews" (1607), Historical Manuscripts Commission and Salisbury Manuscripts, vol. 19 (London: Her Majesty's Stationery Office, 1965), 473–74.

Sir Thomas Sherley to the King

The profit that may be raised to your Majesty out of the Jews is different three manner of ways.[1] First, I entertained them with a promise to become a suitor[2] to your Majesty for privilege for them to inhabit in Ireland, seeking to draw them thither, because doubtless their being there would have made that country very rich, and your Majesty's revenue in Ireland would in short time have risen almost to equal the customs[3] of England. For, first, they were willing to pay your Majesty a yearly tribute of two ducats for every head; and they, being most of them merchants, would have raised great customs where now are none; and they would have brought into the realm great store of bullion of gold and silver by issuing of Irish commodities into Spain, which will be of high esteem there considering their natures, viz.[4] salted salmons, corn, hides, wool, and tallow; of all which there will be great abundance if once the people give themselves to that industry, which doubtless they will do as soon as they find that their labors will procure them money.

2. The second course is to give them privilege to be and inhabit in England, and have synagogues, and for that I suppose I could have drawn them to pay a greater annual tribute for every head, because their chief desire is to be here.

3. But sith[5] your Majesty (like a most zealous and religious Christian prince) is not pleased that they should have any synagogue within any of your dominions, there is a third course to be taken with them, which is this: they must give your Majesty a fine for leave to trade for so much the year,[6] within any of your ports. This I know they will purchase at a high rate when they see that they can obtain no more. I saw a precedent of this at Naples this summer passed: . . . the Jews being banished out of all the King of Spain's dominions, they desired leave[7] to trade for 500,000 ducats the year only within the kingdom of Naples,[8] for 5 years, and to have their bodies and goods secured; and for this they gave to the King 100,000 ducats. Now you may if you please give license for much more and there is no synagogue allowed in this kind.

You shall reap many extraordinary commodities out of the Jews, besides the customs and fines. And the first and greatest is that if the Eastern Jews once find that liking of your countries, which I am sure they will, then many

---

[1] **three manner of ways:** Sherley proposes three different plans. [2] **a suitor:** one who makes requests. [3] **customs:** tax levied by ruler on imported or exported goods. [4] **viz:** for example. [5] **sith:** since. [6] **leave to trade for so much the year:** permission to trade for a stipulated amount of money per year. [7] **leave:** permission. [8] **Naples:** Naples was under Spanish control at this time.

of them of Portugal (which call themselves Morani[9] and yet are Jews) will come fleeing hither. And they will bring more wealth than all the rest, and by them the most part of the trade of Brazil will be converted hither. Wherein your Majesty may give the King of Spain (who is your secret enemy) a greater blow in peace than Queen Elizabeth of glorious memory did with all her long and tedious war. The King of Spain cannot justly "except" against[10] this; and if he do, he knows not how to mend it.[11]

The second commodity will be that if your Majesty shall have any occasion to be at a great extraordinary charge, you may at any time borrow a million of the Jews with great facility, where your merchants of London will hardly be drawn to lend you 10,000 [pounds]. There is experience of both: one in your Majesty, for the Londoners; the other in the Duke of Mantua for the Jews. His estate is one of the least of all Italy, and therefore cannot contain the tenth part of those Jews that may very well be in your dominions, by privilege of trade only, without a synagogue; and yet once in three years he picks 300,000 or 400,000 crowns out of his Jews. The Duke of Savoy were not able to maintain his estate without their help and the benefit he reaps by them.

Daily occasions will be offered to make greater commodities out of them if once you have hold of their persons and goods. But at the first they must be tenderly used, for there is great difference in alluring wild birds and handling them when they are caught; and your agent that treats with them must be a man of credit and acquaintance amongst them, who must know how to manage them, because they are very subtle people. The politique Duke of Florence will not leave his Jews for all other merchants whatsoever.

[9] **Morani:** Marranos or Jews converted to Christianity who continue to practice Judaism in secret. [10] **"except" against:** object to. [11] **how to mend it:** how to alter it for the better.

→ NICOLAS DE NICOLAY

From The Navigations, Peregrinations, and Voyages Made into Turkey

1585

Translated by Thomas Washington

Nicolas de Nicolay (1517–1583), a French explorer, began his career as a soldier, serving in France and throughout Europe. On his return from military service abroad, he was appointed geographer and gentleman of the King's chamber by

---

Nicolas de Nicolay, *The Navigations, Peregrinations, and Voyages Made into Turkey* (1568), trans. Thomas Washington (London, 1585), 130–32.

Henry II. He traveled with ambassador Gabriel d'Aramon to Constantinople, Algiers, and Tripoli. In addition to speaking almost all the European languages, he was a deft artist; he wrote a number of books on his travels and military career and provided the drawings for the engravings that illustrate his works. His *Navigations et Pérégrinations Orientales* . . . was first published in Lyon in 1568 (*Nouvelle Biographie Générale* 37: 1016–17). His account of the Jews of Constantinople, excerpted here in the English translation of *The Navigations* (1585), mentions trade as a source of the Jews' livelihood. In the latter half of the sixteenth century, Jews were increasingly abandoning moneylending in favor of commerce as a profession. As the English began to trade with Turkey in the 1580s, they were more and more likely to encounter Jews as merchants rather than usurers (Vitkus).

CHAPTER 16

Of the merchant Jews dwelling in Constantinople and other places of Turkey and Greece.

The number of the Jews dwelling throughout all the cities of Turkey and Greece, and principally at Constantinople, is so great, that it is a thing marvelous and incredible. For the number of these using trade and traffic of merchandise, like[1] of money at usury, doth there multiply so from day to day, that the great haunt and bringing of merchandises which arrive there of all parts as well by sea as by land is such that it may be said with good reason that at this present day they have in their hands the most and greatest traffic of merchandise and ready money that is in all Levant. And likewise, their shops and warehouses, the best furnished of all rich sorts of merchandises which are in Constantinople, are those of the Jews. Likewise, they have amongst them workmen of all arts and handicrafts most excellent; and specially . . . the Marranos[2] of late banished and driven out of Spain and Portugal[3] . . . have taught the Turks diverse inventions, crafts, and engines of war, as to make artillery, harquebuses,[4] gunpowder, shot, and other munitions. They have also there set up printing, not before seen in those countries, by the which in fair characters they put in light diverse books in diverse languages, as Greek, Latin, Italian, Spanish, and the Hebrew tongue, being to them natural. . . . They have also the commodity and usage

---

[1] like: likewise.   [2] Marranos: Jewish converts to Christianity who secretly maintain Jewish practice.   [3] driven out of Spain and Portugal: the Jews were expelled from Spain in 1492; the Portuguese Inquisition was established in 1521, though forcible conversions began there in the 1490s.   [4] harquebuses: early portable guns of varying size.

to speak and understand all other sorts of languages used in Levant, which serveth them greatly for the communication and traffic, which they have with the other strange nations, to whom oftentimes they serve for Dragomans, or interpreters. Besides, this detestable nation of the Jews, are men full of all malice, fraud, deceit, and subtle dealing, exercising execrable usuries amongst the Christians and other nations without any consciences or reprehension, but have free license, paying the tribute: a thing which is a great ruin unto the country and people where they are conversant. . . .

> ## The Levant Company's Charter   *1581*

The Levant Company was one of the several joint stock companies that began to be organized in the latter half of the sixteenth century to finance the development of new trade by pooling large numbers of small investments. They usually sought a charter from the crown, a document that made shares in the company more easily transferrable by sale and secured a monopoly on a particular branch of commerce. The Mediterranean trade, which included goods from Asia, was mostly in the hands of the Italians and others until England negotiated directly with the Turkish sultan to obtain trading privileges. This resulted in the formation of the Turkey Company (incorrectly identified here in Hakluyt as the Levant Company) and the Venice Company, which shared the highly lucrative monopoly of Mediterranean trade. The two combined to form the Levant Company in 1593 (Clay 2: 129).

The letters patents,[1] or privileges granted by her Majesty to Sir Edward Osborne, Master Richard Staper, and certain other merchants of London for their trade into the dominions of the great Turk, in the year 1581.

Elizabeth by the grace of God Queen of England, France and Ireland, defender of the faith, etc. To all our officers, ministers, and subjects, and to all other people as well within this realm of England, as else where under our obeisance, jurisdiction, or otherwise, unto whom these our letters shall

---

[1] **letters patents:** "An open letter or document, usually from a sovereign or person in authority, issued for various purposes, e.g., to put on record some agreement or contract, to authorize or command something to be done, to confer some right, privilege, title, property, or office" (*OED*).

The Levant Company's Charter (1581), reprinted in *The Principal Navigations, Voyages, and Discoveries of the English Nation,* ed. Richard Hakluyt (London, 1589), 172–74.

be seen, showed or read, greeting. Where our well-beloved subjects Edward Osborne, alderman[2] of our city of London, and Richard Staper, of our said city, merchant, have by great advantage and industry, with their great costs and charges, by the space of sundry late years, travailed,[3] and caused travail to be taken, as well by secret and good means, as by dangerous ways and passages both by land and sea, to find out and set upon a trade of merchandise and traffic into the lands, islands, dominions, and territories of the great Turk, commonly called the Grand Signoir, not heretofore in the memory of any man now living known to be commonly used and frequented by way of merchandise by any the merchants or any subjects of us, or our progenitors. And also have by their like good means and industry, and great charges, procured of the said Grand Signoir (in our name,) amity, safety, and freedom, for trade and traffic of merchandise to be used and continued by our subjects within his said dominions. Whereby there is good and apparent hope and likelihood both that many good offices may be done for the peace of Christendom, and relief of many Christians that be or may happen to be in thralldom[4] or necessity under the said Grand Signoir, his vassals or subjects, and also good and profitable vent and utterance[5] may be had of the commodities of our realm, and sundry other great benefits to the advancement of our honor, and dignity Royal, the increase of the revenues of our crown, and general wealth of our realm. Know ye, that hereupon we greatly tendering the wealth of our people, and the encouragement of our subjects in their good enterprises for the advancement of the commonweal, have of our special grace, certain knowledge and mere motion,[6] given and granted, and by these presents[7] for us, our heirs and successors, do give and grant unto our said trustee, and well-beloved subjects Edward Osborne, and unto Thomas Smith of London Esquire,[8] Richard Staper, and William Garret of London Merchants, their executors, and administrators and to the executives and administrators of them, and of every of them, that they, . . . their servants, factors or deputies, and to such others as shall be nominated according to the tenor of these our letters patents, shall and may during the term of seven years from the date of these patents, freely trade, traffic, and use feats of merchandise into, and from the dominions of the said Grand Signoir, and every of them, in such order, and manner, form, liberties and condition to all intents and purposes as shall be between them limited and agreed, and not

---

[2] **alderman:** magistrate below the rank of mayor. [3] **travailed:** labored or traveled. [4] **thralldom:** captivity. [5] **vent and utterance:** selling of goods. [6] **mere motion:** sole action. [7] **these presents:** this present document. [8] **Esquire:** denotes membership in the gentry.

otherwise, without any molestation, impeachment, or disturbance, any law, statute, usage, diversity of religion or faith, or other cause or matter whatsoever to the contrary notwithstanding. . . .

And further, we of our more ample and abundant grace, mere motion and certain knowledge, have granted, and by these patents for us, our heirs and successors, do grant to the said Edward, Thomas, Richard and William, their executors and administrators, that they . . . shall have the whole trade and traffic, and the whole entire only[9] liberty, use and privilege of trading, and trafficking, and using feat of merchandise, into, and from the said dominions of the said Grand Signoir, and every of them. . . . And that they the said Edward, Thomas, Richard and William, their executors and administrators, . . . shall have full and free authority, liberty, faculty, license and power to trade and traffic into and from all and every the said dominions of the said Grand Signoir, and into, and from all places where, by occasion of the said trade, they shall happen to arrive or come, whether they be Christians, Turks, gentiles, or other, and into, and from all seas, rivers, ports, regions, territories, dominions, coasts and places with their ships, barks,[10] pinnaces[11] and other vessels, and with such mariners and men, as they will lead with them or send for the said trade, as they shall think good at their own proper cost and expenses, any law, statute, usage, or matter whatsoever to the contrary notwithstanding. And that it shall be lawful for the said Edward, Thomas, Richard and William, and the persons aforesaid, and to and for the mariners and seamen to be used and employed in the said trade and voyage to set and place in the tops of their ships and other vessels the arms of England with the red cross[12] over the same, as heretofore they have used the red cross, any matter or thing to the contrary notwithstanding.

And by virtue of our high prerogative royal (which we will not have argued or brought in question) we straightly[13] charge and command, and prohibit for us, our heirs, and successors, all our subjects (of what degree or quality soever they be) that none of them directly, or indirectly, do visit, haunt, frequent or trade, traffic or adventure by way of merchandise into, or from any of the dominions of the said Grand Signoir, or other places above said by water or by land (other than the said Edward, Thomas, Richard and William, their executors and administrators, or such as shall be admitted, and nominated as is aforesaid) without express license, agreement, and consent of the said governor, and company or the more part of them, whereof

---

[9] **whole entire only:** complete and sole. [10] **bark:** small ship or barge. [11] **pinnaces:** small, light, two-masted vessels. [12] **red cross:** English national emblem, St. George's Cross. [13] **straightly:** strictly.

the said governor always to be one, upon pain of our high indignation, and of forfeiture and loss, as well of the ship and ships, with the furniture thereof, as also of the goods, merchandises, and things whatsoever they be of those our subjects which shall attempt, or presume to sail, traffic, or adventure, to or from any the dominions, or places above said, contrary to the prohibition aforesaid. The one half of the same forfeiture to be to the use of us, our heirs and successors, and the other half to the use of the said Edward, Thomas, Richard and William, and the said company, and further to suffer imprisonment during our pleasure, and such other punishment as to us, for so high contempt, shall seem meet[14] and convenient.

[14] **meet:** fitting.

→ **JOHN WHEELER**

From A Treatise of Commerce   *1601*

John Wheeler lived in the late sixteenth and early seventeenth centuries, and was secretary of the Merchant Adventurers' Company, which held the monopoly on cloth exportation to northwest Europe. His *Treatise of Commerce* defends the need for the Merchant Adventurers' Society against a growing popular resistance to monopolies. Wheeler also insists on the dignity and importance of the merchant's vocation, deeming it suitable for the nobility despite the common assumption that only persons of lower class labored in their professions. In sum, the *Treatise* is interested in presenting the value and virtue of the merchant, suggesting that not all held him in high esteem in this period. How does Wheeler's representation of the merchant's profession compare with the way in which Antonio is viewed in *The Merchant of Venice*? Can you tell from Antonio's own speech or behavior, or that of other characters, whether he represents a gentleman or a member of the middle class?

Wherein are showed the commodities[1] *arising by a well ordered and ruled trade, such as that of the Society of Merchants Adventurers is proved to be: written for the better information of those, who doubt of the necessariness of the said Society in the state of the realm of England, by John Wheeler, secretary of the said Society.*

There be two points about the which the royal office and administration of a prince is wholly employed, to wit: about the government of the persons of men; next, of things convenient and fit for the maintenance of humane soci-

[1] *commodities:* conveniences.

John Wheeler, *A Treatise of Commerce* (London, 1601), 5–8, 21–23.

ety, wherein principally the civil life consisteth and hath her being. And therefore the prince that loveth the policy, and rules by sage and good counsel, is to constitute and appoint certain laws, and ordinary rules, both in the one and the other of the abovesaid points, and specially in the first as the chiefest, which is conversant and occupied about the institution of the persons of men in piety, civil conversation in manners and fashion of life, and finally in the mutual duty of equity and charity one towards another. Of the which my purpose is not to entreat,[2] but somewhat of that other point, namely, the government of things convenient and fit for the maintenance of humane society. Whereupon men's action and affections are chiefly directed, and whereabouts they bestow, and employ not only the quickness and industry of their spirits, but also the labor and travail of their hands, and sides,[3] that so they may draw from thence with commodity or pleasure, or at leastwise thereby supply and furnish their several wants and necessities. From hence, as from a root or fountain, first proceedeth the estate of merchandise, and then consequently in a row, so many, diverse, and sundry arts, as we see in the world. At which it should seem that man beginneth the train, or course of his life, and therein first of all discovereth not only the dexterity and sharpness of his wit, but withal that naughtiness and corruption which is naturally in him. For there is nothing in the world so ordinary and natural unto men, as to contract, truck, merchandise, and traffic one with another, so that it is almost impossible for three persons to converse together two hours, but they will fall into talk of one bargain or another, chopping, changing,[4] or some other kind of contract. Children, as soon as ever their tongues are at liberty, do season their sports with some merchandise or other; and when they go to school, nothing is so common among them as to change, and rechange, buy and sell of that, which they bring from home with them. The prince with his subjects, the masters with his servants, one friend and acquaintance with another, the captain with his soldiers, the husband with his wife, women with and among themselves, and in a word, all the world chopeth and changeth, runneth and raveth after marts,[5] markets, and merchandising, so that all things come into commerce, and pass into traffic (in a manner) in all times, and in all places, not only that, which nature bringeth forth, as the fruits of the earth, the beasts and living creatures, with their spoils,[6] skins and cases, the metals, minerals, and such like things, but further also, this man maketh merchandise of the works of his own hands, this man of another man's labor, one sells words, another

---

[2] **to entreat:** to handle.   [3] **sides:** of one's body.   [4] **chopping, changing:** exchanging one thing for another.   [5] **marts:** marketplaces or bargains.   [6] **spoils:** animal skins or other remains.

makes traffic of the skins and blood of other men, yea there are some found so subtle and cunning merchants, that they persuade and induce men to suffer themselves to be bought and sold, and we have seen in our time enough, and too many which have made merchandise of men's souls. To conclude, all that a man worketh with his hand, or discourseth in his spirit, is nothing else but merchandise . . . all negotiations, or traffics whatsoever, are none other thing but mere matter of merchandise and commerce. Now, albeit this affection be in all persons generally both high and low, yet there are of the notablest, and principalest traffickers which are ashamed, and think scorn to be called merchants. Whereas indeed merchandise, which is used by way of proper vocation, being rightly considered of, is not to be despised, or accounted base by men of judgement, but to the contrary, by many reasons and examples it is to be proved that the estate is honorable, and may be exercised not only of those of the third estate[7] (as we term them), but also by the nobles and chiefest men of this realm with commendable profit, and without any derogation to their nobilities, high degrees, and conditions, with what great good to their states, honor, and enriching of themselves and their countries, the Venetians, Florentines, Genoveses, and our neighbors the Hollanders have used this trade of life, who knoweth not? Or having seen the beauty, strength, opulence, and populousness of the abovesaid cities and provinces, wondereth not thereat? . . .

So that it appeareth, that not only a prince may use this kind of men, I mean merchants, to the great benefit, and good of his state, either for foreign intelligence or exploration, or for the opening of an entry and passage unto unknown and fair distant parts, or for the furnishing of money, and other provisions in time of wars, and death, or lastly, for the service and honor of the prince, and country abroad at all times requisite, and expedient. But also, this kind of life may be exercised and used with commendation, and without loss of one jot of honor in those who are honorable, or of eminent degree, as aforesaid. Whereunto I add this further, that without merchandise, no ease or commodious living continueth long in any state, or commonwealth, no not loyalty, or equity itself, or upright dealing. Therefore herein also, as in the former point, good order and rule is to be set where it is wanting, or where it is already established, there it ought to be preserved, for the maintenance of so necessary and beneficial an estate in the commonwealth, by constituting meet and well proportioned ordinances over the same and over those things, which are thereupon depending between the

[7] **the third estate:** commoners; aristocrats were thought to be above work in this period.

merchants, and those things which are merchandised, or handled likewise with convenable[8] and well appropriated magistrates and overseers for the maintenance and execution of the said ordinances. . . .

[S]uch is the value, profit, and goodness of the English commodities, that all nations of these parts of Europe, and elsewhere, desire them: and on the other side, the English merchants buy up, and carry into England so great a quantity of foreign wares, that for the sale thereof all strange merchants do, and will repair unto them. Now what these English commodities are and how they be so profitable, may appear by the particulars following.

First, there is shipped out yearly by the abovesaid company, at least sixty thousand white clothes,[9] besides colored clothes of all sorts, kerseys[10] short, and long, baize,[11] cottons, northern dozen,[12] and diverse other kind of coarse clothes. The just value of the sixty thousand white cloths cannot well be calculated, or set down, but they are not less worth (in mine opinion) than six hundred thousand pounds sterling, or English money. The colored clothes of all sorts, baize, kerseys, northern dozens, and other coarse clothes, I reckon to arise to the number of forty thousand clothes, at least, and they be worth one with another four hundred thousand pounds sterling, or English money.

There goeth also out of England, besides these woolen clothes, into the low countries,[13] wool, fell,[14] lead, tin, saffron, conyskins,[15] leather, tallow, alabaster stone, corn, beer, and diverse other things, amounting unto great sums of money: By all which commodities, a number of laboring men are set on work, and gain much money, besides that which the merchants gaineth which is no small matter. Hereunto add the money which shippers, and men that live upon the water get by freight and portage of the aforesaid commodities from place to place; which would amount to a great sum, if the particulars thereof were, or could be, exactly gathered. Hereby in short may be seen, how great and profitable the Company of the Merchant Adventurers trade has been, and is in the places where they hold their residence: besides, the profit raised upon the chambers, sellers and packhouses, which they must have for four or five hundred merchants, whereby rents are maintained and kept up; and the great expenses otherwise, which the said merchants are at for their diet, apparel, etc.; to say nothing of the prince's or generalities' profit and revenues by their tolls, convoys, imports, excises, and

---

[8] convenable: suitable. [9] clothes: cloths. [10] kerseys: coarse woolen cloth, usually ribbed. [11] baize: a fine, light woolen fabric. [12] northern dozen: coarse woolen cloth. [13] low countries: Belgium, Luxembourg, and the Netherlands. [14] fell: animal skin. [15] conyskins: rabbit skins.

other duties, whereof there can be no certain notice had. But to show the greatness thereof, let this one sign so long ago serve for all, that Philip the Good, Duke of Burgundy, and first founder of the order of the Golden Fleece, gave the aforesaid Fleece for a livery or badge of the order, for that he had his chiefest tolls, revenues, and incomes by wool and woolen cloth.

Thus you have seen what profit is raised by strangers, upon the English trade, it followeth to show, what the Merchant Adventurers buy for return of strange nations, and people frequenting their mart towns and bringing their country commodities thither.

Of the Dutch and German merchants, they buy Rhenish Wine,[16] fustians,[17] copper, steel, hemp, onion seed, copper and iron wire, latten,[18] kettles, and pannes,[19] linen cloth, harness,[20] saltpeter, gunpowder, all things made at Nurenberg, and in sum, there is no kind of ware that Germany yields but generally the Merchant Adventurers buy as much, or more thereof than any other nation.

Of the Italians, they buy all kind of silk wares, velvets, wrought[21] and unwrought, taffetas, satins, damasks, sarsenets,[22] Milan fustians, cloth of gold and silver, grosgrains,[23] camlets,[24] satin, and sowing silk,[25] organza, . . . and all other kinds of wares either made or to be had in Italy.

Of the Easterlings[26] they buy flax, hemp, wax, pitch, tar, wainscot,[27] deal-boards,[28] oars, corn, furs, cables, and cable yarn,[29] tallow, ropes, masts for ships, soap-ashes,[30] . . . and almost whatsoever is made or groweth in the East Countries.

Of the Portugese, they buy all kind of spices and drugs: with the Spanish and French, they had not much to do, by reason that our English merchants have had a great trade in France and Spain, and so serve England directly from thence with the commodities of those countries.

Of the Low Country merchants, or Netherlanders, they buy all kind of manufacture, or band work not made in England, tapestry, buckrams,[31] white thread, inkle,[32] linen cloths of all sorts, cambrics, lawns,[33] . . . and an

---

[16] **Rhenish Wine:** wine from the area near the Rhine river.   [17] **fustians:** coarse cloth of cotton and flax.   [18] **latten:** yellow metal, brass.   [19] **pannes:** soft cloth with long nap, like velvet.   [20] **harness:** possibly household and personal equipment, or military equipment (armor, mail, etc.).   [21] **wrought:** embroidered.   [22] **sarsenets:** fine, soft silk.   [23] **grosgrains:** corded fabric.   [24] **camlets:** fabric made of camel hair, wool, or Angora goat hair and silk.   [25] **sowing silk:** i.e., sewing silk, or silk thread.   [26] **Easterlings:** inhabitants of eastern Germany or the Baltic Coast.   [27] **wainscot:** oak planks or logs.   [28] **deal-boards:** thin boards of pine or fir.   [29] **cables and cable yarn:** hemp ropes and fibers.   [30] **soap-ashes:** ashes of certain kinds of woods used in forming lye for soap-making.   [31] **buckrams:** fine linen or cotton.   [32] **inkle:** narrow woven strip of linen.   [33] **cambrics, lawns:** fine linens.

infinite number of other things, too long to rehearse in particular, but hereby I hope it sufficiently appeareth, that it is of an exceeding value, which the Merchant Adventurers buy and carry into England.

➔ **DANIEL PRICE**

The Merchant: A Sermon Preached at Paul's Cross
1607

Daniel Price (1581–1631) received a bachelor's and a master's degree from Oxford and became a minister in 1604. He was a distinguished preacher, often speaking against Catholic practice, and served as chaplain to the sons of James I and to the king himself; he preached frequently at court. In the excerpt below, Price takes as the text of his sermon two verses from Matthew: "Again, the kingdom of heaven is like unto a merchant man, seeking goodly pearls: Who, when he had found one pearl of great price, went and sold all that he had, and bought it" (13:45–46). Dedicated to the London Company of Merchants, this speech uses the biblical text to defend merchants from popular criticism of their vocation. As Thomas Wilson intimated in his *Discourse upon Usury* (p. 194), merchants were often associated with worldliness and materialism, qualities contrasted negatively in Christian doctrine to spirituality and generosity. Price, however, praises the merchant for the hard work he undertakes and risks he runs, representing him as a figure for the exemplary Christian. In this respect, Price's merchant resembles Antonio, who puts his wealth in peril both in trade and in a free loan to his friend Bassanio.

The king of heaven is like to a merchant etc. Beloved,[1] the action of this merchant is not for any small, but for great gain, not for any carnal, but for spiritual glory, not for any transitory but for an eternal treasure. *The king of heaven is like to a merchant etc.* What trade more honorable than the merchant, what merchandise more honorable than the kingdom of heaven? Ye are many of you come hither as buyers, as sellers, as merchants, and therefore at this time what argument more feasible, more plausible, more forceable,[2] more available than this *the king of heaven is like to a merchant?* like to a seeking, finding, buying, selling, exchanging merchant.

---

[1] **Beloved:** Price is addressing his auditors. [2] **forceable:** carrying more force.

Daniel Price, *The Merchant: A Sermon Preached at Paul's Cross on Sunday the 24th of August Being the Day before Bartholomew Fair, 1607* (Oxford, 1608), 5–6, 12–20, 22–24.

In these words I will observe these two general points: 1. the difficulty of obtaining the kingdom of heaven, intimated in that it is compared to a merchant, the most diligent, careful, assiduous, industrious, laborious, and indefatigable of all other kinds of life; 2. the earnestness required in pursuing this kingdom, expressed in the seeking, finding, buying, selling, exchanging all. . . .

The kingdom of heaven is like to a merchant man.

If that complaint were true, which Erasmus[3] took up in his time against merchants, it is a marvel why I should compare the kingdom of heaven to a merchant, when so few merchants are like to the kingdom of heaven. His words are these . . . : "The trade of merchants account nothing good or holy, but only the only lucre[4] of money; for the attaining of which they have dedicated, and consecrated themselves as unto God. By this they measure piety, amity, honesty, credit, and fame, and all humane and divine things." I am sure he spoke by the figure[5] of some, in the name of all; for the stories and customs of Jews and Gentiles, Grecians and barbarians, infidels and Christians, do acknowledge the necessity, dignity, and excellence of merchants, and they have approved the merchant of all men[6] to be the most diligent for his life, the most assiduous in his labor, the most adventurous on the sea, the most beneficial to the land, the glory of his country, and the best pillar of his commonwealth. The word in the original is εμποροσ[7] translated by some *mercator*,[8] by some *negotiator*.[9] . . . The words differ in this, that *mercator* hath a house and family; *negotiator* is he that still travels, voyages, ventures, changing his seats like the true Christian, who is ever traveling to change his country, knowing that here he hath no abiding city, but doth seek one to come.

The common gloss[10] showeth why this kingdom is for our instruction compared to the merchant: . . . because we ought by right understanding and by good practicing to negotiate. There be two things to be observed in the merchant: the profit, and the danger of the trade. Of the profit we shall find what great commodity came of[11] Solomon by the triennial coming of the navy of Tharssis, that brought unto him gold, and silver, ivory, and apes, and peacocks, even all things, for profit, and for pleasure (1 Kings 10:32). Where the Holy Ghost doth show, that this trade was the occasion of the enriching of Solomon, and surely it doth mutually enrich all kingdoms, making the proper commodities of one country[12] common to another. Wit-

---

[3] **Erasmus:** Desiderius Erasmus (1469–1536), preeminent Renaissance humanist.   [4] **only the only lucre:** only the sole gain.   [5] **figure:** example.   [6] **of all men:** above all men.   [7] *εμποροσ:* "*emporos*," a merchant.   [8] *mercator:* merchant.   [9] *negotiator:* businessman.   [10] **gloss:** interpretation.   [11] **of:** to.   [12] **making the proper commodities of one country:** the goods belonging to one country.

ness our gold from India, our spices from Arabia, our silks from Spain, our wines from France, and so many other commodities from other countries, whereby the merchant is the key of the land, the treasurer of the kingdom, the venter[13] of his soil's surplus, the combiner of nations, and the adamantine[14] chain of countries. Of the danger David speaketh in the Psalm, "They that go down to the sea in ships and merchandise[15] in great waters, these men see the works of the Lord, and his wonders in the deep, for at his word, the stormy winds arise, which lift up the waves thereof, they are carried up to heaven, and down again to the deep, their soul melteth within them all, and all their cunning[16] is gone" (Psalm 107:23–27). . . . So that of all men I may say with David, these men see the works of the Lord and his wonders in the deep. And surely so it is with the state of the godly, in this life in most danger, subject to the greatest affliction; they are in the waves of the world, yet they above all others see the works of the Lord and his wonders, in the deepest of their misery his power in delivering them, his favor in preserving them, his mercy in comforting them, his love in caring for them, his care in protecting them, though with Paul they be in perils in the City, perils in the wilderness, perils among false brethren, perils among his own, perils in the sea,[17] as he was most fearfully, the winds being contrary, the sailing jeopardous.

The doctrine I observe out of the word *merchant,* is this; that the state of a Christian is not an idle vain speculation, but must be a careful, painful, diligent, walking in his vocation. The reason of this doctrine, is proved a *contrario*[18] by the antithesis between the state of the godly, and ungodly, under the name of the fool.[19] The fool foldeth his hands, and eateth up his own flesh. Better is one handful, saith he, with quietness, than two handfuls with labor and vexation of spirit.[20] But contrariwise, the wise merchant, the true Christian, he seeketh, he taketh pains, he laboreth, he endeavoreth to follow hard to the mark, . . . no peril no danger, no cost, no temptation, no opposition can confront[21] him. . . .

The second use of this doctrine is more particular, belonging only to those that be merchants, that seeing the merchant here is so studious, careful, diligent, and earnest in good pearls, that every one of them seek by all means to become heavenly merchants to seek, and labor, and endeavor to obtain this merchandise, to lay up his treasure in heaven, where neither rust

---

[13] **venter:** vender.   [14] **adamantine:** incapable of being broken.   [15] **merchandise:** trade.   [16] **cunning:** wisdom.   [17] **in perils in the City . . . sea:** paraphrase of 2 Corinthians 11:26.   [18] **a contrario:** in the contrary.   [19] **under the name of the fool:** the ungodly are represented by the fool in this example.   [20] **Better . . . spirit:** Ecclesiastes 4:5–6.   [21] **confront:** stand against.

nor moth doth corrupt, and where thieves break not through and steal, that as their trade of life is more honorable than others among men, so God should be more honored of them than of other men. . . .

If many about this honorable city should be asked this question [Jonah 1:8: "What is thy occupation?"], you should find an infinite number that walk in the counsel of the ungodly, and stand in the way of sinners, and sit in the seat of scorners, able to answer no otherwise than Satan did to God, that they live by compassing the earth to, and fro, and by walking in it.[22] They are vagrants, I may say vagabonds, wandering persons, as the planets in the Zodiac, never keeping a fixed place. Of no endowment, employment, art or trade, or calling, or mystery, unless they profess the mystery of iniquity. . . .

Another sort there be as bad as these conjurers, charmers, tellers of fortune, robbers by land, pirates by sea, cozeners,[23] harlots, brokers, usurers, who by cozenage, imposture, frauds, tricks, and circumventions, do set at sale honesty, truth, conscience, oaths, their soul and own salvation. The sum of that I would speak in this point if the time did serve is this, that every man ought to look well, to the lawfulness of his calling. God hath given diverse gifts but the same spirit, he hath given diversities of spirit but the same Lord,[24] he hath given the warriors a spirit to fight, to counselors a spirit to direct, to Judges a spirit to discern, to magistrates a spirit to govern, to ministers a spirit to convince, to instruct, to reprove, to direct, to merchants a spirit to trade, to traffic, to buy, to sell, to exchange, but of everyone of them in their vocation, he requireth that as he is holy, so they should be holy. Wherefore beloved, seeing the merchant in my text, is by the finger of God particularly pointed at in this place, give me leave in one word, to remember[25] you, that as your calling is honorable, and here your comparison is honorable, in that you are compared to the kingdom of heaven. O remember beloved, that if ye so much care, and labor, and travel for earthly things, how much more ought ye to care for spiritual things? I know not what reason many learned men have to condemn merchants, and merchandise so much, that Tully[26] in his book *De Republica*, should affirm of the Phoenicians, that being merchants, they by their merchandise brought in covetousness, pride, luxury, and all kind of wickedness into Greece. That St. Hierome on the third of Jeremy,[27] called the Arabians, who much traded in merchandise, the thieves of the world. That the Carthaginians would not

---

[22] able to answer . . . by walking in it: a reference to Job 1:7.   [23] cozeners: cheaters.   [24] God hath given . . . but the same Lord: 1 Corinthians 12:4–5.   [25] remember: remind.   [26] Tully: Marcus Tullius Cicero.   [27] St. Hierome on the third of Jeremy: St. Jerome, one of the Church fathers, in his commentary on the third chapter of Jeremiah.

suffer them to be common with[28] their citizens. That the Greeks would not let them enter their city, but caused them to keep their markets without the suburbs, as Cornelius Agrippa observeth, that Plato admitted them not into his commonwealth. That Aristotle detested them and their life, that the ancient laws did not admit any merchant to bear any office, or to be admitted into the Counsel or Senate, that Cicero affirmeth their getting of money to be most odious, giving this reason . . . that they get their living by lying. I hope the merchants of our time deserve not to be so thought of. Many of these merchants were Jews, gentiles, heathens, infidels, pirates, robbers; I hope none such are to be found among you, for you are Christians. I hope there be some such merchants amongst you, . . . some such as Apolonius,[29] who having long used merchandise at the last became a physician of the poor and needy and bestowing all his time and store in providing necessities for poor, aged, lame, blind people. . . .

Our merchant seeketh good pearls. I might here stand upon the color, splendor, luster, nature, effect and form of pearls: the learned know their color to be diverse, their splendor to be gracious, their luster glorious, their nature and effect miraculous, their orbicular form most perfect, and surely many great wonders hath God made known unto man, in precious pearls, but the time I have to spend is precious and I must not linger longer in these. By good pearls in my text, Avendanus[30] understandeth *virtutes animi, Albertus*[31] *legam & prophetas, Hierome calestia dona,*[32] and other diversely do expound these words, but the proper doctrine arising naturally is this, that it is a Christian's part to seek the best things. The reason of this doctrine is drawn *ab universali,*[33] all things do desire that which is good and therefore of all, a Christian ought to desire it, and endeavor to obtain it. Not that which is good in opinion only, for so good may be evil, and evil good, not good in imagination only, for so light may be darkness and darkness light, sweet may be sour, and sour sweet. But good indeed *bonum qua bonum.*[34] The difference that men have made of good is infinite. St. Austen in his 15th

---

[28] **would not suffer them to be common with:** would not allow merchants to associate with.
[29] **Apolonius:** possibly Appolonius of Citium, an Alexandrian physician of about 50 B.C.E. (*Oxford Classical Dictionary*).   [30] **Avendanus:** probably an author of Bible commentaries.
[31] *Albertus:* Albertus Magnus (c. 1200–1280), German philosopher, teacher, bishop, and saint; in addition to other scholarly works, he wrote commentaries on the Bible (*EBO*, s.v. Albertus Magnus).   [32] *virtutes animi, Albertus legam & prophetas, Hierome calestia dona:* Avendanus interprets "good pearls" to mean virtues of the soul; Albertus interprets them as the law and prophets; while St. Jerome understands them to refer to heavenly gifts.   [33] *ab universali:* from the general.   [34] *bonum qua bonum:* good which is good.

book *de Civitate Dei*[35] out of Varro[36] collecteth 288 opinions that men had concerning the sum of goodness. So much men have differed, some in missing of their good in reason, some in religion, some impropriating[37] the name of good to bad, of bad to good.

The philosophers' vanity excepted, none were more vain and vile in this than the heretics of all ages who professed that their religion was only true and good, the rest were most abhorrent and false. The Arians[38] as Socrates recordeth, affirmed that their religion was only good and all other false. . . . The Turks as Sozomen[39] recordeth, coming lineally from Hagar[40] will be called Sarazens of Sara.[41] The scripture hath taught us that there be some that cry *Templum Domini Templum Domini,* and yet would destroy *Dominum Templi* (Jeremiah 7:4[42]). That there be some that call themselves the seed of Abraham that be malicious Pharisees (John 8:39). That there be some that call themselves Jews, that are but the Synagogue of Satan (Apocalypse 3:9). That there be some that had Abraham to their father and yet are not Israelites (Romans 9:7). That there are some that have the name of Jesuits, yet have no part of the faith of truth or profession of Jesus. Some also I fear that are called Christians that have no part or portion in Christ. Thus hath Satan masked folly in the habit of wisdom, falsehood in the habit of truth, vice in the habit of virtue, sin in the habit of godliness, lewdness in the habit of goodness and as Polidor Virgil[43] hath observed of the Romish Church in electing their Popes if any were fearful they would call him Leo, if any cruel Clemens, if any wicked Pius, if any covetous Bonifactus, if any most vile Innocentius.[44] So hath good been esteemed bad, and bad good, and so many have deceived themselves in the seeking of good pearls.

---

[35] **St. Austen in his . . . *de Civitate Dei*:** St. Augustine, a Church father, in his book *The City of God;* the reference is erroneous: Varro is mentioned in book nineteen.   [36] **Varro:** Marcus Terentius (116–27 B.C.E.), Roman scholar, author of numerous works, including the *Menippean Satires* (*EBO,* s.v. Marcus Terentius Varro).   [37] **impropriating:** appropriating.   [38] **Arians:** followers of "the doctrine of Arius, a presbyter of Alexandria in the 4th c., who denied that Jesus Christ was consubstantial, or of the same essence or substance with God" (*OED*).   [39] **Sozomen:** fifth-century church historian Salamanes Hermeios Sozomen (c. 400–450) (*EBO,* s.v. Sozomen).   [40] **Hagar:** concubine of Abraham; see Genesis 14 and following.   [41] **Sara:** According to Sozomen, the Turks, though actually descendents of the concubine Hagar, pretend to be descended from Sara, Abraham's official wife, and hence take their name "Sarazens" from "Sara."   [42] **Jeremiah 7:4:** in Jeremiah 7:4 the Jews say, "The Temple of the Lord, the Temple of the Lord"; Price accuses the Jews of seeking to destroy "the Lord of the Temple," or Jesus in Christian thought.   [43] **Polidor Virgil:** (1470–1555) Italian-born humanist and author of a history of England (*EBO,* s.v. Polydore Vergil).   [44] **fearful they would call him Leo . . . Innocentius:** the fearful is named "lion," cruel is named "kind," wicked is named "pious," greedy is named "generous," and vile "innocent."

CHAPTER 3

Religion

————————————— ✣ —————————————

The Reformation in England led to enormous changes in religious structure, practice, and belief — changes complicated by the varying ecclesiastical allegiances of the Tudor monarchs. While Henry VIII broke away from the Catholic Church and established the Protestant Church of England, he maintained many of the doctrines, ceremonies, and institutions of Catholicism, much to the dismay of some of his more radical prelates. After Henry, Edward VI, under the influence of the duke of Somerset, the *de facto* ruler during the king's minority, sought to make the country more rigorously Protestant and sped up the pace of reform which had slackened in the final years of Henry's reign. Upon the accession of Mary Tudor, the country reverted to Catholicism, which was reinforced by her marriage to the most powerful Catholic ruler of the period, King Philip II of Spain. When Elizabeth took the throne in 1558, she reestablished England as a Protestant country, though one that tried to find a middle way between the Catholics and the radical Protestants. In the space of little more than thirty years, the country had witnessed several changes of religion, numerous uprisings, and executions of both Protestants and Catholics for heresy. The contest between the two branches of Christianity ran bitter and deep, with each side protesting its persecution and martyrdoms, asserting the authenticity of its own doctrine, and denouncing the authority of the other. The presses in

England and abroad produced numerous opposing religious texts that attempted to convince readers of their validity while denying that of the opposition. A wider understanding of the religious controversies of the period both clarifies and complicates the representations of relations between Christians and Jews in *The Merchant of Venice*. While the play clearly draws on a tradition of theological conflicts between Christians and Jews, it also reflects issues of contemporary concern within Christianity.

Catholics versus Protestants

While England was officially a Protestant country by the end of the six-teenth century when *The Merchant of Venice* was being written and per-formed, it was home to a significant Catholic minority. The state sought to secure the allegiance of its Catholic subjects by forcing them to conform to the Protestant Church of England. The stringent laws that were passed to force Catholics to conform or convert to Anglicanism[1] from the mid-1500s onward were justified as a political necessity: in the wake of Pope Pius V's bull of 1570, which excommunicated Elizabeth and nullified her legitimacy as a queen for Catholics, the state feared both domestic and foreign attempts to overthrow the government. Yet, while some Catholics may have sought to return their country to Catholicism, many loyal Catholic subjects felt the laws unjustly persecuted them for their religious beliefs. This state of affairs created anxieties and suspicions on both sides of the religious question. John Foxe voiced an influential Protestant perspective in his enormous *Acts and Monuments* (1563), which focused on the history of Catholic persecution of Protestants in England, and condemned Catholic practice generally (see p. 250 in this volume). In his introduction, he questions why Protestants should experience persecution at the hands of their fellow Christians, and contrasts that persecution with the (allegedly) merciful treatment offered to Catholics in England. Protestant viewers of *The Merchant of Venice* who identified its Italian setting with the seat of Catholic power (as the author of *A Discovery* does in an earlier excerpt [p. 150]), might have related the Vene-tians' treatment of Shylock to the Spanish Inquisition as represented by Foxe. This Catholic institution — first directed against the Jews — allowed greed and cruelty to cloak themselves in the robes of sanctity. Would the audience of the play be able to imagine the "good" Christian characters as similarly persecuting Shylock?

---

[1] **Anglicanism:** the form of Protestantism practiced by the Church of England, the official state church.

A Catholic viewer of the play might be more sensitive to the experience of a religious minority in a society dominated by Protestants. William Allen challenges Foxe's claim that Catholics are well used in England by demonstrating the cruelty of inflicting Protestant beliefs on "their infinitely distressed consciences" (p. 257) and contrasts this to the religious toleration of Catholics and other religious minorities in different countries. If they were better used in England, he argues, fears of Catholic rebellion would have never arisen. Robert Parsons argues that English Catholics are loyal subjects to the queen and also argues for religious tolerance, noting that radical Protestant sects live with greater impunity in England than do Catholics (see p. 258). He further emphasizes the sufferings of the Catholics by focusing on the physical and material punishments they experience; nonconforming Catholics could be fined, suffer the confiscation of their property, and face imprisonment or even death in cases where treason was alleged. Two examples of state rhetoric about Catholic criminality appear in the excerpts of proclamations aimed at Jesuits and other Catholic priests, whose mission is represented as advancing the aims of England's Catholic enemies, especially the pope and the king of Spain, rather than ministering to English Catholics (p. 263). While seeking to justify her suspicions of the priests' treason, Elizabeth argues that her judicial action against them is meant to circumscribe their political ambitions, rather than persecute them on religious grounds. Nevertheless, her proclamation firmly rejects a policy of religious toleration because it "would not only disturb the peace of the church but bring this our state into confusion" (p. 267). As a Jew and an alien in Venice, Shylock is susceptible to laws that apply only to a religious minority and carry both financial and penal threats. How would Catholics who lived under analogous circumstances respond to the trial scene in act 4 of *The Merchant of Venice*?

The sufferings of both Protestants and Catholics brought about the articulation of a discourse of religious toleration, which may also have influenced viewers of the play on both sides of the religious divide. However, these ideas were extremely limited in scope and applicability. While alive to the sufferings of his co-religionists, and seeking compassion from his opponents, Foxe nevertheless displays intolerance toward Catholics and religious others, such as Jews and Muslims. Parsons also summons the idea of religious tolerance in support of his argument, but imposes a condition, the logic of which would transform English Protestants into heretics susceptible to the same secular punishments he denounces earlier:

when a man has received once the Christian Catholic religion, and will by new devices and singularity corrupt the same by running out and making dissension in Christ his body (as all heretics do), then, for the conservation of unity in the Church, and for restraint for this man's fury and pride, the Church hath always from the beginning allowed that the civil magistrate should recall such a fellow by temporal punishment to the unity of the whole body again. (p. 262)

Any Christian who leaves Catholicism would therefore be liable to punishment by the secular authorities. In what ways could such restrictive perspectives on religious freedom influence interpretations of the play's representation of Jews?

Jews as Other

Early Christianity formulated itself both within and against Judaism. The Jew was the primary religious other Christians needed to separate themselves from in order to establish an alternative and autonomous religion; this struggle is worked out in part in the pages of the New Testament. One important way in which Judaism differed from Christianity was its continued commitment to practicing the laws given in the Torah, the Hebrew Bible (the Christian Old Testament). St. Paul argues in the excerpt taken from Galatians that Jesus's fulfillment of the law releases Christians from its observance (p. 269). Salvation now depended on having faith in God's merciful grace rather than in performing the law.[2] Hence, Christians seem to judge Jewish observance of the law as, at best, ignorance or, at worst, a stubborn rejection of God's will and mercy. Paul's juxtaposition of Jewish law and Christian mercy is taken up by the play's association of Shylock with justice and Portia with mercy.

In addition to their place in the theological tradition of Christianity, Jews also figured frequently in contemporary controversies. In the early modern period, the Jew as religious other and opponent of God's will becomes a figure by which Christian religious opponents can denigrate and delegitimize each other. Both Protestants and Catholics attempted to undermine the authority of their religious adversaries by representing them as Jews. Andrew Willet articulates part of his criticism of Catholicism in terms of the similarities of its rites and teachings with Jewish ceremonial law. He begins his analysis by alleging that Catholics rely on the "beggarly ceremonies" and "unwritten traditions of the Jews" which have been condemned by Jesus (p. 270). However, as he proceeds, he shifts his accusation to suggest that Catholics are *worse* than Jews: "And yet never did the Jews use half of

[2] Specifically, the so-called ceremonial and civil categories of Jewish law are abrogated in Christianity; the moral laws are still considered binding, but one cannot effect one's own salvation through their performance.

those ceremonies . . . . as papists do. . . . The Jews themselves would blush to behold such things" (p. 272). He concludes that the Jews, though they reject the salvation offered by Jesus, are better off than Catholics, since the former, at least, were subject to divinely ordained "legal ceremonies, not to the inventions of men" (p. 273). From a Catholic perspective, Richard Bristow likens Protestants to Jews insofar as the former, in effect, deny that Jesus is God by rejecting the authenticity of his Catholic Church. He attacks the Protestant claim that their religion is based on Scripture by insisting that the Catholic Church and the Bible are not at odds with each other; he dismisses the Protestant Church and its leaders as a "synagogue" and "Rabbis" (p. 276). Hence polemics that represented disagreements between Christians and "Jews" might in fact refer to controversies within Christian doctrine and practice. Is Shylock's conflict with the Venetian community best understood as a clash between Jews and Christians, or might it reflect a dispute between Catholics and Protestants?

Attitudes about the Jews themselves were frequently but not exclusively negative; the Hebrews of the Bible were sometimes viewed as the locus of authentic practice or knowledge of God's word. In breaking away from hundreds of years of Catholic authority, Protestants turned to the Jews, especially as the preservers of the Hebrew Bible, as a source of legitimacy by which they could justify their practices. Basing their arguments on close readings of the Bible, Protestants "proved" the divine authenticity of their reforms and "exposed" the human artificiality of Catholic beliefs and rites. English Protestants in particular "believed themselves to be the second Israel" (Greenfield 52), and could be called upon to identify with Jews. William Perkins urges his auditors to do just this in his sermon preached on the first two verses of chapter 2 of Zephaniah in which the prophet chastises the children of Israel for provoking God's wrath with their disobedience (p. 277). Perkins begins by citing contemporary views of Jews as "a most vile and wicked people, a froward generation, . . . worthy to taste deeply of all God's plagues, who so far abused his love and mercy" (p. 278). However, he forces his audience to acknowledge that England is equally unworthy of God's love for failing to appreciate his actions on their behalf. As he enumerates God's blessings on the English and their ingratitude toward him, he is forced to conclude that the English people might be more sinful than the Jews. While Jews could be conceptualized as archetypal sinners and heretics, they also could be seen as less disobedient than Christian opponents and even as better servants of God than Christians themselves. In what ways does *The Merchant of Venice* emphasize the resemblances between Jews and Christians in Venice? If Shylock is guilty of various transgressions, are his Christian opponents equally or even more guilty of similar crimes?

Conversion

Conversion was an issue of central importance given the number of changes in religion experienced by the population in sixteenth-century England. In establishing a state church that was headed by the monarch, Henry VIII effectively tied his subjects' political allegiance to their religious faith. Every time a ruler ascended the throne, the people were required to pledge loyalty to both the sovereign and to his or her religion. As England settled into Protestantism during the reign of Elizabeth, concerns about conversion were still dominant in the minds of both Catholic and Protestant clerics; each side was interested in bringing the people to the true faith. While the state was particularly interested in converting Catholics and establishing a unified political and religious realm, it tended to focus on the political aspect of conformity, which entailed allegiance to the queen. Although stringent laws were legislated and enforced to promote conformity to Anglican practices and hence political obedience, the state stopped short of endorsing coercion of belief (Questier 158–59, 168–70). Both Catholic and Protestant clerics who sought to convert the other side were more concerned with religious belief than political affiliation, and they felt even more strongly that true conversion could be effected only by "an arousal of the will under the influence of grace" and not by force (Questier 186). Foxe's account of Archbishop Cranmer's forced renunciation of and subsequent return to Protestantism during the reign of Queen Mary illustrates the limits of force and suggests that the torments of coerced belief are worse than the agonies of martyrdom. Catholic writers in the period echoed these views. How do they apply to the process Shylock undergoes in act 4? Since religious beliefs aren't explicitly at stake in the enforcement of the law against him, is Shylock expected merely to conform to an appearance of Christian observance? What do we make of the fact that his conversion is stipulated by Antonio as a condition of mercy, and not in fact required by law? Would an audience view the conversion as a forcing of conscience or as a free embracing of Christian beliefs?

In a biblical context, Jews could be viewed positively; however, toleration of contemporary Jews frequently depended on their willingness to convert. The calling of all the Jews to Christianity was understood to be a necessary precursor to Jesus's second coming; this concept achieved a certain urgency in a period where millenarian expectations were rising. However, in spite of the sense of divine imperative driving this phenomenon, those who write about converting actual Jews nevertheless reject compulsion as a legitimate or effective strategy. Gregory Martin articulates this view in his consideration of the organized attempt to convert Jews in Rome. Although the Jews' refusal to acknowledge the divinity of Jesus renders them, in his view, active enemies of Christianity who deserve God's punishment, he nevertheless sees them as less problematic than Christians who have turned away from Catholicism. Like Parsons, he notes that only those who have been confirmed in the Church can be compelled to return if they have left it. Since Jews were never "children of the Church," they cannot be forced to convert, although they may be "by all charitable means . . . invited and persuaded to forsake obstinate Judaism and to become Christians" (pp. 281–82). Although Roman Jews are subject to a number of legal restrictions, they have some rights within this constraint, including "justice against Christians . . . in all lawful causes" and some measure of freedom in the practice of their religion. Nevertheless they are required to listen to sermons against Judaism, which are preached on the Jewish Sabbath by a priest learned in Hebrew. While Martin implies this method meets with steady success in Rome, Coryate complained that the rate of conversion among Venetian Jews was very low (see p. 142). Even those Christian authors who consider the conversion of the Jews in the larger theological context of the second coming of Jesus — such as Thomas Draxe, who suggests it is close at hand — eschew force and perhaps even active proselytizing. While Draxe encourages severe laws to curb Jews from practicing usury with Christians,[3] and encourages Christian rulers to force Jews to listen to readings of Christian scripture, he also cautions Christians against prideful and cruel treatment of the Jews, which might serve to hinder their conversion. Because this conversion is a "mystery, . . . not common or ordinary, let us not be curious to dive and descend further into particulars than God's word, . . . but rest in expectation, until the time come, and in the interim help [the Jews] by our prayers, and further them by our zealous and holy example" (p. 287). Here he argues that setting good examples of Christian behavior would be more effective in bringing about conversions than proselytizing. How would these views inform an audience's evaluation of Jessica's and Shylock's conversions in *The Merchant of Venice*? How do these two models of conversion comment on each other in the play?

While the forced conversion of Jews was not advocated in many of the writings on the subject circulating in early modern England, historically Jews had been the victims of persecution and forced conversions. In some instances, such as the case of the 1492 expulsion of the Jews from Spain

---

[3] By this period, very few Jews in Europe made a living lending money at high rates.

where they were given the option of leaving or converting, the choice was not entirely free owing to the hardships of emigration. In his *Consolation for the Tribulations of Israel*, which circulated in the sixteenth-century English New Christian community,[4] Samuel Usque presents a chronicle of Jewish history that includes a reaction to the experience of conversion attempts (see p. 288). In his consideration of the Jews of medieval England, Usque imaginatively represents Christians acknowledging the inefficacy of both persuasive and coercive attempts to convert the Jews, and recommending that they be killed. The choice that the Jews are ostensibly offered in the narrative, either death (disguised as religious freedom) or conversion, amounts to the same thing, as Jewish tradition considers apostasy a kind of death. While Jewish law sets out a principle requiring that one choose death rather than a forced conversion, in practice, rabbis of the medieval and early modern periods were more lenient on this issue, and victims of forced conversion were expected to return to Jewish practice as soon as possible, or continue to practice Judaism in secret, rather than give up their lives (Friedman 13–14). While early modern audiences of *The Merchant of Venice* would probably not be familiar with Usque's text, it nevertheless provides an important context for the play by encouraging us to imagine it from a contemporary Jewish perspective. Would a Jewish viewer in the period sympathize with Shylock's plight or condemn him for relinquishing his faith to save his life and goods? How would Jessica's conversion be viewed from this perspective?

Jews in England

While Usque's tales of English attempts to expel the Jews may have little basis in historical fact, the event he chronicles — the 1290 expulsion of 2,000 to 3,000 Jews from England — actually did occur. Edward I agreed to expel the Jews from England on the condition that Parliament levy a tax to provide much-needed revenue. Members of Parliament wanted to take action against the Jews because they were frequently in the Jews' debt and the king could rely on them for funds when Parliament refused to authorize

---

[4] **English New Christian community:** Jewish converts to Christianity were often referred to as New Christians or marranos. While some of these converts did change faiths when they converted, others who converted under duress simply professed Christianity publicly while continuing to practice Judaism in secret.

a tax. A small community of converts from Judaism remained, but there was no significant Jewish presence in England again until the end of the seventeenth century, although occasional Jewish visitors and a number of recently converted New Christians and crypto-Jews from other countries came to sojourn or settle in England. However, attitudes and narratives about the medieval Jewish community circulated in the law as well as in chronicles of English history, keeping them alive in the early modern public discourse. A medieval law recorded in *Fleta* prescribes live burial as the punishment for those Christians who "have connection with Jews and Jewesses or are guilty of bestiality or sodomy" (p. 304); Edward Coke cites this "ancient law" in the *Third Part of the Institutes,* suggesting that it would still pertain (p. 303). Foxe recounts a number of incidents involving Jews in medieval England; even though he is careful to justify the cruel treatment they received at the hands of the English, he vividly records the suffering and mass suicide of the Jews resulting from the escalation of a minor scuffle with Christians during the coronation of Richard I (see p. 30). Holinshed recasts the common association of Jews with usury by representing them rebuking Henry III for his "excessive taking of money, as well of his Christian subjects as of them"; the king is reported as "fleec[ing] the Jews to the quick," burdening them as well as his other subjects with "exactions and impositions" (pp. 298–99). He also gives a sympathetic account of a deception that cost many Jews their goods and their lives during the expulsion; in his version of the tale, the wicked perpetrators were punished for their fraud by hanging. However, he, like Foxe, also gives voice to the specious charge of ritual murder directed at Jews throughout medieval and early modern Europe. In this widespread fantasy, the Jews kidnap and crucify a Christian child, with the aim, apparently, of parodying and ridiculing Christianity. Early modern Christian doctrine frequently characterizes the Jews as being responsible for the crucifixion of Jesus and hence bearing malice toward all Christians. All Jews were suspected of haboring this hatred of Christians, even Jewish converts to Christianity. When the New Christian Roderigo Lopez is convicted of plotting with Spain to poison Queen Elizabeth, his accusers identify his Jewishness as the basis of the crime, comparing his assassination attempt with Judas's betrayal of Jesus and scoffing at his profession of Christianity (see pp. 307, 310). While early modern audiences of *The Merchant of Venice* would have had few opportunities to meet with practicing Jews, they were exposed to a number of sources about Jews in England. How do these negative and positive representations of Jews relate to the depiction of Shylock and Jessica in the play and the attitudes of the Venetians toward them?

→ JOHN FOXE

From Acts and Monuments   *1570*

John Foxe (1516–1587) was an activist Protestant who, even in his college days at Oxford, refused to conform to certain requirements of the Church of England. When Mary Tudor succeeded to the throne in 1553, Foxe protested her intention to return England to Catholicism; he joined other Marian exiles (Protestants who fled under Mary's reign) in Germany and Switzerland before returning to England in 1559 after Elizabeth I became queen. He is the author of a number of theological treatises, but his most famous and influential work is *Acts and Monuments,* a history of Catholic persecutions of Protestants in England. A canon, or church law, was proposed to have copies of Foxe's martyrology placed in cathedral churches. While Parliament never confirmed this canon, the book spread throughout England, and became an authoritative source for Puritans on the subjects of church history and anti-Catholic polemic. This huge work, totaling well over two thousand pages in the two-volume 1570 edition and covering hundreds of years of history, presents an avowedly biased view of relations between Catholics and Protestants. Despite his zeal and piety, he was considered a man of "cheerful temperament" and was capable of friendships with Catholics. For example, he maintained a lifelong friendship with a former student, the Catholic-leaning duke of Norfolk; when Norfolk was executed for conspiring with Catholics against Queen Elizabeth, Foxe walked with him to the scaffold. In the excerpts from *Acts and Monuments* printed below, Foxe describes various persecutions fomented by Catholics. In his diatribe against the Spanish Inquisition, his tone is frequently sarcastic, referring to the justification of interrogation and torture as "holy doctrine." The following sections question Catholic persecution of Protestants, in contrast to the allegedly fair treatment of Catholics in England, and raise issues of freedom of religious belief, or conscience. Given the centrality for English culture of these stories of religious persecution, what context do they provide for an early modern audience viewing the attitudes of intolerance against Judaism expressed in *The Merchant of Venice*?

THE FORM AND MANNER OF THE EXECRABLE[1] INQUISITION OF SPAIN

The cruel and barbarous inquisition of Spain first began by King Ferdinand and Isabella his wife, and was instituted against the Jews, who, after their baptism, maintained again their own ceremonies. But now it is practiced against them that be ever so little suspected to favor the verity of the

---

[1] **Execrable:** deplorable.

John Foxe, *Acts and Monuments* (1570; London: Seeleys, 1857), vol. 4, part 2, 451–52.

Lord. The Spaniards, and especially the great divines there, do hold that this holy and sacred inquisition cannot err, and that the holy fathers, the inquisitors, cannot be deceived.

Three sorts of men most principally be in danger of these inquisitors: they that be greatly rich, for the spoil of their goods; they that be learned, because they [the inquisitors] will not have their misdealings and secret abuses to be espied and detected; [and] they that begin to increase in honor and dignity, lest they, being in authority, should work them [the inquisitors] some shame or dishonor.

The abuse of this inquisition is most execrable. If any word shall pass out of the mouth of any, which may be taken in evil part; yea, though no word be spoken, yet if they bear any grudge or evil will against the party, incontinent[2] they command him to be taken, and put in a horrible prison, and then find out crimes against him at leisure, and in the meantime no man living is so hardy[3] as once to open his mouth for him. If the father speak one word for his child, he is also taken and cast into prison as a favorer of heretics, neither is it permitted to any person to enter in to the prisoner. But there he is alone, in a place where he cannot see so much as the ground where he is, and is not suffered either to read or write, but there endure in darkness palpable, in horrors infinite, in fear miserable, wrestling with the assaults of death.

By this it may be esteemed what trouble and sorrow, what pensive sighs and cogitations they sustain, who are not thoroughly instructed in holy doctrine.[4] Add, moreover, to these distresses and horrors of the prison, the injuries, threats, whippings, and scourgings, irons, tortures, and racks which they endure. Sometimes also they are brought out, and showed forth in some higher place to the people, as a spectacle of rebuke and infamy. And thus are they detained there, some many years, and murdered by long torments, and whole days together treated much more cruelly, out of all comparison, than if they were in the hangman's hands to be slain at once. During all this time, what is done in the process, no person knows, but only the holy fathers and the tormentors, who are sworn to execute the torments. All this is done in secret, and (as great mysteries[5]) pass not the hands of these holy ones. And after all these torments so many years endured in the prison, if any man shall be saved, it must be by guessing; for all the proceedings of the court of that execrable inquisition are open to no man, but all is done in hugger-mugger[6] and in close corners, by ambages,[7] by covert ways,

---

[2] **incontinent:** immediately. [3] **hardy:** courageous. [4] **holy doctrine:** meant sarcastically. [5] **great mysteries:** meant sarcastically. [6] **hugger-mugger:** secrecy. [7] **ambages:** mysterious ways of action.

and secret counsels. The accuser is secret, the crime secret, the witness secret, whatsoever is done is secret, neither is the poor prisoner ever advertised of any thing. If he can guess who accused him, whereof and wherefore, he may be pardoned peradventure of his life; but this is very seldom, and yet he shall not incontinent be set at liberty before he hath long time endured infinite torments. And this is called their 'Penitence,' and so is he let go: and yet not so but that he is enjoined, before he pass the inquisitor's hands, that he shall wear a garment of yellow colors for a note of public infamy to him and his whole race. And if he cannot guess right, showing to the inquisitors by whom he was accused, whereof and wherefore (as is before touched), incontinent the horrible sentence of condemnation is pronounced against him, that he shall be burned for an obstinate heretic. And so yet the sentence is not executed by and by, but after he hath endured imprisonment in some heinous prison.

And thus have ye heard the form of the Spanish inquisition. By the vigor and rigor of this inquisition many good true servants of Jesus Christ have been brought to death, especially in these latter years, since the royal and peaceable reign of this our Queen Elizabeth.

The Second Question

My second question is this, to demand of you, Catholic professors of the pope's sect, who so deadly malign and persecute the Protestants professing the gospel of Christ: what just or reasonable cause have you to allege for this your extreme hatred ye bear unto them, that neither you yourselves can abide to live with them, nor yet will suffer the others to live amongst you? If they were Jews, Turks, or infidels, or in their doctrine were any idolatrous impiety, or detestable iniquity in their lives; if they went about any deadly destruction, or privy conspiracies to oppress your lives, or by fraudulent dealing to circumvent you; then had you some cause to complain, and also to revenge. Now seeing in their doctrine ye have neither blasphemy, idolatry, superstition, nor misbelief to object unto them — seeing they are baptized in the same belief, and believe the same articles of the creed as ye do, having the same God, the same Christ and Savior, the same baptism, and are ready to confer with you in all kind of Christian doctrine, neither do refuse to be tried by any place of the Scripture — how then riseth this mortal malice of you against them? If you think them to be heretics, then bring forth, if ye can, any one sentence which they arrogantly hold, contrary to the mind of holy Scripture, expounded by the censure of most ancient doctors.[8] Or what

---

[8] **doctors:** those proficient in knowledge of theology.

is there in all the Scripture to be required, but they acknowledge and confess the same? See and try the order of their lives and doings; what great fault find you? They serve God, they walk under his fear, they obey his law, as men may do; and though they be transgressors towards him, as other men are, yet toward *you* what have they done, what have they committed or deserved, why you should be so bitter against them? . . .

Let us turn now to the peaceable government in this realm of England, under this our so mild and gracious queen now presently reigning. Under whom you see how gently you are suffered, what mercy is showed unto you, how quietly ye live. What lack you that you would have, having almost the best rooms and offices in all the realm, not only without any loss of life, but also without any fear of death? And though a few of your arch-clerks[9] be in custody, yet in that custody so shrewdly are they hurt, that many a good Protestant in the realm would be glad with all their hearts to change rooms and diet with them, if they might.[10] And albeit some other for their pleasure have slipped over the seas,[11] if their courage to see countries abroad did so allure them, who could let[12] them? Yet this is certain, no dread there was of death that drove them. For what papist[13] have you seen in all this land to lose either life or limb for papistry, during all these twelve years hitherto since this queen's reign?[14] And yet, all this notwithstanding, having no cause to complain, so many causes to give God thanks, ye are not yet content, ye fret and fume, ye grudge and mutter, and are not pleased with peace, nor satisfied with safety, but hope for a day, and fain[15] would have a change. And to prevent[16] your desired day, ye have conspired, and risen up in open rebellion against your prince, whom the Lord hath set up to be your governor.

And as you have since that now of late disturbed the quiet and peaceable state of Scotland, in murdering most traitorously the gentle and godly regent of Scotland (who, in sparing the queen's life there, when he had her in his hands, hath now therefore lost his own),[17] so, with like fury, as by your rebellion appears, would you disturb the golden quiet and tranquility of this realm of England, if ye might have your wills. Which the merciful grace of

---

[9] **arch-clerks:** higher-ranking clerics.   [10] **And though . . . if they might:** Foxe is saying that imprisoned Catholics live better than some free Protestants.   [11] **over the seas:** gone abroad.   [12] **let:** prevent.   [13] **papist:** derogatory term for a Catholic.   [14] **this queen's reign:** Elizabeth I, who reigned from 1558 to 1603.   [15] **fain:** gladly.   [16] **prevent:** in anticipation of.   [17] Lord James Stewart, earl of Moray (1531?–1570) was the Protestant half-brother of the Catholic Mary, Queen of Scots, and regent of Scotland from 1567 until his assassination in 1570. When Mary was in his power, he treated her leniently; his murder, engineered by a Catholic faction, was "approved of, if not instigated by Mary, who liberally rewarded the assassin" (*DNB;* see entries for Stewart and James Douglas, earl of Morton).

the Almighty, for Christ his Son's sake our Lord, forefend[18] and utterly disappoint. Amen!

Wherefore, these premises considered, my question is to ask of you and know, what just or reasonable cause ye have of these your unreasonable doings, of this your so mortal and deadly hatred, fury, and malice, you bear against these your even-christened;[19] of these your tumults, conjurations, gaping and hoping, rebellions, mutterings, and murders, wherewith you trouble and disquiet the whole world? Of all which mischiefs, if the true cause were well known, the truth would be found doubtless to be none other but only the private cause of the bishop of Rome, that he is not received,[20] and the dignity of his church exalted.

ADMONITION TO THE READER, CONCERNING THE EXAMPLES ABOVE MENTIONED

It has been a long persuasion,[21] engendered in the heads of many men these many years, that to ground a man's faith upon God's word alone, and not upon the see[22] and church of Rome, following all the ordinances and constitutions of the same, was damnable heresy, and to persecute such men to death, was high service done to God. Whereupon have risen so great persecutions, slaughters and murders, with such effusion of Christian blood through all parts of Christendom by the space of these seventy years, as hath not before been seen. And of these men Christ himself does full well warn us long before, truly prophesying of such times to come, when they that slay his ministers and servants should think themselves to do good service unto God. Now what wicked service, and how detestable before God this is, which they falsely persuade themselves to be godly, what more evident demonstrations can we require, than these so many, so manifest, and so terrible examples of God's wrath pouring down from heaven upon these persecutors. Whereof a part we have already set forth; for to comprehend all (which in number are infinite), it is impossible. Wherefore, although there be many which will neither hear, see, nor understand, what is for their profit, yet let all moderate and well-disposed natures take warning in time. And if the plain word of God will not suffice them, nor the blood of so many martyrs will move them to embrace the truth, and forsake error, yet let the desperate deaths and horrible punishments of their own papists persuade them, how perilous is the end of this damnable doctrine of papistry. For if these

---

[18] **forefend:** forbid.   [19] **even-christened:** fellow Christians.   [20] **he is not received:** he is not accepted as supreme head of the Church.   [21] **persuasion:** opinion.   [22] **see:** jurisdiction.

papists, which make so much of their painted[23] antiquity, do think their proceedings to be so catholic, and service so acceptable to God, let them . . . tell us how come then their proceedings to be so accursed of God, and their end so miserably plagued, as by these examples above specified is here notoriously to be seen? Again, if the doctrine of them be such heresy, whom they have hitherto persecuted for heretics unto death, how then is Almighty God become a maintainer of heretics, who hath revenged their blood so grievously upon their enemies and persecutors?

[23] painted: alleged.

⇥ **WILLIAM ALLEN**

From A True, Sincere, and Modest Defense of English Catholics
1584

William Allen (1532–1594) studied at Oxford during the reigns of Edward VI and Mary, maintaining his Catholicism throughout, even while teaching as a fellow during the early years of Elizabeth's reign. He fled England in 1561 because of his ardent Catholicism, but having fallen ill, returned a year later on the recommendation of a doctor. While in England he urged his fellow Catholics to disobey the law requiring them to attend Protestant services, recruited converts, and generally promoted the practice of Catholicism. In 1565 Allen left England for the last time and went to Belgium, where he was ordained a priest. In 1568 he established an English college at Douay, France, to train English students abroad, prepare priests to restore the Catholic faith to England if circumstances permitted, and teach Catholicism to students sent from England. Allen set up another English college in Rome in 1579, which was placed under Jesuit control; in the following year, he encouraged the Jesuits to help promote Catholicism in England, which resulted in Robert Parsons's mission there. Allen sought to return England to Catholicism, first through religious means by persuading the seminary priests. However, he later supported intrigue and force, becoming, along with Parsons, an advocate for the Spanish claim to the English throne. He was named cardinal by Pope Sixtus V in preparation for a Spanish armed victory against England. As the Armada sailed, Allen published a treatise, *An Admonition to the Nobility and People of England* (1588), which declared Elizabeth a heretic and usurper and called for her deposition. The loyalty of English Catholics to Elizabeth brought Allen's political plans to nought, but his support of Catholic education in this period probably

---

William Allen, *A True, Sincere, and Modest Defense of English Catholics That Suffer for Their Faith Both at Home and Abroad against a False, Seditious, and Slanderous Libel, Entitled "The Execution of Justice in England"* (Ingolstadt, Germany, 1584), 209–12, 215–17.

prevented Catholicism from being destroyed in England. The work excerpted below was written in response to William Cecil, Lord Burghley's *Execution of Justice in England* (1583), in which Burghley, lord high treasurer of England and influential advisor to Elizabeth, argues that Catholics are executed only for treason, not religion. Allen's treatise calls for religious tolerance, requesting that Catholics be permitted places of worship, and citing the liberty enjoyed by Catholics in other Protestant countries as well as the "courtesy [that] the Christians find among the very Turks, or very Jews among Christians" (p. 257). However, even while he calls for a kind of religious freedom in England, Allen's preference would be to return the entire country to Catholicism (p. 258).

CHAPTER 9

It cometh often to our minds, that if anything avert God's ire from our Prince[1] and country, it is the abundance of holy blood shed these late years and ever since the first revolt.[2] Which though by justice, it might cry rather to God for vengeance (and so it doth in respect of the impenitent, and the clamor thereof shall never be void[3]) yet we trust it sueth[4] for mercy, specially in respect of the infinite number of all estates[5] that never consented to this iniquity. It is the heroic endeavor of a great many zealous priests and worthy gentlemen, that continually offer not only their prayers, and other devout and religious offices, but themselves in sacrifice, for the salvation of their best beloved country. It is the ardent and incessant care of his holiness,[6] seeking our reconcilement with charity unspeakable. It is the general conjunction of all Christian minds in the whole world, towards our recovery. No church, no company, monastery or college of name in Christendom, that with earnest devotion and public fasts and prayers, labors not to God for mercy towards us. Finally, even those things and persons, that the adversaries account to be the cause of all their troubles and fears, are indeed the only hope of God's mercy, their own pardon, and our country's salvation. . . .

But (alas) if any mercy, just or tolerable treaty were meant, or ever had been offered to Catholics upon any reasonable conditions whatsoever, our adversaries had never needed to have fallen to such extreme proceedings with their own flesh and blood. Nor ever had any such troubles, fears, or dangers been thought upon, whereof now they have so deep apprehension if any pitiful ear had ever been given by the superiors to the incessant groans, cries, tears, and supplications of the Catholic subjects, desiring but relief of

---

[1] **Prince:** Elizabeth I; *prince* is commonly used for both genders.   [2] **the first revolt:** when England left the Catholic Church.   [3] **void:** annulled.   [4] **sueth:** requests.   [5] **estates:** classes.
[6] **his holiness:** the pope.

their infinitely distressed consciences, tormented by damnable oaths, articles, and exercises of Calvinism, that were forced upon them; if they might have had either by license or convenience, in never so few places of the realm, never so secretly, never so inoffensively, the exercise of that faith and religion, which all their forefathers since our country was converted, lived and died in, and in which themselves were baptized and from which by no law of God nor man they can be compelled to any sect or rite of religion which they nor their forefathers ever voluntarily accepted or admitted; if of all the noble churches, colleges, and other inestimable provisions of the realm, founded and made only by Catholics for Catholics, and for no Protestants nor any [of] their sacrilegious ministries at all, some few had been permitted to the true owners, and to that true worship of God, for which they were instituted; if they might have obtained any piece of that liberty, which Catholics enjoy in Germany, Switzerland, or other places among Protestants, or half the freedom that the Huguenots[7] have in France and other countries; yea, or but so much courtesy as the Christians find among the very Turks, or very Jews among Christians; upon any reasonable or unreasonable tribute (which has been often in most humble and lamentable sort offered and urged). Or (to be short) if any respect, care, or compassion in the world has been had, either of Catholic men's souls, bodies, or goods, our adversaries should never have been troubled nor put in jealousy of so many men's malcontentment at home, nor stand in doubt of the departure and absence of so great a number of nobility and principal gentlemen abroad. . . . There should have been no cause of writing so many books for defense of our innocence, and the faith of our forefathers, and for our just complaint to the Christian world, of the intolerable rigor or cruelty used against us. . . .

But to return to our purpose and to the libeler's[8] proffer[9] of mitigation or ceasing this persecution, upon condition we would deal no more in secret, but openly. We protest before God and all his Saints, that we will (upon any reasonable security of our persons, liberty of conscience, permission to exercise Christian Catholic offices, to the salvation of our own souls and our brethren) do the same things publicly which we now do secretly, in all peaceable and priestly sort as hitherto we have accustomed: and that so, those things which now you suspect to be done against the state (for that they be done in covert[10]) may plainly appear unto you nothing else indeed, but mere matter of conscience and religion, as in verity they are.

---

[7] **Huguenots:** Protestant sect.   [8] **libeler:** Lord William Burghley, author of the treatise to which Allen is replying.   [9] **proffer:** offer.   [10] **in covert:** in secret.

Therefore, if such as govern our state under her Majesty at this day, cannot be induced to revoke themselves and the whole realm (which were absolutely the best) to the former Catholic state and condition wherein their ancestors left it . . . yet at the least, let their wisdom consider, that their principal worldly error was, that in the beginning, or long since, they gave not liberty of conscience to Catholics (being far the greater and more respective part of the realm) as other [countries] of their religion and profession have done, to their own great advantage. . . .

But sure we are, that the first best for our English nation, as well Prince as people, were, both in respect of God and the world, of themselves, and other men; to restore the state again to the obedience of God's church, and to the happy fellowship of all their forefathers, and other faithful people and Princes now living. The next best were in respect of their own security and perpetuity (if the first may not take place) to desist from persecuting their Catholic subjects and brethren, and to grant some liberty for exercise of their consciences, divine offices and holy devotions; that so they may pray for her Majesty and counselors as their patrons, whom now they pray for only as their persecutors.

If to none of these conditions they can be brought, but will have our bodies, goods, life and souls, then let our Lord God the just arbiter of all things, and judge of Princes as well as poor men, and the only comforter of the afflicted, discern our cause. In whose holy name, word and promise, we confidently tell them, and humbly even in Christ's blood pray them, to consider of it: that by no humane[11] force or wisdom they shall ever extinguish the Catholic party, overcome the holy Church, or prevail against God.

[11] humane: *humane* carries the possible sense of *human* as well.

→ ROBERT PARSONS

From A Brief Discourse Containing Certain Reasons Why Catholics Refuse to Go to Church   *1580*

Robert Parsons or Persons (1546–1610) was a Jesuit missionary and author of numerous tracts and books advancing the Catholic faith. He received his bachelor's and master's degrees from Oxford, where he remained as tutor, dean, and ultimately chaplain-fellow of his college, but never took orders in the Anglican

Robert Parsons, *A Brief Discourse Containing Certain Reasons Why Catholics Refuse to Go to Church* (London, 1580), ‡2v–‡7r, ‡‡2v–‡‡4v.

Church. He left England in 1574 to study medicine at Padua, but subsequently went to Rome, where he became a Jesuit. In 1580 he went on a mission to England, approved by the pope, to convert ministers and lay Protestants to Catholicism; he had strict instructions not to involve himself in or even discuss political matters. While in England Parsons set up a secret printing press, which published and distributed polemical works. In response to an anonymous treatise arguing that Catholics should comply with the English law requiring them to attend the Protestant service, Parsons wrote *A Brief Discourse Containing Certain Reasons Why Catholics Refuse to Go to Church* (1580), excerpted below. Later in his career Parsons became involved with Spain in political maneuvers against the English, and encouraged Philip II to attack England with the Armada. At the time of Elizabeth's 1591 proclamation in which he is mentioned (see p. 265), Parsons was encouraging Philip to attack England once more and was rallying support among English Catholics. Parsons believed in the power of the pope to depose rulers who had left the Catholic religion; prior to the attack of the Armada he also debated whether to justify a Spanish claim to the English throne by right of conquest or inheritance. In 1594 he published his *Conference about the Next Succession*, which advances a "historical and legal argument to prove the right of the people to alter the direct line of succession for just causes, especially for religion; and . . . a genealogical argument, [which] balances the various claims, and points to the infanta of Spain . . . as the fittest successor to Elizabeth" (*DNB*). However, in his *Brief Discourse,* Parsons writes in a less political mood, focusing on freedom of conscience and requesting tolerance for Elizabeth's loyal Catholic subjects. While his plea for religious freedom seems more inclusive than Foxe's, he in effect precludes tolerance for Protestants at the end of this excerpt by stating that anyone who has left the Catholic church can be forced "by temporal punishment to the unity of the whole . . . again" (p. 262). With Foxe, Allen, and Parsons we see the development of a discourse of religious tolerance in response to religious persecution; how do they imagine the extent and limits of religious freedom? How applicable are their definitions to faiths other than their own?

From *A Brief Discourse*

[T]he state of many a thousand of your grace's most loving, faithful, and dutiful subjects, who being now afflicted for their consciences, and brought to such extremity as never was heard of in England before, have no other means to redress and ease their miseries, but only as confident children to run unto the mercy and clemency of your Highness their Mother and born sovereign Princess. Before whom, as before the substitute and Angel of God, they lay down their griefs, disclose their miseries, and unfold their pitiful afflicted case, brought into such distress at this time, as either they must renounce God by doing that, which in judgment and conscience they do condemn, or else sustain such intolerable molestations, as they cannot bear. Which your majesty by that which followeth, more at large, may please to understand.

There are at this day in this your Majesty's realm, four known religions, and the professors thereof, distinct both in name, spirit and doctrine: that is to say, the Catholics, the Protestants, the Puritans,[1] and the Householders of Love,[2] besides all other petty sects, newly born and yet groveling on the ground. Of these four sorts of men, as the Catholics are the first, the most ancient, the more in number, and the most beneficial to all the rest (having begotten and bred up the other, and delivered to them this realm, conserved by Catholic religion, these thousand years and more) so did they always hope to receive more favor than the rest, or at least wise, equal toleration with other religions disallowed by the state. But God knoweth, it has fallen out quite contrary. For other religions have been permitted to put out their heads, to grow, to advance themselves in common speech, to mount to pulpits, with little or no control. But the Catholic religion hath been so beaten in with the terror of laws and the rigorous execution of the same, as the very suspicion thereof hath not escaped unpunished. . . .

To this now if we add the extreme penalties laid upon the practice of certain particulars in the Catholic religion; as imprisonment perpetual, loss of goods and lands, and life also, for refusal of an oath against my religion; death for reconciling myself to God by my ghostly father;[3] death for giving the supreme Pastor supreme authority in causes of the Church; death for bringing in a crucifix in remembrance of the crucified: death, for bringing in a silly pair of beads, a medal, or an agnus dei,[4] in devotion of the lamb that took away my sins, which penalties have not been laid upon the practice of other religions. Your Majesty that shall easily find to be true, so much as I have said, which is, that the Catholic religion, wherein we were born, baptized, and bred up, and our forefathers lived and died most holily in the same, hath found less favor and toleration than any other newer sect or religion whatsoever. . . .

But now, these afflictions, how grievous and heavy soever they were, yet were they hitherto more tolerable, because they were not common, nor fell not out upon every man: and if there were any common cross laid upon them, (as there wanted[5] not) they bear it out with patience. . . .

---

[1] **Puritans**: radical Protestants.   [2] **Householders of Love**: a Christian group that followed the teachings of Dutch mystic Hendrick Niclaes.   [3] **reconciling . . . ghostly father**: making confession to a Catholic priest.   [4] **silly pair of beads, a medal, or an agnus dei**: Catholic religious objects: a rosary, a saint's medal, and a figure of a lamb bearing a cross or flag (sometimes stamped in wax and blessed by the pope).   [5] **wanted**: lacked.

But at this time present, and for certain months past, the tempest hath been so terrible upon these kind of men, and their persecutions so universal, as the like was never felt, nor feared before. For besides the general molestation, and casting into jails, both of men, women, and children of that religion throughout all parts of your Majesty's realms, there are certain particulars reported here, which make the matter more afflictive. As the disjoining of man and wife in sundry prisons, the compelling of such to die in prison, which could not stand or go in their own houses; the sending of virgins to Bridewell[6] for their consciences; the racking and tormenting of diverse,[7] which was never heard of before in any country for religion. And that which above all other things is most grievous, injurious, and intolerable, is, the giving out publicly[8] that all Catholics are enemies and traitors to your Royal Majesty; and this not only to utter in speech, but also to let it pass in print, to the view of the world, and to the rending of Catholic hearts, which are privy[9] of their own truth and dutiful affection towards your Highness, estate, and person. . . .

They [i.e., Catholics] are as ready to spend their goods, lands, livings, and life, with all other worldly commodities whatsoever, in the service of your Majesty and their country, as their ancestors have been to your noble progenitors before this, and as dutiful subjects are bound to do unto their sovereign Princess and Queen; only craving pardon, for not yielding to such conformity in matters of religion, as is demanded at their hands, which they cannot do, but by offence of their consciences. . . .

In consideration of which goodwill and service, they cannot imagine to ask of your Majesty any so great gift, recompense, or benefit in this world, as should be to them, some favorable toleration with their consciences in religion. The which consciences, depending on judgment and understanding, and not of affect and will; cannot be framed by them at their pleasures, nor consequently reduced always to such conformity, as is prescribed to them by their superiors. And yet this nothing diminisheth their dutiful love towards the same superiors, seeing conscience (as I have said) dependeth of judgment and not of will.

Now because as the Philosopher[10] sayeth, that is only good unto every man, which each man's understanding telleth him to be good, unto the which the Scripture and Divines agree, when they say that we shall be judged at the last day, according to the testimony of our conscience (2 Corinthians 1 and 1 John 3), here of it followeth, that whatsoever we do

---

[6] **Bridewell:** a prison in London. [7] **diverse:** various people. [8] **the giving out publicly:** spreading the word. [9] **privy:** aware. [10] **Philosopher:** the marginal note in the original cites the first book of Aristotle's *Rhetoric* here.

contrary to our judgment and conscience, is (according to the apostle) damnable.[11] *Because we discern it* (to be evil) *and yet do it* (Romans 14). So that, how good soever the action in itself were (as for example, if a gentile should for fear say or swear that there were a Messiah) yet unto the doer it should be a damnable sin, because it seemed nought[12] in his judgment and conscience; and therefore to him, it shall be so accounted at the last day. Which thing hath made all good men, from time to time, to stand very scrupulously in defense of their conscience, and not to commit anything against the sentence and approbation of the same. All Princes also and Potentates of the world have abstained from the beginning, for the very same consideration, from enforcing men to acts against their conscience, especially in religion: as the histories both before Christ and since, do declare. And amongst the very Turks at this day, no man is compelled to any act of their religion, except he renounce first his own. And in the Indies and other far parts of the world, where infinite infidels are under the government of Christian Princes, it was never yet practiced, nor ever thought lawful by the Catholic Church, that such men should be enforced to any one act of our religion. And the reason is: for that, if the doing of such acts should be sin unto the doers: because they do them against their conscience, then must needs the enforcement of such acts be much more grievous and damnable sin, to the enforcers. Marry[13] notwithstanding this, when a man has received once the Christian Catholic religion, and will by new devices and singularity corrupt the same by running out and making dissension in Christ his body[14] (as all heretics do), then, for the conservation of unity in the Church, and of restraint for this man's fury and pride, the Church hath always from the beginning allowed that the civil magistrate should recall such a fellow by temporal punishment to the unity of the whole body again.[15]

So that we, keeping still our old religion, and having not gone out from the Protestants, but they from us, we cannot be enforced by any justice to do any act of their religion.

---

[11] **whatsoever . . . damnable:** marginal note to St. Augustine's *Of Christian Doctrine,* book 3, chapter 10, and book 1, chapter 40.   [12] **nought:** nothing.   [13] **Marry:** however.   [14] **Christ his body:** the body of Christ, here denoting the Christian community.   [15] **the Church . . . again:** the church allowed state authority to punish heretics to preserve the unity and conformity of the religion.

→ QUEEN ELIZABETH I

Proclamations on Priests   *1591, 1602*

With the succeeding reigns of Edward VI, Mary Tudor, and Elizabeth I, the official religion of England switched from Protestant to Catholic and back again to Protestant. Because the faith of the ruler determined the national religion in England, laws requiring loyalty both to the monarch and to his or her religion were passed at the beginning of each reign. Elizabeth, in conjunction with Parliament, legislated statutes on her accession which stipulated adherence to her and to the Protestant Church of England; Catholics who resisted these laws were subject to fines, imprisonment, and death in extreme cases. With parliamentary approval, she passed similar anti-Catholic statutes throughout her reign, imposing increasingly severe penalties both on lay Catholics and priests. Over 150 Catholics, many of them priests, were executed under these statutes *(Catholic Encyclopedia On-line).* In addition to these parliamentary laws, Elizabeth also issued proclamations against Catholics. The ones excerpted below set out some of the restrictions on Catholic action within her realm. The first, published in 1591, calls for the establishment of commissions of inquiry to investigate individuals suspected of supporting the pope or the Catholic king of Spain. This was only three years after England defeated Spain, her most formidable Catholic adversary, in the famous battle of the Spanish Armada. Philip II, the King of Spain (1556–1598) was the former brother-in-law of Elizabeth and had served as co-regent of England with his wife, Mary, from 1553 to 1558. Elizabeth refused Philip's offer of marriage after Mary's death, and for the remainder of their respective reigns the rulers engaged in overt or covert battles over national, international, and religious issues. The pope, who had excommunicated Elizabeth in 1570, posed a similar political threat to her reign by releasing her Catholic subjects from allegiance to her. English policy attempted to distinguish between Catholics who remained faithful to their queen in spite of the pope's excommunication of her, and those Catholics who sought to foment treason against England by joining with her enemies.

While practicing Catholics were permitted to live in England, Elizabeth firmly rejects the distressing prospect of "toleration" in the second proclamation, which calls for the banishment of Catholic clergy. The priests and Jesuits targeted in these proclamations, although clearly pursuing some religious aims in coming to Elizabeth's realm, were nevertheless suspected, sometimes correctly, of seeking to overthrow Protestant England. During this period the concept of freedom of conscience was increasingly used to argue for religious tolerance; a powerful counterargument was the political threat posed by religious differences.

Elizabeth I, Proclamation 738, "Establishing Commissions against Seminary Priests and Jesuits" (Richmond, October 18, 1591), and Proclamation 817, "Banishing All Jesuit and Secular Priests" (Richmond, November 5, 1602), in *Tudor Royal Proclamations,* ed. Paul L. Hughes and James F. Larkin (New Haven: Yale University Press, 1964–69), 3: 86–91, 250–51, 253–54.

Proclamation 738: Establishing Commissions against Seminary Priests and Jesuits

Although we have had probable cause to have thought that now towards the end of 33 years, being the time wherein Almighty God hath continually preserved us in a peaceable possession of our kingdoms, the former violence and rigor of the malice of our enemies, specially of the King of Spain, would, after his continuance in seeking to trouble our state without any just cause so many years, have waxed faint and decayed in him and all others depending on him, and been altered into some peaceable humor meet[1] to have disposed him to live in concord with us and other Christian princes his neighbors, and by such good means to establish an universal peace in Christendom, now by his wars only, and no otherwise, disturbed: yet to the contrary we find it, by his present mighty actions, so great as be never before this time attempted the like.

[The proclamation lists a number of allegations of Spanish and Catholic aggressions.]

And though these manner of popish[2] attempts have been of long time used, yet in some sort also they have been impeached[3] by direct execution of laws against such traitors for mere treasons, and not for any points of religion as their fautors[4] would color falsely their actions, which are most manifestly seen and heard at their arraignments how they are neither executed, condemned, nor indicted but for high treasons, affirming amongst other things that they will take part with any army sent by the pope against us and our realm. And of this that none do suffer death for matter of religion there is manifest proof in that a number of men of wealth in our realm professing contrary religion are known not to be impeached for the same either in their lives, lands, or goods or in their liberties, but only by payment of a pecuniary sum[5] as a penalty for the time that they do refuse to come to church, which is a most manifest course to falsify the slanderous speeches and libels of the fugitives abroad. Yet now it is certainly understood that these heads of these dens and receptacles, which are by the traitors called seminaries and colleges of Jesuits, have very lately assured the King of Spain that though heretofore he had no good success with his great forces against our realm,[6] yet if now he will once again renew his war this next year, there shall be found ready secretly within our dominions many thousands (as they make their account

---

[1] **meet:** fitting.   [2] **popish:** derogatory term for Catholic.   [3] **impeached:** prevented or impeded.
[4] **fautors:** adherents.   [5] **pecuniary sum:** an amount of money.   [6] **though . . . realm:** England repulsed the Spanish Armada in 1588, three years before this proclamation.

for their purpose) of able people that will be ready to assist such power as he shall set on land. And by their vain vaunting they do tempt the King hereto who otherwise ought, in wisdom and by his late experience, conceive no hope of any safe landing here: showing to him in Spain by the special information of a schoolman named Parsons,[7] arrogating to himself the name of the King Catholic's Councilor, and to the pope at Rome by another scholar named Allen,[8] now for his treasons honored with a cardinal's hat, certain scrolls or beadrolls[9] of names of men dwelling in sundry parts of our countries, as they have imagined them, but specially in the maritimes, with assurance that these their headmen[10] named seminaries, priests, and Jesuits are in the sundry parts of the realm secretly harbored, having a great part of them been sent within these 10 or 12 months, and shall be ready to continue their reconciled people in their lewd constancy to serve their purpose both with their forces and with other traitorous enterprises when the Spanish power shall be ready to land. Upon which their impudent assertions to the pope and to the King of Spain (though they know a great part thereof to be false) they have now very lately advertised into diverse parts by their secret messengers, whereof some are also very lately taken[11] and have confessed the same, that the King upon their informations and requests hath promised to employ all his forces that he can by sea this next year to attempt once again the invasion of this realm. . . .

Wherefore considering that these the intentions of the King of Spain are to us in this sort made very manifest, and although we doubt not but Almighty God, the defender of all just causes, will (as always hitherto He hath) make the same void, yet it is our duty as being the supreme governor under [God's] almighty hand to use all such just and reasonable means as are given to us, and therewith to concur or rather attend upon his most gracious favor by the help of our faithful subjects both to increase our forces to the uttermost of their powers and by execution of laws and by all other politic ordinances to impeach the foresaid practices of these seditions and treasons. . . .

And lastly, to withstand and provide speedy remedy against the other fraudulent attempts of the seminarians, Jesuits, and traitors, without the which (as it appeareth) [our] forces should not be now used, the same [attempts] being wrought only by falsehood, by hypocrisy, and by underminings of our good subjects under a false color and face of holiness to make breaches in men's and women's consciences and so to train[12] them to their treasons, and that with such a secrecy by the harboring of the said traitorous

---

[7] **Parsons:** Robert Parsons; see p. 258.   [8] **Allen:** William Allen; see p. 255.   [9] **beadrolls:** lists.
[10] **headmen:** leaders.   [11] **lately taken:** recently captured.   [12] **train:** drag.

messengers in obscure places as without very diligent and continual search to be made, and severe orders executed, the same will remain and spread itself as a secret infection of treasons in the bowels of our realm, most dangerous; . . . therefore we have determined, by advice of our council, to have speedily certain commissioners, men of honesty, fidelity, and good reputation to be appointed in every shire, city, and port towns within our realm, to inquire by all good means what persons are, by their behaviors or otherwise, worthy to be suspected to be any such person as have been sent, or that are employed in any such persuading of our people, or of any residing within our realm, to treason, or to move any to relinquish their allegiance to us or to acknowledge any kind of obedience to the pope or to the King of Spain, and also of all other persons that have been thereto induced and that have thereto yielded; and further to proceed in the execution of such their commission as they shall be more particularly directed by instructions annexed to their said commission.

Proclamation 817: Banishing All Jesuit and Secular Priests

As the clemency wherewith we have ever found our heart possessed towards our subjects of all sorts and our desire to avoid all occasions of drawing blood, though never so justly grounded upon the rules of policy and vigor of our laws, have been a great cause that of late[13] years we have used greater forbearance from the execution of some ordinances established by advice of our parliaments for the conservation of the true religion now professed in our kingdoms, and for the resisting of all disturbers and corrupters of the same, especially from foreign parts, and the receivers and harborers of them, than the just consideration of the safety of our estate may well endure, or the examples of some other princes, where one form of religion hath ever been only allowed, do regularly approve. So do we truly confess that our hope was that those Romish[14] priests, who were sent into this realm by foreign authority to seduce our people from their affection to religion and so by consequence from the constancy of their obedience to us, . . . would either by our clemency have been moved, or out of their own judgment have learned, to forbear to provoke us to any sharper course of proceeding and not so notoriously have abused our mercy as they have done.

For whilst we in our princely commiseration and pity of their seduced blindness held this so mild and merciful a hand over them, they in the

---

[13] late: recent. [14] Romish: from Rome, i.e., Catholic.

meantime, greatly forgetting our patience and lenity,[15] have sought like unthankful subjects the utter ruin both of us and of our kingdoms to the uttermost of their abilities.

Furthermore, we cannot conjecture, but do wonder upon what grounds they proceed (except it be our sufferance and benignity which is greatly neglected by them), in that they carry themselves in so great and insolent animosity as they do almost insinuate thereby into the minds of all sorts of people (as well the good that grieve at it as the bad that thirst after it) that we have some purpose to grant toleration of two religions within our realm, where God (we thank him for it, who seeth into the secret corners of all hearts) doth not only know our own innocence from such imagination, but how far it hath been from any about us, once to offer to our ears the persuasion of such a course as would not only disturb the peace of the church but bring this our state into confusion.

And to the further aggravating of this their audacious boldness, we find that their said concept of a toleration is accompanied with very great liberty and intolerable presumption in that they dare adventure to walk in the streets at noondays, to resort to prisons publicly, and execute their functions in contempt of our laws, never ceasing, the one side as well as the other, by these and many more their intolerable proceedings to waken our justice, which for the respect before mentioned hath lain in a slumber, where in all good policy it had been their parts (if ever) by a far contrary course to have prescribed to themselves the strictest rules and cautions of giving any such notorious scandals to so notable clemency,[16] never moved but by constraint to think upon any severity. . . .

So to avoid in some sort all these inconveniences, mischiefs, murmurings, and heart burnings in this realm (the government whereof hath been and is as well in temporal as ecclesiastical things most firmly established by general consent in parliament) . . .

And therefore we have resolved to publish this our admonition and commandment, whereby we first require and charge all Jesuits and secular priests combined together, as is before expressed, who are at liberty within this our realm (by whose sole act of their very coming into this kingdom they are within the danger of our laws), that they do forthwith depart out of our dominions and territories, and not by their abode any longer provoke us to extend the rigor of our laws upon them.

---

[15] lenity: lenience, mercifulness.　[16] in all good policy . . . notable clemency: in all good sense they should be taking a course, contrary to their present one, of avoiding any occasion of transgressing against the merciful English state.

→ ST. PAUL

On Law and Grace

1611

The trial in act 4, scene 1 frames a debate in which Shylock demands justice while Portia advocates for mercy, a debate reflecting the New Testament's view of the relation of Judaism to Christianity. For Jews, justice and mercy are both attributes of God; performance of the law constitutes obedience to God's will, but if humans sin through neglect of the commandments, their true repentance is answered by divine mercy and forgiveness. St. Paul, the Jewish antagonist of Christianity who subsequently converted and joined the disciples of Jesus (Acts 9:1–31), explains in his letters to the Romans and Galatians that the Mosaic law was a temporary measure that discouraged sin, but did not eradicate it. Jesus established a new dispensation in which adherents do not justify themselves through obedience to the Law, but believe in the saving grace offered by a merciful God. The doctrine of Grace, attained through faith on the adherents' part and mercy on God's part, is set against Law, or Judaism, in order to nullify or supercede it. As Paul writes in Galatians 3:28, "There is neither Jew nor Greek, there is neither bond nor free, there is neither male nor female: for ye are all one in Christ Jesus" (p. 269). In the course of act 4 of *The Merchant of Venice*, Portia's mercy ostensibly prevails over Shylock's justice, and his Judaism is nullified in a forced conversion; at the play's end, the characters "are all one in Christ Jesus."

ROMANS 9

31. But Israel, which followed after the law of righteousness, hath not attained the law of righteousness. 32. Wherefore? Because they sought it not by faith, but as it were by the works of the law. For they stumbled at that stumblingstone; 33. As it is written, Behold, I lay in Zion a stumblingstone and rock of offence: and whosoever believeth in him shall not be ashamed.

ROMANS 10

1. Brethren, my heart's desire and prayer to God for Israel is, that they might be saved. 2. For I bear them record that they have a zeal of God, but not according to knowledge. 3. For they being ignorant of God's righteousness, and going about to establish their own righteousness, have not submitted themselves unto the righteousness of God. 4. For Christ is the end of the law for righteousness to every one that believeth. 5. For Moses describeth the righteousness which is of the law, That the man which doeth those things shall live by them. 6. But the righteousness which is of faith speaketh

King James Bible (London, 1611).

on this wise, Say not in thine heart, Who shall ascend into heaven? (that is, to bring Christ down from above). 7. Or, Who shall descend into the deep? (that is, to bring up Christ again from the dead). 8. But what saith it? The word is nigh thee, even in thy mouth, and in thy heart: that is, the word of faith, which we preach. 9. That if thou shalt confess with thy mouth the Lord Jesus, and shalt believe in thine heart that God hath raised him from the dead, thou shalt be saved. 10. For with the heart man believeth unto righteousness; and with the mouth confession is made unto salvation.

GALATIANS 2

16. Knowing that a man is not justified by the works of the law, but by the faith of Jesus Christ, even we have believed in Jesus Christ, that we might be justified by the faith of Christ, and not by the works of the law: for by the works of the law shall no flesh be justified. 17. But if, while we seek to be justified by Christ, we ourselves also are found sinners, is therefore Christ the minister of sin? God forbid. 18. For if I build again the things which I destroyed, I make myself a transgressor. 19. For I through the law am dead to the law, that I might live unto God. 20. I am crucified with Christ: nevertheless I live; yet not I, but Christ liveth in me; and the life which I now live in the flesh I live by the faith of the Son of God, who loved me, and gave himself for me. 21. I do not frustrate the grace of God: for if righteousness come by the law, then Christ is dead in vain.

GALATIANS 3

21. Is the law then against the promises of God? God forbid: for if there had been a law given which could have given life, verily righteousness should have been by the law. 22. But the scripture hath concluded all under sin, that the promise by faith of Jesus Christ might be given to them that believe. 23. But before faith came, we were kept under the law, shut up unto the faith which should afterwards be revealed. 24. Wherefore the law was our schoolmaster to bring us unto Christ, that we might be justified by faith. 25. But after that faith is come, we are no longer under a schoolmaster. 26. For ye are all the children of God by faith in Christ Jesus. 27. For as many of you as have been baptized into Christ have put on Christ. 28. There is neither Jew nor Greek, there is neither bond nor free, there is neither male nor female: for ye are all one in Christ Jesus. 29. And if ye be Christ's, then are ye Abraham's seed, and heirs according to the promise.

→ ANDREW WILLET

From *Tetrastylon Papisticum* 1593

Andrew Willet (1562–1621) was an outspoken critic of Catholicism; his major work, *Synopsis Papismi* (1592), written in response to the work of Catholic theologian Bellarmine, presented a catalogue of criticism of papal theory. He added a supplement, the *Tetrastylon Papisticum*, one year later; these two volumes (which were bound together in subsequent printings) were so popular that eight more editions came out in quick succession. Although Willet's writings were strongly polemical and his personal views tended toward Calvinism, he strongly supported the Church of England and criticized Catholics and Puritans who rejected Anglicanism. This excerpt from the third book of the *Tetrastylon Papisticum*, which attacks "loose arguments, weak solutions and subtle distinctions," focuses on those elements of Catholicism allegedly drawn from Jewish practice. Willet both implies that Catholicism is derived from Judaism, and argues that it's actually worse than Judaism. In either case, Willet uses *Judaism* as a derogatory term to defame Catholic practice.

The Third Pillar of Papistry

All the world beginneth to see their nakedness and beggary, what sleight arguments, what loose conjectures, what poor shifts they use; and how in most of their chief questions, they are fain to beg some help of the Jews, and run to their beggarly ceremonies, as St. Paul called them, for succor. It shall not be amiss to see a few examples of this matter.

To prove their traditions beside scripture, they allege the unwritten traditions of the Jews. And yet we read of no such authentical traditions which they had, but those which were unlawful and superstitious, condemned by our Savior Christ. *Ye reject the commandments of God, to observe your own traditions* (Mark 7:9).

They ground the usurped Monarchy of the Pope over the whole Church upon the example of the high priesthood in the law, which was a type and figure of Christ, and in him accomplished, Hebrews 4:15 and 9:24.

The name of clerks or clergymen Bellarmine deriveth from the Jews: amongst whom the Levites were said to be the Lord's lot and inheritance, Numbers 18. And thus he would bring in a legal and Judaical difference between the ministers of the gospel and the people, as there was between

Andrew Willet, *Tetrastylon Papisticum, That Is the Four Principal Pillars of Papistry* (London, 1593), 106–09.

the Levites and the other tribes: Whereas before the Lord there is no difference between them in that respect: for they are all, both people and pastors,
the Lord's inheritance and lot, 1 Peter 5:3.

The single life of the clergy he would prove, by the example of the priests
in the law, who when their course came to sacrifice, separated themselves
from their wives. But who seeth not, that this was a legal sanctity only, representing the integrity and purity of the true and perfect priest Christ Jesus?
For otherwise by the same reason, they may exact the life abstinence of all
Christian people, because all Israel was commanded to keep from their
wives three days, before the Lord appeared in Sinai, Exodus 19.

Vows and Monkery proved by the example of the Nazarites and Rech-
abites, amongst the Israelites.

Their lenten fast warranted by Moses and Elias forty days fast, Matthew 4:2.

They say, the temples of Christians ought to be built, *ad similtudinem
templi Salominus,* after the similitude and pattern of Solomon's Temple.

Their consecrated oil, salt, water, ashes, and such trumpery, they would
warrant by the like ceremonies used in the law, as the salt water, Numbers 5.
The water mingled with ashes, Numbers 19.

Their Chrisme,[1] which is used in the Popish Church in Confirmation
brought in by superstitious imitation of the holy ointment, whereby the
Priests and the tabernacle were annointed in the law.

The mass, a sacrifice propitiatory,[2] because the Jews had sacrifices for sin,
Leviticus 4–5. As though all those sacrifices were not types and figures of
one only propitiatory sacrifice of Christ upon the cross.

Their private Masses, wherein the Priest receives alone, the people
standing by, authorized by the practice of the priests in the law, who sacrificed within, and the people waited without. As though the veil of the
temple was not rent asunder in the passion of Christ, which before kept the
people from the sight of the holy things: and now the Papists would draw
the curtain before their eyes still.

Popish Massing garments invented to resemble the costly attire of the
high Priest in the law, whose rich ornaments, and beautiful attire, were evident types and figures of the spiritual beauty and excellency of the kingdom
and Priesthood to Christ, Psalms 45:8, 13, 14.

The superstitious dedication of their churches they borrow from the
practice of the people under the law:[3] as the church was solemnly dedicated

---

[1] **Chrisme:** consecrated scented oil used for annointing. [2] **propitiatory:** in order to appease.
[3] **people under the law:** the Jews.

in Solomon's time, and in Ezra his days, 2 Chronicles 7; Ezra 6. And by the Maccabees, 1 Maccabees 4. And therefore Christian churches ought to be dedicated in like sort. Whereas it is certain, that the solemn dedication of the Jewish temple, was a lively resemblance of the dedication of the true tabernacle, which was the body of Christ, Hebrews 8:2. And of the new and living way, dedicated unto us by the veil, that is his flesh, Hebrews 10:20. And yet never did the Jews use half of those ceremonies (but such toys none at all) in dedicating the temple, as papists do in hallowing of theirs: Such are the burning of tapers, picturing of Crosses upon the walls, sprinkling of water, and of ashes upon the pavement, making of characters. The Jews themselves would blush to behold such things.

The adorning of their churches, with images, crosses, vestments, of silver, gold, silk, precious stone, their gorgeous and sumptuous buildings warranted by the example of Solomon's temple, which was beautiful both within and without, yet they might have known this, that the beauty and glory of the first house did shadow forth unto the Jews the spiritual comeliness and excellence of the last house, that is, the Church of God under Christ, Haggai 2:10. And so saith the prophet Isaiah, speaking of the spiritual temple: I will lay thy stones with the carbuncle,[4] and thy foundations with Sapphires, Isaiah 54:11.

Yet we deny not, but that the external churches also of Christians ought to be built, and adorned, with moderate cost after a decent and comely sort.

The priests in the law, were to judge of leprous persons, and to discern their leprosy, Leviticus 13:14. Therefore Christians are now also bound to make particular confession and enumeration of their sins in the ears of the priests. Yet the priests were not to take knowledge of every infirmity or disease, but of this contagious and infectious kind: how then can they hence conclude, that the knowledge of all sins both great and small belongeth to the priest? And who knoweth not, that herein the priesthood of the law, did decipher the priesthood of Christ, by whom our spiritual leprosies are discerned and cured?

Nay they do not content themselves with an apish imitation of Jewish ceremonies, but they also belie them, and father upon them[5] such things, as they never used: as that they pray for the dead, which it is certain the Jews to this day do not, and whereas the fact of Judah Maccabee is commended for praying for the dead, 2 Maccabees 12:44. It seemeth to be put into the story

---

[4] **carbuncle:** precious stone, possibly a ruby.   [5] **father upon them:** pretend that they originated.

by the author (whosoever he was)[6] of his own: for Josephus,[7] who wrote five books of those matters, and out of whose works this story seemeth to be abridged, entreating of this place, makes no mention at all of prayer for the dead. . . . And again, in this place prayer is made for open idolaters, which is contrary to the practice of the popish Church, who deny prayer to be made for those which die in deadly sin.

In like manner they burden the Jews with authentic unwritten traditions besides scripture: such they had none as we have showed before.

Lastly, the papists do far exceed the Church of the Jews, in number of ceremonies, but in lightness and vanity of such childish toys, the Jews never came near them, or were once like unto them. We will give one influence of the ceremonies used in baptism: First they touch the ears and the nostrils with spittle of the party baptized, that his ears may be opened to hear the word, and his nostrils do discern between the smell of good and evil. Secondly, the priest signeth his eyes, ears, mouth, breast, forehead with the sign of the cross, that all his senses thereby may be defended. Thirdly, hallowed salt is put into his mouth, that he may be seasoned with wisdom, and kept from putrifying in sin. Fourthly, the party is anointed with oil in his breast, that he may be safe from evil suggestions; and between the shoulders, which signifieth the receiving of spiritual strength. Fifthly, he is anointed with the Chrisme in the top of his head, and thereby is become a Christian. Sixthly, a white garment is put upon him to betoken his regeneration. Seventhly, a veil is put upon his head, in token that he is now crowned with a royal diadem. Eighthly, a burning taper is put into his hand, to fulfill that saying in the gospel, *Let your light shine before men* [Matthew 5:16]. I report now to the Jews, if ever they used such toys, or do to this day in any of their rites and ceremonies. So that we may justly say with Augustine, *ipsem religionem oneribus premunt, ut tolerabilior sit conditio Judeorum, qui legalib sarcinis, non humanis prasumptionibus subijcuntur:* They cumber religion with their burdensome inventions: so that the Jews case was more tolerable (than theirs that live under Popery) who were subject only to legal ceremonies, not to the inventions of men. And thus we see the weakness of Popish religion, and feebleness of their cause, who through very beggary are constrained to patch up their tattered garments with Jew's rags.

---

[6] author (whosoever he was): the book of Maccabees is apocryphal and its authorship therefore uncertain.   [7] Josephus: Flavius Josephus was a Jewish historian of the first century C.E. whose compendium of Jewish history, *Jewish Antiquities*, covers the battles of the Maccabees.

→ RICHARD BRISTOW

From Demands to Be Proponed of Catholics
to the Heretics   *1596*

Richard Bristow (1538–1581) received his bachelor's and master's degrees from Oxford University; having a reputation for public speaking, he was selected in 1566 to present a public disputation to Queen Elizabeth during a visit to the university. In 1569 he left England for Louvain, France, where he met William Allen (see p. 255). He became Allen's "right hand," serving as a teacher at Douay, where Bristow continued his studies and became the first member of the college to enter the priesthood. When Allen had to move the college from Douay to Reims, he put Bristow in charge; there he taught and wrote polemics against Protestantism. Like Allen, he fell ill and was advised to return to England to be strengthened by "his native air"; however, the cure failed and he died there. In his *Demands to Be Proponed* [proposed for consideration] *of Catholics to the Heretics* (1596), he identifies Protestants as heretics on the level of Jews and pagans in their rejection of the Catholic Church.

THE SECOND DEMAND

Whereas Christ and his Christians have, besides schismatics[1] and heretics, two other kinds of enemies, to wit, Paynims[2] and Jews: and whereas the ancient writers have made many goodly books against those enemies, either to confound them, or to persuade them, that Christ is God; as it was then, in the first beginning of Christians, very necessary for them so to do: let the learned Protestants be likewise demanded, whether those Christian writers in those books have not made, amongst others, this argument, to prove that Christ is God, namely St. Chrysostom,[3] both against the Paynims, in his book named, *Contra Gentiles demonstratio, quod Christus sit Deus, A Plain Demonstration against the Gentiles, That Christ Is God,* and also against the Jews, in the second of his five orations that he made against them. That Christ (I say) is God, because his church, although it had but a small and poor beginning, and even then very many, very mighty, and very fierce enemies, yet could not, nor cannot ever profitably be suppressed. But contrariwise, being in the beginning, as it were but one little sparkle of fire, and whole floods, yea seas of persecutions being poured out upon it; yet could it

---

[1] schismatics: those who have broken away from the Church.   [2] Paynims: pagans.   [3] St. Chrysostom: (347–407 C.E.) Church father.

Richard Bristow, *Demands to Be Proponed of Catholics to the Heretics* (London, 1596), 8–10, 30–32, 78–79, 100–02, 106–07.

not be extinguished; but contrariwise (I say) partly hath, and partly shall set all the world on fire, first or last bringing all to Christ: according to Christ's own prediction, which he also doth there allege: . . . *I will build up my church, and Hell Gates shall not prevail against it* (Matthew 16:18).

Now let it be considered whether their argument do hold, if it be true which they [the Protestants] say, to wit, that the Church of Christ was not invincible, but that it hath been these many hundred years quite suppressed; yea, and in Chrysostom's own time no Church of Christ at all. For they know, if they grant that then to have been the true Church, that they must grant also ours now to be the true Church, as being alone with that. If then they will say, that this is not a good argument, let them be further demanded, whether they dare take part also with the very Jews and Paynims against the Christians, yea and against the Godhead[4] of Christ himself; and whether they will go about or whether they be able (which the Jews and Paynims were never able, nor never shall be able) to answer this argument. . . .

Insomuch that the ancient Christians writing against the Jews and Paynims for the Godhead of Christ, do first show, that the miracles of Christ himself and his Apostles were wrought to set up a visible church, that should continue forever (whereof it follows, that all heretics do rise in vain). And secondly they do by such miracles of the church, as I have said, daily to be seen, prove unto those infidels the miracles of Christ and his Apostles, recorded in the holy scripture, which those infidels did deny, because they did not see them. Let the Protestants therefore be asked, whether they dare join herein also with the infidels against the Christians, and answer for them, that the argument is not good, because these miracles of the cross, and of relics, and such like, are not miracles, but illusions. And then, when they shall by and by hear the infidels say even so also of Christ and his Apostles' miracles, let us see a[5] God's name, how wisely the Protestants will reply, and show them a plain difference between Christ's miracles, and our miracles.[6] Or rather shall we not see them hereby not only confirm the infidels in their incredulity, but also prepare weak Christians to infidelity, yea and themselves also ready to give over thereunto: as in our country (God help) who sees not. . . .

Because (I say) they openly renounce the claim of authority, confessing thereby that it is not of them,[7] that the creed saith, *I believe the Church,* insomuch that they have suffered of late *an unlearned Christian* (as he is called)

---

[4] **Godhead:** divinity.   [5] **a:** in.   [6] **hear the infidels . . . our miracles:** Protestants challenged the primacy of saints, the efficacy of relics, and the validity of later miracles confirmed by the Catholic Church.   [7] **it is not of them:** it does not refer to them.

to set out in print a vain libel[8] against the *authority of the Church of God*, comparing and opponing[9] unto it the *authority of the word of God*, as though the word of God and the Church were one against the other: it being yet so plainly written, that as the Father said of his Son, *ipsum audite*, Hear him, (Matthew 17) so the Son said of his Church, . . . *If he will not hear the Church, do thou use him, as* (the Jews did) *an heathen and a publican* (Matthew 18). And yet this fellow *trusts so much in his own folly* (Proverbs 17), that he is bold to provoke all Catholics to answer his childishness, or else they must be accompted[10] (sayeth he) no less than very *murderers*. It were good for him poor man, that he had in him no more pride than learning. My best counsel to him for his salvation is, that he read humbly these demands and look whether any of his great masters will answer them. And if after this his stomach serve him still, let him set out his libel more orderly with his name, and with approbation of their Rabbis, and with privilege, that we labor not in vain: and with the grace of God he shall quickly see it answered, as unworthy as it is. . . .

The Forty-eighth Demand

Whether it be not our Church only, which now, and ever, so blessed of God, and so imbued with Christ's blood, that she hath grace in her sacraments (as well for remission of sins after baptism, as of sins before baptism; to the unspeakable comfort of all that be heavy laden), merit in her works, force in her word, power in her teachings, so that she breedeth the devotion, turneth to the religion, and to the search of salvation strangely altereth the hearts of men. Her children therefore being the saddest[11] sort of people, men of best order in all families, towns, and cities, for all goodness best beloved both of God and man. And whether with the Protestants, all be not clean contrary: no preaching of penance, no grace in sacraments, nothing but sin in Good works although they be done in Christ, no power to bring under devils, no blessing, no comfort, and their followers therefore easy to be noted by their ill conditions. All persons as they fall from order and godliness, more near they become to their religion: a general observation, that all men as they return to our Church, bettered and amended, as they all to their synagogue, much worsed and more than afore corrupted. . . .

---

[8] **libel**: literally a defamatory tract, in this case probably more polemical than truly libelous. I have not been able to identify the text referred to here. [9] **opponing**: opposing. [10] **accompted**: considered. [11] **saddest**: most sober.

THE FORTY-NINTH DEMAND

Whether it be not our Church only, which all the enemies of Christ do fight against, conspiring all against us as the company that only stands in their way, and that only bears their brunt: specially all sects and heretics for that cause bearing intolerably with one another's blasphemies, and (as it is called) syncretizing[12] and tied together by the tails (like Samson's foxes)[13] their heads being most far asunder, and counting Turks, and Jews, and very Atheists, for their friends, and all that be not Papists. And therefore, whether our Church be not the true Church and our Church only; as which only both now, and ever, have been of all maligned, and by Hell-gates impugned.

[12] syncretizing: reconciling diverse beliefs. [13] like Samson's foxes: see Judges 15:4–5.

⤷ **WILLIAM PERKINS**

From A Faithful and Plain Exposition upon the Two First Verses of the Second Chapter of Zephaniah *1606*

William Perkins (1558–1602), who received his master's degree from Cambridge, was a popular teacher and preacher as well as author of numerous texts that advanced his Puritan views. His writings were well received and were translated into a number of languages. In the sermon excerpted below, Perkins, like Willet (p. 270) and Bristow (p. 274), uses the Jewish people to represent the ungrateful sinner. However, he goes on to accuse his fellow English men and women of exceeding the Jews in their transgressions and ingratitude, thus encouraging his audience both to identify with the Jews and to acknowledge their own greater offenses. In what ways would Shakespeare's audience have been able to identify with Shylock?

Thus for soul and body, [Israel] were every way a nation, blessed of God, a people beloved of God above all others. Now, how did this people (thus beloved of their God) requite this his love, which they had not more deserved, than any other nation? Certainly, as they deserved it not afore they had it, so they requited it not when they had it: but requited this love of God with sin,

William Perkins, *A Faithful and Plain Exposition upon the Two First Verses of the Second Chapter of Zephaniah Containing a Powerful Exhortation to Repentance* (London, 1606), 13–16.

with rebellion, and with disobedience. They tempted him, they provoked him to wrath, they presumed of his mercy, and proved a most stubborn and stiff-necked people, a froward[1] generation: Moses partly saw this in his own experience, and better discerned it in the spirit of prophesy: and therefore, wondering at this their wickedness, he cried out, "Do you thus requite the Lord: O foolish people, and unwise thus," that is, with sin, and disobedience, which is the only means to displease the Lord and to provoke him to wrath: for this cause, they are worthily called a foolish and unkind people by Moses, and here, by the prophet. A nation not worthy to be beloved: namely, for their unthankfulness, and unkindness: which was such, as they not only were slack, and careless in performance of such duties as God required, but even multiplied their sins and committed those foul rebellions, which his soul hated.

And amongst many, the prophet here in this chapter, noteth three of their great sins, for which they were a nation not worthy to be beloved. Covetousness,[2] cruelty, and deceit: all which were the more heinous and intolerable, because they were the sins of their princes, their rulers, and their priests: who should have been lights and examples of the rest. . . .

And therefore there is no man, but if he be asked what he thinks of this nation of the Jews, he will answer, that they are a most vile and wicked people, a froward generation, and that they are worthy to taste deeply[3] of all God's plagues, who so far abused his love and mercy.

But what does this belong to them alone? And is Israel only a nation not worthy to be beloved? Nay, I may cry out with as good cause: O England, a nation not worthy to be beloved. For, God hath been as good a God to us, as he was to them, and we have been as unkind a people to him. But that I may be free from discrediting our nation, and from defiling my own nest:[4] let us prove both these points, and lay them open to the view of the world.

First therefore, the same mercies, and far greater, have been poured and heaped upon us: He hath called us out of darkness, first of heathenism and then of popery; his covenant of grace and salvation he hath confirmed with us; his treasures of his word and sacraments he hath imparted to us; his holy word never better preached; and the mysteries thereof never more plainly opened, since the time of the Apostles. And as we have religion, so we have it under a religious prince, whereby it comes to pass, that these blessings of salvation we enjoy not in secret or by stealth, but we have it countenanced by authority, so that religion is not barely allowed, but even as it were thrust upon men. Besides all this, we have a land also, that flows with milk and

---

[1] froward: contrary.   [2] Covetousness: greed.   [3] taste deeply: experience fully.   [4] defiling my own nest: figuratively speaking, to do wrong to one's own.

honey; it is plentiful in all good things: we have liberty and peace under a peaceable prince, and the companions of peace, prosperity, plenty, health, wealth, corn, wool, gold, silver, abundance of all things, that may please the heart of man. Thus hath God deserved the love of England.

But now England, how hast thou requited this kindness of the Lord? Certainly even with a great measure of unkindness: that is, with more and greater sins than ever Israel did. So that if Moses spake true of them, then may our Moses much more truly cry out against England: "Do thou thus requite the Lord thou foolish people?" And if this Prophet said thus of Israel for three sins, then may it be said of England, for 300 sins (O England) a nation not worthy to be beloved: for thou hast multiplied thy transgressions, above theirs of Israel, even as though thou had resolved with thyself, the more God's kindness is heaped on thee, the more to multiply thy sins against him. For thou England, as thou hast requited the Lord with sins, so not with a few sins or small sins, or sins which hardly could have been prevented: for that had been a matter of some excuse, or not of so great complaint. But thy sins are many, and grievous, and capital. And which is worst of all, willful and affected, even as though God had deserved evil of us, and that therefore we ought maliciously to requite him.

If any man make doubt of this, and therefore think I speak too hardly of our Church, I will then deal plainly and particularly, and rip up the sores of our nation, that so they may be healed to the bottom.[5]

The common sins of England, wherewith the Lord is requited, are these. First, ignorance of God's will and worship (I speak not of that compelled ignorance in many corners of our land, which is to be pitied because they want the means) but willful and affected ignorance. . . .

The second main sin of England is contempt of Christian religion. Religion hath been among us these five and thirty years; but the more it is published, the more it is contemned,[6] and reproached of many; insomuch, as there is not the simplest fellow in a country town, who although he knows not one point of religion, yet he can mock and scorn such as are more religious than himself. . . .

O what a cursed sin is this? To contemn the greatest favor that God can give us, that is, his holy religion: for the which we should rather praise him all the days of our lives. All that God can give a man in this world, is his gospel, what then can God give to be regarded, when his gospel is contemned?

This sin was never amongst the Jews: they indeed regarded it not so as it deserved, but who did ever make a mock and a scorn of it but England? O

---

[5] **sores . . . healed to the bottom:** Perkins uses the idea of lancing an infected sore as a metaphor for attacking the religious transgressions of the English.   [6] **contemned:** despised.

England, how can thou answer this? . . . The third common sin of England is blasphemy, many ways, but especially in vain swearing, false swearing, and forswearing, and the abuse of all the names and titles of the Lord God. This sin is general, even over the whole land, especially, in fairs and markets, where men for a little gain, will not care to call the Lord of hosts to be witness to a lie, and the God of truth, to testify an untruth.

The fourth general and great sin is profanation of the Sabbath. . . .

The fifth sin of our nation is unjust dealing in bargaining betwixt man and man. How hard is it to find an honest, simple, plain dealing man. . . . [T]herefore labor to approve yourselves sincere hearted men; remember the counsel of the holy Ghost. Let no man oppress nor defraud his brother in bargaining: for the Lord is the avenger of all such things (1 Thessalonians 4:6). These sins are general and universal as a canker. And so are the sins of the 6, 7, and 8 Commandments (though they be not altogether so common as these be), murders, adulteries, usuries, briberies, extortions, cozenages,[7] they are a burden, under which our earth groans, and they cry against us to heaven, so that upon as good or much better cause may it be said to us, as to the Jews: O nation not worthy to be beloved.

[7] cozenages: deceptions.

GREGORY MARTIN

→ *From* Roma Sancta   *1581*

Gregory Martin (d. 1582) received his bachelor's and master's degrees from Oxford and was active in English Catholic circles in the 1560s. Unwilling to conform to the Church of England, he fled to the English College at Douay, France, where he was ordained as a priest in 1573. He was subsequently hired by William Allen (p. 255), founder of the college, to teach Hebrew and give lectures on the Bible. He went to Rome in 1576 to help organize the English College there; according to his address to the reader in *Roma Sancta*, he remained there until 1578. He followed Allen to Rheims when the Douay college was forced to move, and he spent the rest of his life translating the Vulgate (Latin bible) into English. He was assisted in this task by William Allen and Richard Bristow (p. 274), among others; the New Testament was published in 1582; the Hebrew Bible (Christian Old Testament) appeared in 1609–10. *Roma Sancta* praises the capital of western Catholicism, and reflects Martin's impressions dur-

Gregory Martin, *Roma Sancta: The Holy City of Rome, So Called, and So Declared to Be, First for Devotion, Secondly for Charity; in Two Books* (1581), ed. George Bruner Parks (Rome: Edizioni di Storia e Letteratura, 1969), 75–78, 81–82.

ing his stay there. The excerpt that follows describes the organized effort of the Christian community of Rome to convert the Roman Jewish community. Martin insists that the Jews are not to be "forced" to convert, but "by all charitable means they are invited and persuaded to forsake obstinate Judaism" (p. 282). How do his views toward the Jews resonate with attitudes toward Jews expressed in *The Merchant of Venice?* In what ways does his discussion of the conversion of the Jews reinforce or contradict the representation of Shylock's conversion?

BOOK I: THE DEVOTION OF THE CITY OF ROME

Chapter 26: Preaching to the Jews for Their Conversion

What need I stand here in the commendation of this exercise? It concerneth the conversion of them that were always the greatest enemies to Christ and Christian Religion. Myself was never delighted with any one thing in Rome more than this. But to omit all prefaces, hear only the manner and circumstance thereof, as I will briefly put it down according as I saw and heard for the space of one year and a half.

As in many great cities of the world (specially under the Turk) this nation of the Jews is dispersed for their horrible sin of crucifying our Savior and their Messiah, according to the prophecies of them: so in Rome also since the sacking of their city and temple by Titus the Emperor, forty years after the death of Christ, there have been always very many of the captivity.[1] And during the time of infidelity they were permitted to live (as now under the Turk) for paying their tributes and other duties, but since the reign of Christianity they have been always suffered to live and shall be for sundry great causes; namely, for the confirmation of our faith (as Saint Augustine sayeth) not only by their persons, who protest that they are the children of them that put Christ to death, but also by their books of the Old Testament in their Hebrew tongue, wherein they carry our testimonies against themselves. And this is it which the psalm speaketh of, *Ne occidas eos* etc. *Kill them not, lest my people forget. Disperse them O Lord in thy power* [Psalm 58]. But principally for that God would have many of them also saved from time to time by few and few, which in scripture are called *reliquiae* (the remnant) [Romans 11] and in the end all together that are left by the preaching of Enoch and Elias.[2] In these respects the Holy Church by God's providence spared their lives, as also for that they were never children of the Church,

---

[1] **captivity:** refers to the exile of the Jews from Israel initiated at the destruction of the Second Temple in 70 C.E.   [2] **Enoch and Elias:** the marginal note here cites Malachi 4; in verse 5, Elias, or the prophet Elijah, is mentioned here as being sent "before the coming of the great and dreadful day of the Lord." He and Enoch, who first appears in Genesis 5:23, are identified as the "two witnesses" of Revelations 11:3 who will prophesy at the end of days (Jeffrey 237).

nor made profession in baptism, and therefore may not be compelled to the faith, as St. Paul said of them and all pagans: *Quid mihi de his qui fortis sunt judicare?* What have I to judge of them that are without?

This people therefore thus hitherto preserved in the world as they are not forced, so by all charitable means they are invited and persuaded to forsake obstinate Judaism and to become Christians, as every year they do. But it shall not be amiss, before I come to the preaching, to tell their whole usage and state in that city. They are known by their yellow cap or hat which everyone wears, and must wear under a pain. They live altogether in that which is called the Jews' street; and of late, because of the number, there is another corner added. Where they have their synagogue and in it reading of the Old Testament in the Hebrew tongue, where no Christian may enter in at service time under pain of excommunication. In the night time the gates of their streets are shut upon them, as well for their own safety, as for inconvenience that might ensue perforce by them. Likewise in Passion Week,[3] two or three days together all is shut, because the people's stomach by occasion of that time and the preaching then riseth against them, and some or other might upon indiscreet zeal do them a displeasure. Otherwise they have their market there in their shops, and once a week a solemn sale called the Jew's market, in a place called Agona:[4] but yet restrained to one kind of merchandise, namely apparel ready made and that for the most part, old clothes new dressed, hangings also and other implements: so that they may live, and only live, except a few which grow to some wealth. And so they are able upon a sudden to furnish a stranger for his body, or for his house. And when he departs, they will take it again, whether he buy it, or borrow, as the manner is: but if they can possibly, they will deceive him. For all is well won (they think) of a gentile, that is, a Christian.

They have all justice against Christians, as these against them, and in all lawful causes, a cardinal their protector to deal for them, as all other nations. Their own festival days they keep among themselves, and ours they may not break in any open show. They must not in word or deed impeach Christian religion, nor deal by any means to the hurt thereof. Their usury, which they think lawful to exact of a Christian and not of their brethren, and to the which that covetous nation is given exceedingly, always talking of their Quadrines,[5] that (I say) is so defeated, that for the help of poor Christians, who otherwise might be driven to borrow of them, there are in Rome *montes*

---

[3] **Passion Week:** the week between Passion Sunday (two Sundays before Easter) and Palm Sunday (the Sunday before Easter). [4] **Agona:** Piazza Navona, a plaza in Rome. [5] **Quadrines:** small coins.

pietatis,[6] where for a pledge, they may borrow, and never hinder themselves, as I will more at large declare hereafter. This people therefore so far from good life and Christian faith, the exceeding charity of his Holiness[7] and of that virtuous city draweth little by little into the church of Christ, and maketh them of that one fold which our Savior speaketh of in the Gospel, and so fulfilleth the prophecies both of the old and new testament, in this order and by this means as followeth.

Upon every Saturday, which is their Sabbath day and wherein they work not, because they shall have no excuse of other businesses, they are bound under a penalty to be present in the church of the company of the Blessed Trinity there to hear what may be said for Christianity against their Judaism. The audience being thus assembled every Saturday about two o'clock after dinner, there come up into the pulpit two excellent men, one after another, for the space of two hours. The one and the first, a Jesuit or some other of great skill and good spirit, to move; the other, a great Rabbi sometime of their own, but now these many years a zealous and learned Christian, named master Andreas. Of the which Christian Rabbis (by the way) they are in Rome four very famous, one a Dominican friar, another a Jesuit, the third Reader of the Hebrew in . . . the university: and the fourth, this M. Andreas. Whose zeal for his brethren the Hebrews (for so they are and would be called) not unlike to St. Paul's in the like case, his manner of utterance to teach and convince and confound, his knowledge and readiness in the Hebrew Bible and all the Hebrew commentaries and Chaldee paraphrases and the Syriac[8] and Arabic tongue, who can conceive that hath not heard him, and who can express that hath heard him? Well, this man is chosen of purpose to confute them out of their own books and doctors, and to confound them by their own peevish opinions and absurd imaginations and foolish practices, which he knoweth as well as the best Rabbis, and can disclose all their ridiculous mysteries, himself having been sometime one of them, and knowing the greatest points that then blinded himself, and marveling now that he could be so sotted and bewitched. Of all these things he telleth them in goodly order and after the best manner, to their great confusion, and to the angering of their Rabbis and the rest that are obstinate and malicious among them, as by their gesture they declare. . . .

---

[6] *montes pietatis:* in Italy in 1452 the Franciscans established these "charitable funds for the granting of interest-free loans secured by pledges to the poor. . . . Later, in order to prevent the premature exhaustion of funds, *montes pietatis* were compelled to charge interest and to sell by auction any pledges that became forfeit" *(EBO).*   [7] **his Holiness:** the pope.   [8] **Chaldee . . . Syriac:** both refer to Aramaic, the Semitic vernacular language in which the Talmud is written.

And this much of the Principal Preacher, which is this Christian Rabbi, principal (I say) in this kind and to this people. The other that preacheth before him immediately and in the first place, useth all other kind of reasons and proofs out of the old Scriptures and the fathers that have written of purpose against the Jews.

And especially the zeal and compassion that they declare toward the saving of their souls do much move and persuade, God principally working withal in their hearts. Father Possivino the Jesuit occupied the place til he was sent to convert Suetia. After him Lupus a Capuchine. And after him Francisco Maria, all famous men and full of zeal and charity. So by this means it comes to pass, that now one, and now another, and sometime a whole household, sometime of the Rabbines themselves, feel compunction and remorse, and say as it were with their forefathers in the acts of the apostles upon the preaching of Peter, *Quid faciemus viri fratres?* What shall we do brethren? And so signifying their mind, they are received and baptized.

→ THOMAS DRAXE

From The World's Resurrection   *1608*

Thomas Draxe (d. 1618) received a bachelor's degree in divinity from Cambridge and began a career as a Protestant minister. Among his religious writings he published the works of William Perkins (p. 277), which he translated into Latin. In *The World's Resurrection,* he handles a very sensitive subject: the conversion of the Jews during the messianic period. This event was seen by Christians as presaging Jesus's second coming; such millennial hopes gained popularity in seventeenth-century England, but were also seen by the government as potentially subversive. Draxe's text takes the form of a commentary on Romans 11; he quotes verses and offers explanations of their meanings. He includes a series of what might be the early modern equivalent of FAQs — frequently asked questions; these queries allow him to address issues that are not directly alluded to in Romans. He also occasionally considers the "use" of a passage in which he develops an interpretation of Romans and presents it as a point of doctrine or commentary on particular events. As does Gregory Martin (p. 280), Draxe urges the conversion of the Jews through persuasion, not coercion, and understands this conversion as divinely ordained, and therefore inevitable.

Thomas Draxe, *The World's Resurrection* (London, 1608), 1–2, 49–50, 88–89, 92–94.

ROMANS 11

[V. 1.] *I demand then, hath God cast away his people? God forbid: for I also am an Israelite, of the seed of Abraham, of the tribe of Benjamin.*
[V. 2.] *God hath not cast away his people which he knew before? Know ye not what the Scripture saith of Elias, how he communeth with God against Israel, saying.*
[V. 3.] *Lord they have killed thy Prophets, and digged down thine Altars: and I am left alone, and they seek my life etc.*

What then (may some repining Jew object) *hath God,* that is unchangeable in his decree and covenant, and whose compassions fail not, *cast away* hath he wholly and universally cast off and excluded from righteousness and everlasting life *his people?* That is the Israelites or Jews, for whose faith and preservation he hath wrought[1] so many miracles, whom he hath fastened and affianced[2] unto himself by so solemn a covenant and by so many precious promises and whom he hath ennobled and renowned by so many illustrious patriarchs? *God forbid,* far be that from any man's thought and imagination for that cannot be. *For I am also an Israelite, of the seed of Abraham of the tribe of Benjamin,* I by my own example can testify the contrary, for I (notwithstanding I formerly was a Pharisee, a Blasphemer, a persecutor, and an oppressor) am not cast off, but *I am an Israelite,* therefore God hath not cast off all. . . .

Albeit, God hath generally rejected and cast off the body and greatest number of the Jews that were called, and so deemed in their own eyes, and in the estimation of the world, by reason of the tenor of God's covenant and the ceremonies and outward exercises of religion, called (I say) and reputed to be God's people; yet those amongst them *whom he knew before,* whom he predestined to salvation, whom he acknowledged and approved for his own, and whom he prevented[3] by his special favor, this people he never failed nor ever will renounce and relinquish.

V. 15. *For if the casting away of them be the reconciling of the world, what shall their receiving be but life from death?*
V. 16. *For if the first fruits be holy, so is the whole lump, and if the root be holy so are the branches.*

If the casting away of them, the greatest part of the Jews, *be the reconciling of the world,* serve and tend to the calling of the Gentiles whereby they are reconciled unto God, *what shall the receiving be,* the calling of the fullness of the Jews,[4] by which they that before were cast off, shall again be admitted

---

[1] **wrought:** performed.   [2] **affianced:** linked as in marriage.   [3] **prevented:** provided beforehand.
[4] **the fullness of the Jews:** all the Jews.

and received into the Church? [It will be] *life from death*, a recovery and bringing of spiritual life again to the Jews that were so many hundred years dead in their sins, and also their restitution and fullness[5] shall give an occasion of quickening to the Gentiles, and of enriching many with the knowledge of Christ and salvation, and so of enlarging God's kingdom, both amongst Jews and Gentiles, and hence by reason of the common felicity shall be the true and perfect joy of the world. . . .

Even so, if Abraham, Isaac, and Jacob their stock, fathers and founders of their Nation, were (especially) by reason of God's covenant holy and accepted with God: so shall the elect of their posterity be (in some sort) favored for their Fathers' sake. And if the root be holy, full of juice and sap of grace, so shall the branches. The holy remainers by force of God's covenant shall receive and draw juice, grace and goodness from it.

QUESTION: When is likely to be the time of the Jews conversion, before the sacking and burning of Rome or afterward?

ANSWER: In all probability it is like to follow the burning and destruction of Rome, for then the stumbling blocks [that] the Papists[6] offer them, by their imagery, invocation of saints, Latin service, and abominable and most senseless transubstantiation,[7] shall be removed and taken away.

Secondly (as it appears in the 18th and 20th chapters of John's Apocalypse),[8] there shall be some reasonable distance of time between the burning of Rome and the end of the world, in which it is most consonant to truth that the Jews shall be called, for their conversion [is] in the last general sign and forerunner of Christ's second coming so far forth as the scripture reveals unto us.

QUESTION TWO: Whether shall the Jews recover the holy land again or not, and be all converted and dwell there; seeing that it is said their deliverer shall come out of Zion, or must we think rather they shall be converted in the countries in which they dwell, and into which they are dispersed or shall then be found inhabiting?

ANSWER: They are likely never to recover it, for they have no such promise, neither have they any possibility of means to compass it.[9] Secondly, Christ's coming unto them shall not be *visible but spiritual*, not from the earthly *Zion*, which long since has been made desolate, but from his spiritual *Zion* of his Catholic Church.[10]

---

[5] **fullness:** completeness.   [6] **Papists:** derogatory term for Catholics.   [7] **imagery . . . transubstantiation:** a list of Catholic practices rejected by Protestants: images of Jesus and saints were considered idolatrous, and the intercessory power of saints, the Latin text of the mass, and the doctrine of transubstantiation of the host were all denounced.   [8] **Apocalypse:** the Book of Revelation.   [9] **to compass it:** to bring it about.   [10] **Catholic Church:** given Draxe's antipathy toward Catholicism elsewhere, I assume he means the true Christian church.

Lastly it is most probable and likely that they shall be converted in those countries into which they are dispersed and in which they have their residence.

For first we have some small beginnings (here and there) of it. Secondly they shall better and sooner by their *zeal and example revive* the faith of the Gentiles being mixed and conversant with them, and living amongst them there, than if they should dwell and be contained all in one country.

[V. 25.] *That you should not be wise in your own conceit.*

The cause of stubbornness and obstinacy, which breeds and begets absurd and rebellious opinions in us, is that men will be wise of themselves, and will not seek to understand and know heavenly mysteries by diligent search and examination of the Scriptures and by revelation of God's spirit, which is the only means to understand them. But they either wholly neglect them, or else measure them by their sense and imagination which is shallow and will deceive them.

The reason hereof is, because they want[11] God's spirit and humility to guide and direct them, without which all other means are vain and without force.

Use 1. The first use hereof serves to condemn the badness and madness of many not only Papists, but others in many countries, who because they will not be thought to err, will stiffly maintain gross, false and absurd opinions, as we have many Lutherans, Schismatics, Sectaries,[12] temporizing flatterers[13] for instance.

Use 2. Let us not measure this mystery of the Jews' conversion by sense and reason, but by faith. And seeing it is a mystery, yet that it is (in respect of the form and manner of their conversion) not common or ordinary, let us not be curious to dive and descend further into particulars than God's word, or (at least) very probable arguments, (not contrarying, the same will warrant)[14] but rest in expectation, until the time come, and in the interim help them by our prayers, and further them by our zealous and holy example.

Obstinacy is partly come to Israel, that is, it is not universal nor perpetual, and *so all Israel shall be saved* [Romans 11:25]. Seeing that the Jews are not altogether rejected, but that there is always a remnant remaining. . . . And especially seeing that all Israel, the greatest part and number of that nation, are to be called and converted, we must not rashly either contemn,[15] much less condemn the Jews, nor expel them out of our coasts and countries, but

---

[11] **want:** lack.   [12] **Schismatics, Sectaries:** those who have broken away from the Church; here referring to radical Protestants who separate themselves from the Church of England. [13] **temporizing flatterers:** self-serving sycophants.   [14] **very probable arguments, (not contrarying, the same will warrant):** very probable arguments, which by not contradicting God's word will instead be supported by it.   [15] **contemn:** despise.

hope well of them, pray for them, and labor to win them by our holy zeal and Christian example.

The first reason hereof is, there are some of them [Jews] called and converted in all ages, which are a preludium and forerunner of the conversion of the rest.

Secondly they are the faithful keepers and preservers of the Old Testament.

Thirdly, they, in respect of the time past, that is, since Christ's ascension until this day, do (in a sort) confirm the Christian faith, seeing that the judgment of God is come upon many of them to the full, and that they suffer those things which the Prophets threatened to the enemies of the Messiah.

Fourthly, amongst us Christians scarce one of a hundred answers his holy profession, and therefore we have little reason to insult over the Jews, that are so faulty ourselves.

Lastly, the great plenty and exceeding number of them, for in Asia and Africa (to omit diverse places of Europe), there are infinite numbers of them, who when they shall be converted, shall both in respect of themselves and us Gentiles be the reviving and resurrection of the world must keep us from rash censuring of them.

Use [1]. Christian Princes and Potentates, must take order that the Jews amongst them, may by degrees be taught true religion, yea, they should force them to hear the gospel, and not leave the miserable souls in perpetual darkness, that they every day grow worse, and more willful in their error.

Secondly they must by severity of laws and punishment curb and moderate their unmeasurable usuries, whereby they much damnify[16] and impoverish Christian men.

Use 2. Let the people amongst whom these Jews live and dwell beware lest by their pride and cruelty they do not hinder their [the Jews] conversion, for were it not for this, doubtlessly in many places many more Jews than now are, would be moved and drawn to embrace the Gospel.

[16] **damnify:** cause loss to.

→ SAMUEL USQUE

From Consolation for the Tribulations of Israel   *1553*

Samuel Usque was born in the late fifteenth century in Portugal to parents who had left Spain when the Jews were expelled in 1492. The time of his birth coincided with the forcible conversions of Portuguese Jews to New Christians. Some

Samuel Usque, *Consolation for the Tribulations of Israel* (Ferrara, 1553), trans. Martin Cohen (Philadelphia: The Jewish Publication Society of America, 1965), 182–85.

of these converts, including Usque, secretly continued their practice of Judaism. He probably left Portugal after the Inquisition was instituted there in 1521 and traveled extensively through Europe and the east, even visiting Israel. By 1551 he had settled in Ferrara, Italy, and published his *Consolation for the Tribulations of Israel* there in 1553 (Martin Cohen 12–14, 16). We know from documents of the Portuguese Inquisition that shortly after the publication of the *Consolation* it was circulating in the crypto-Jewish New Christian community in Bristol, England (Baron 15: 126). This text records the oppression of the Jews throughout history and situates it in the context of divine providence, which will ultimately redeem Israel and punish its enemies. The excerpt below, taken from a modern transla-tion, combines fiction and fact to provide a fantastical history of the Jews in medieval England. In addition, it offers a trenchant condemnation, similar in tone to that of Catholic critics, of Henry VIII and the English Reformation. Usque provides a counterpoint to Christian accounts of the Jews' sojourn in and expul-sion from England in 1290, focusing specifically on the issue of compulsory con-version. As one who had first-hand experience in the matter, he presents a Jewish perspective on the experience of Christian attempts to convert Jews. Are there any parallels between this view and Shylock's reaction to his required conversion?

England

In this same country, England, I witnessed another fierce and terrible calamity. After that king had left this life he was succeeded by another,[1] who disregarded the past expulsion and recalled all the Jews who had left the kingdom, offering to take them back and let them live in peace.

The Jews took counsel among themselves in France, Flanders, Spain, and other lands where they had dispersed. They decided that for no reason, including ease or material gain, would they re-enter a place where they had suffered so great a calamity except to see the children they had left behind, to talk with them and convince them to return to the faith of their fathers and to leave England. This reason seemed sufficient for their return: they desired to win back their lost children.[2]

When they arrived, the populace received them kindly because of the good will the king showed them.

After a few years, the plague broke out all over that island; it was so severe that a large number of people died every day. A terrible famine also hit the land at this time, and not long thereafter these misfortunes were increased by a bitter war against the English by many peoples, including

---

[1] **king . . . another:** Henry III (1207–1272) was succeeded by his son, Edward I (1239–1307).
[2] **lost children:** in the anecdote that precedes this one, Usque relates the story of a forced mass conversion of the children of the Jews in retaliation for the voluntary conversion of a monk to Judaism.

Scotland, a kingdom close by. Clearly this was a punishment for the sin which that people[3] had committed against Israel, so manifest that even fools came to recognize it. But the English, blind and ignorant of God's secrets, did not perceive it. The king, astonished at these many simultaneous tribulations, began to look for counsel concerning a way to check their causes.

After many possible causes had been considered, including the events in their kingdom and in the history of other peoples, the king and his advisors resolved to strike a blow in the weakest spot, where it would hurt them least: all agreed that the sins of the Jews were responsible for the calamities the kingdom was suffering. For this they found no better remedy to assuage the wrath of the heavens than to make all the Jews Christians by force, since they could not convert them through love.

The king forthwith proclaimed that any Jew attempting to leave the kingdom would be put to death. Then he assembled all the Jews. He admonished them to change their religion and promised them many favors and benefits, but they refused to convert. He finally baptized them by force, against their will, hoping that the famine, pestilence, and wars would then abate.

But after the Jews were forcibly converted to Christianity, these calamities multiplied to such an extent that the land was almost desolate. The people's understanding was confused, and they marveled at the opposite effect which their action had accomplished. The king again took counsel, and there were many who voiced the opinion that the calamities had doubled because of the force which had been used to bring the Jews to the Christian faith; spiritual matters, they asserted, should be freely decided, since our Lord had granted freedom to the human will. For this reason, they advised the king to restore the Jews to their former estate. If any Jew should wish to come to the Christian faith through love, he alone was worth more than all the forced converts.

The council would have unanimously favored this way of thinking, had not the Adversary[4] intervened, dressed in human clothing, to condemn it. Another point of view arose: "It is obvious that if the Jews are given their liberty, not one of them will choose the Christian faith over the Jewish law. Their law is deeply rooted in them; the resistance which they showed to becoming Christians is proof of this. Thus, if they become Jews again, the sin which is the cause of the misfortunes our kingdom is suffering will again be present. Our troubles have now doubled because of the force used against

---

[3] **that people:** the English. [4] **Adversary:** in Jewish thought, השטן, the Satan or adversary, is not the devil, but an accusing angel who argues against the Jews. For an example, see the first two books of Job.

the Jews. No other remedy remains for us to consider except to uproot the cause from the world; then the sin and its punishment will cease."

This argument pleased the king, and his counselors approved it. But the Jewish people were numerous and, if killed, their bodies might contaminate the air. The king therefore ordered two pavilions to be set up by the seashore, one distant from the other. In one he placed the Law of Moses received from the Lord on Mount Sinai, and in the other the cross of Christianity. In the middle an elaborate scaffold was erected. Here the king sat down. He ordered brought before him all the Jews whom he had made Christians by force. With a cheerful countenance and feigned joy he spoke these words to them:

"It is true that I believed that if I made you Christians, it would remedy the adversities my kingdom has suffered. Now I see that not only did they not abate, but they are increasing daily, and I realize that the force I used against you has been responsible for their having doubled. Since this is so, I would now restore to you your freedom of religion that you may freely choose which of these two religions you desire. There in that pavilion near the sea is the Law of Moses, and in this one the Christian's faith. Let each one run to the one he likes, and I shall let him live in that faith without any hurt whatsoever."

The Hebrew people greatly rejoiced at the liberty which the king was granting them, not realizing the deceit which lay beneath his cogent arguments. With their wives, and their children in their arms, they ran to the pavilion where the Law of Moses had been placed. There they fell into the snare which the enemy had set for them. They were able to enter only one at a time because the entrance was very narrow, especially prepared for this purpose; and when an Israelite lamb entered, he was decapitated by a hidden English butcher and cast into the sea. Each one knew not what happened to his brother; their lives were taken by the sword, and their bodies fed to the fish.[5]

O cruel Englishmen, were you so righteous and holy a people that you could not presume that your misfortunes were a punishment for your own deeds? . . . O people so quick to hurt me and so blind to the reason for which you harmed me! How is it that you have not become inured to crime from of old? Are not the ways of your princesses adulterous? Is not the garb of your masses woven of robberies, hatreds, and killing? I do not need to cite ancient histories to tell this, for modern and contemporary records proclaim and testify to it.

Consider in the few years that King Henry reigned, how many acts of adultery were committed by his own queens; how many treacheries were

---

[5] **fed to the fish:** I have found no evidence that this event occurred.

attempted by the king's noblest and closest associates; how many heads were placed on London Bridge because of these and other ghastly crimes; and how many queens were killed by the sword, and other deprived of dominion. The churches where you prayed were demolished by your own hands or converted to stables. Your priests were shamefully expelled. The gold and silver images to which you bowed were broken apart. The wooden ones were burned in the fire. Others were strewn on dunghills and filthy places. And your pope, cardinals and bishops became a laughingstock and byword[6] among you.

According to your religion, all these acts, singly and collectively, are regarded as sins. They manifest extreme wickedness, and you deserved punishment for them. Misfortunes did not come to your land because of the Jews' sins alone, as you say. When your sins made supplication, you received an answer from heaven. Why, you even refused to acknowledge that you had been like the passengers in the ship that carried the prophet Jonah. Heaven thought little of them, except for the righteous Israelite; and when he was cast into the sea, the storm abated. Yet the others considered themselves righteous and innocent. You tried hard to make your case bear a resemblance to this one.[7]

But what shall I say, O wretched me? Who blinded me to return again to this England, where I had fared so ill in former times? How many devices sin employs to bring me punishment. It made my children remain on that island so that they might be the means by which these prophetic words were fulfilled against me:

"I will lay stumbling blocks before these people, that the fathers and their sons may fall in them" (Jeremiah 6:21). Thus, "calamity shall come upon calamity" (Ezekiel 7:26). "You feared a sword: I will draw out a sword upon you, says the Lord, and none will escape by flight, though he strive to escape" (from Amos 9:1–4).

These judgments, O Lord, were cruel. And, moreover, "with treacherous talk have they led me (to slaughter)" (from Psalms 109:2–3). "And (my children) did not . . . understand the Lord's counsel" (from Micah 4:12), in order for this prophecy of yours to be carried out, Isaiah: "The wisdom of their wise men shall perish and the understanding of their prudent men shall be hid at that time" (Isaiah 29:14).

---

[6] **byword**: object of ridicule.   [7] **the passengers . . . this one**: see Jonah 1.

⇢ JOHN FOXE

From Acts and Monuments   *1570*

John Foxe's account of Archbishop Thomas Cranmer in *Acts and Monuments* emphasizes the necessity of freedom of conscience in religious belief. Thomas Cranmer (1489–1556), named Archbishop of Canterbury by Henry VIII, was condemned for heresy and deprived of his archbishopric under Mary Tudor. He was forced to convert to Catholicism, but recanted his conversion shortly before his execution. Foxe not only condemns coercive conversions but illustrates their ultimate ineffectiveness in his triumphant portrait of Cranmer's final return to Protestantism before his death. Would an early modern understanding of religious persecution and forced conversion drawn from this account cast light on Shylock's plight in the play?

THE ARCHBISHOP CONTENT TO RECANT

With these and like provocations, these fair flatterers ceased not to solicit and urge him, using all means they could to draw him to their side, whose force his manly constancy did a great while resist. But at last, when they made no end of calling and crying upon him, the archbishop, being overcome, whether through their importunity,[1] or by his own imbecility,[2] or of what mind I cannot tell, at length gave his hand.

It might be supposed that it was done for the hope of life, and better days to come. . . . But howsoever it was, plain it was, to be against his conscience. But so it pleases God, that so great virtues in this archbishop should not be had in too much admiration of us without some blemish, or else that the falsehood of the popish generation, by this means, might be made more evident, or else to diminish the confidence of our own strength, that in him should appear an example of man's weak imbecility.

HIS MISERABLE CASE AFTER HIS RECANTATION

This recantation of the archbishop was not so soon conceived, but the doctors and prelates without delay caused the same to be printed, and set abroad in all men's hands. . . . All this while Cranmer was in uncertain assurance of his life, although the same was faithfully promised to him by the doctors; but after that they had their purpose, the rest they committed to

---

[1] **importunity:** ceaseless requests.   [2] **imbecility:** weakness.

John Foxe, *Acts and Monuments* (1570; London: Seeleys, 1857), vol. 8, part 2, 81–84, 86–88, 90.

all adventure, as became men of that religion to do.[3] The queen . . . received his recantation very gladly; but of her purpose to put him to death, she would nothing relent.

Now was Cranmer's cause in a miserable taking, who neither inwardly had any quietness in his own conscience, nor yet outwardly help in his adversaries. . . .

There came to him the Spanish friar, witness of his recantation, bringing a paper with articles, which Cranmer should openly profess in his recantation before the people, earnestly desiring him that he would write the said instrument[4] with the articles with his own hand, and sign it with his name; which when he had done, the said friar desired that he would write another copy thereof which should remain with him; and that he did also. But yet the archbishop being not ignorant whereunto their secret devices tended, and thinking that the time was at hand in which he could no longer dissemble the profession of his faith with Christ's people, he put secretly in his bosom[5] his prayer with his exhortation written in another paper, which he minded[6] to recite to the people, before he should make the last profession of his faith.

Cranmer at length came from the prison of Bocardo unto St. Mary's Church. . . . The lamentable case and sight of that man gave a sorrowful spectacle to all Christian eyes that beheld him. He that late[7] was archbishop, metropolitan, and primate of England, and the king's privy counselor, being now in a bare and ragged gown, and ill-favoredly clothed, with an old square cap, exposed to the contempt of all men, did admonish men not only of his own calamity, but also of their state and fortune. For who would not pity his case, and bewail his fortune, and might not fear his own chance, to see such a prelate,[8] so grave a counselor, and of so long continued honor, after so may dignities, in his old years to be deprived of his estate, adjudged to die, and in so painful a death to end his life?

Cole,[9] [who preached the sermon before Cranmer's execution] called back the people that were ready to depart, to prayers. "Brethren," said he, "lest any man should doubt of this man's earnest conversion and repentance, you shall hear him speak before you; and therefore I pray you, master Cranmer, that you will now perform what you promised not long ago, namely, that you would openly express the true and undoubted profession of your faith, that you may take away all suspicion from men, and that all men may

---

[3] **after that they had . . . to do:** once they achieved their goal of securing Cranmer's recantation, they didn't care what followed; note again Foxe's sarcasm. [4] **instrument:** text. [5] **bosom:** pocket. [6] **minded:** intended. [7] **late:** recently. [8] **prelate:** high-ranking church official. [9] **Cole:** Henry Cole (1500–1580) was a prelate and later Dean of St. Paul's; Queen Mary instructed him to preach at Cranmer's execution (*DNB*).

❡The defcription of Doct.Cranmer,how he was plucked downe from the
ftage by Friers and Papiftes,for the true confeffion hys fayth.

FIGURE 13 *Cranmer Recanting His Conversion, from John Foxe,* Acts and Monuments *(1583).*

understand that you are a Catholic indeed." "I will do it," said the arch-bishop, "and that with a good will"; who by and by rising up and putting off his cap, began to speak thus unto the people. "I desire you, well-beloved brethren in the Lord, that you will pray to God for me, to forgive me my sins, which above all mean, both in number and greatness, I have committed. But among all the rest, there is one offense which most of all at this time doth vex and trouble me, whereof in process of my talk you shall hear more in its proper place." And then, putting his hand into his bosom, he drew forth his prayer, which he recited to the people:

"Good Christian people, my dearly beloved brethren and sisters in Christ, I beseech you most heartily to pray for me to the Almighty God, that he will forgive me all my sins and offenses, which be many without number, and great above measure. But yet one thing grieves my conscience more than all the rest, whereof God willing, I intend to speak more here-after. But how great and how many soever my sins be, I beseech you to pray God of his mercy to pardon and forgive them all."

FIGURE 14 *Cranmer Burning the Hand That Signed His Profession of Catholicism, from John Foxe,* Acts and Monuments *(1583).*

[Cranmer enumerates his other sins before returning to the most important one.]

"And now I come to the great thing, which so much troubles my conscience, more than anything that ever I did or said in my whole life, and that is the setting abroad of a writing contrary to the truth; which now here I renounce and refuse, as things written with my hand, contrary to the truth which I thought in my heart, and written for fear of death, and to save my life if it might be; and that is, all such bills and papers which I have written or signed with my hand since my degradation; wherein I have written many things untrue. And forasmuch as my hand offended, writing contrary to my heart, my hand shall first be punished therefore; for, may I come to the fire, it shall be first burned. . . ."

Here the standers-by were all astonished, marveled, were amazed, did look one upon another, whose expectation he had so notably deceived. Some began to admonish him of his recantation, and to accuse him of falsehood. Briefly, it was a world to see the doctors beguiled of so great a hope. I think there was never cruelty more notably or better in time deluded and deceived; for it was not to be doubted but they looked for a glorious victory and a perpetual triumph by this man's retraction. As soon as they heard these things, [they] began to . . . rage, fret, and fume, and so much the more, because they could not revenge their grief — for they could now no longer threaten or hurt him.

And when the wood was kindled, and the fire began to burn near him, stretching out his arm, he put his right hand into the flame, which he held so steadfast and immovable (saving that once with the same hand he wiped his face), that all men might see his hand burned before his body was touched. His body did so abide the burning of the flame with such constancy and steadfastness, that standing always in one place without moving his body, he seemed to move no more than the stake to which he was bound; his eyes were lifted up into heaven, and oftentimes he repeated "his unworthy right hand," so long as his voice would suffer him. And using often the words of Stephen,[10] "Lord Jesus, receive my spirit," in the greatness of the flame he gave up the ghost. . . . And this was the end of this learned archbishop, whom, lest by evil subscribing he should have perished, by well-recanting God preserved;[11] and lest he should have lived longer with shame and reproof, it pleased God rather to take him away, to the glory of his name and profit of his church. So good was the Lord both to his church, in fortifying the same with the testimony and blood of such a martyr, and so good also to the man with this cross of tribulation, to purge his offenses in this world.

[10] words of Stephen: Acts 7:59. [11] lest by . . . preserved: instead of dying for writing a false confession of faith, God preserves him for recanting. Here Foxe inverts notions of life and death.

RAPHAEL HOLINSHED

> *From* Chronicles of England, Scotland, and Ireland *1587*

Raphael Holinshed (d. 1580) was educated at Cambridge and, at the beginning of Elizabeth's reign, served as a translator for the London publisher Reginald Wolfe. Wolfe had been working on a universal history for over a decade using another author's notes, but compiling his own material for the English, Irish, and Scottish sections. Holinshed continued work on the project, and after Wolfe's death was assisted by others in completing the first edition in 1577, which was limited to England, Ireland, and Scotland. After Holinshed's death, another edition was prepared with numerous contributors and printed in 1587.

Raphael Holinshed, *Chronicles of England, Scotland, and Ireland* (1587), ed. Henry Ellis (London, 1807), 2: 435, 437, 492.

The work was popular and influential, providing Shakespeare and other early modern playwrights with material. This excerpt demonstrates the currency in early modern England of a range of stories about medieval Jews. In his discussion of Henry III, Holinshed offers a critique of greedy Christians who extort money from the Jews, even as he recounts the charge of ritual murder, also mentioned by Foxe in the subsequent excerpt. This myth, which probably originated in twelfth-century England and circulated throughout Europe well beyond the early modern period, imagined that Jews would kidnap a Christian child and crucify him in a kind of mocking, murderous imitation of the crucifixion of Jesus. The anger that this myth aroused among Christians resulted in numerous attacks on the Jewish communities in Europe. The narrative from the reign of Edward I describes the agreement between the Parliament and the king to expel all the Jews from England. The powerful men who controlled Parliament disliked the Jews both because they were often in excessive debt to Jewish moneylenders and because Jews in effect limited the leverage Parliament held over the king by providing a source of funding that would allow the king to avoid negotiating with the legislature for tax revenues. In 1289 Edward found himself burdened with such considerable debt, that "he was compelled to bargain with his subjects in parliament. The Expulsion of the Jews from England was the price the [House of C]ommons' representatives demanded for their consent to this voluntary grant of taxation" (Stacey, quoted in Shapiro 54). This act, which resulted in the exodus of the Jewish community, essentially prevented Jews from living in England for the next three or four centuries.

HENRY III, CIRCA 1255

The King Demands Money of the Jews

Moreover, whereas he stood in great need of money, he required by way of a tallage[1] eight thousand marks of the Jews, charging them on pain of hanging, not to defer that payment. The Jews sore impoverished with grievous and often[2] payments, excused themselves by the Pope's usurers,[3] and reproved plainly the king's excessive taking of money, as well of his Christian subjects as of them. The king on the other side, to let it be known that he taxed not his people without just occasion, and upon necessity that drove him thereto, confessed openly, that he was indebted by his bonds obligatory,

---

[1] **tallage:** an arbitrary tax.   [2] **often:** frequent.   [3] **the Pope's usurers:** Foxe, citing the medieval chronicler Matthew of Paris, reports that in 1229 "Stephen, the pope's chaplain . . . brought with him into England . . . bankers and usurers; who, lending out their money upon great usury, did unreasonably pinch the English people" (2: 389). It is not clear how the Jews excuse themselves by the pope's usurers here; perhaps they have no money because of competition with the papal usurers or because they are indebted to them.

in three hundred thousand marks. And again, the yearly revenues assigned to his son Prince Edward, arose to the sum of fifteen thousand marks and above, where the revenues that belonged unto the crown were greatly diminished, in such wise, that without the aid of his subjects, he should never be able to come out of debt. To be short, when he had fleeced the Jews to the quick, he set them to farm[4] unto his brother Earl Richard, that he might pull off skin and all; but yet considering their poverty, . . . [Richard] spared them, and nevertheless, to relieve his brother's necessity, upon a pawn he lent him an huge mass of money. These shifts did the king use from time to time, not caring with what exactions and impositions he burdened the inhabitants of his land, whereby he procured unto himself the name of an oppressor and covetous scraper. But what wonder is it in a king, since *Maxima pars hominum morbo iactatur eodem?*[5]

Also, upon the two and twentieth of November, were brought unto Westminster a hundred and two Jews from Lincoln, that were accused for the crucifying of a child in the last summer, in despite of Christ's religion.[6] They were upon their examination sent to the Tower. The child which they had so crucified was named Hugh, about . . . eight years of age. They kept him ten days after they got him into their hands, sending in the meantime unto diverse other places of the realm, for the other of their nation to be present at the crucifying of him. The murder came out, by the diligent search made by the mother of the child, who found his body in a well, on the back side of the Jew's house, where he was crucified. For she had learned, that her son was lastly seen playing with certain Jew's children of like age to him, before the door of the same Jew. The Jew that was owner of the house was apprehended, and being brought before Sir John de Lexington, upon promise of pardon, confessed the whole matter. For they used yearly (if they could come by their prey) to crucify one Christian child or other. The king, upon knowledge had hereof, would not pardon this Jew that had so confessed the matter, but caused him to be executed at Lincoln, who coming to the place where he should die, opened more matter concerning such as were of counsel and present at the crucifying of the poor innocent. Whereupon at length also eighteen of them that were so brought to London, were convinced,[7] adjudged, and hanged, the other remained long in prison.

---

[4] **set them to farm:** the king allowed his brother to make further profit from them.   [5] *Maxima . . . eodem:* the greater part of men build on this same vice.   [6] **in despite of Christ's religion:** to mock Christianity.   [7] **convinced:** convicted.

EDWARD I, 1290

In the same year [1290] was a Parliament holden at Westminister, wherein . . .
it was also decreed, that all the Jews should avoid out of[8] the land, in consideration whereof, a fifteenth[9] was granted to the king, and so hereupon were
the Jews banished out of all the king's dominions, and never since could they
obtain any privilege to return hither again. All their goods not moveable[10]
were confiscated, with their tallies and obligations;[11] but all other their
goods that were moveable, together with their coin of gold and silver, the
king licensed them to have and convey with them. A sort of the richest of
them, being shipped with their treasure in a mighty tall ship which they had
hired, when the same was under sail, and got down the Thames towards the
mouth of the river beyond Quinborough, the master mariner bethought
him of a wile,[12] and caused his men to cast anchor, . . . till the ship by
ebbing of the stream remained on the dry sands. The master herewith
enticed the Jews to walk out with him on land for recreation. And at length,
when he understood the tide to be coming in, he got him back to the ship,
whither he was drawn up by a cord. The Jews made not so much haste as he
did, because they were not aware of the danger. But when they perceived
how the matter stood, they cried to him for help. Howbeit he told them,
that they ought to cry rather unto Moses, by whose conduct their fathers
passed through the Red Sea, and therefore, if they would call to him for
help, he was able enough to help them out of those raging floods which now
came in upon them. They cried indeed, but no succor appeared, and so they
were swallowed up in water. The master returned with the ship, and told the
king how he used the matter, and had both thanks and reward, as some have
written. But others affirm (and more truly, as should seem) that diverse of
those mariners, which dealt so wickedly against the Jews were hanged for
their wicked practice, and so received a just reward of their fraudulent and
mischievous dealing.

---

[8] **avoid out of:** leave.   [9] **a fifteenth:** a tax of one fifteenth was imposed on personal property.
[10] **goods not moveable:** real estate.   [11] **tallies and obligations:** receipts of payment and contracts of outstanding debts owed.   [12] **wile:** deception.

↦ JOHN FOXE

From Acts and Monuments   *1570*

While John Foxe's *Acts and Monuments* focuses primarily on the history of
Protestant martyrs in England, it also records other historical events. In this
excerpt, which narrates the events surrounding Richard I's reign, Foxe records
the random brutality against Jews that could be stirred up with apparently little
provocation in the twelfth century. In spite of his prohibition of the Jews' atten-
dance at his coronation, Richard seeks to protect the Jews from the increasing
violence against them. The history of England and European countries repeats
this pattern of the ruler often protecting the Jews for his own purposes, against
the hatred of the nobility or the people. Foxe disapproves of the decision to
allow Benedict to return to his Judaism, but seems shaken by the awful account
of Jewish mass suicide and murder to escape Christian violence. However, he
recalls the myth of ritual murder as a way of preventing his empathy and justify-
ing the tragedy. Although these events occurred hundreds of years before the
publication of *Acts and Monuments,* the wide distribution of this text meant that
these old stories were still circulating in the early modern public imagination.

RICHARD THE FIRST

Sunday, September 3rd, 1189

[I]t befell that . . . the day before his coronation [the king], by public edict
commanded both the Jews, and their wives, not to presume to enter either
the church or his palace, during the solemnization of his coronation,
amongst his nobles and barons; yet, while the king was at dinner, the chief
men of the Jews, with diverse others of the Jewish affinity and superstitious
sect, against the king's prohibition, together with other press,[1] entered the
court gates. Whereat a Christian man being offended, struck one of them
with his hand or fist, and bade him stand further from the court gate, as the
king had given commandment; whose example others also following, being
displeased with the Jews, offered them the like contumely.[2] Others also,
supposing that the king had so commanded indeed, as using the authority
of the king, fell upon[3] all the Jews that stood by without the court gate. And
first they beat them with their fists, but afterwards they took up stones and
such other things as they could get, and threw at them, and beat them therewith. And thus driving them from the court gates, some of them they
wounded, some they slew, and some they left for dead.

---

[1] **press:** throngs.   [2] **contumely:** contemptuous treatment.   [3] **fell upon:** attacked.

John Foxe, *Acts and Monuments* (1570; London: Seeleys, 1857), vol. 2, part 1, 276–77.

There was amongst this number of the Jews one called Benedict, a Jew of York, who was so sorely wounded and beaten with the rest, that, for fear of his life, he said he would become a Christian, and was indeed of William, the prior of the church of St. Mary of York, baptized; whereby he escaped the great peril of death he was in, and the persecutors' hands. In the meanwhile there was a great rumor spread throughout all the city of London, that the king had commanded to destroy all the Jews. Whereupon, as well the citizens, as innumerable people more, being assembled to see the king's coronation, armed themselves and came together. The Jews thus being for the most part slain, the rest fled into their houses, where for a time, through the strong and sure building of them, they were defended. But at length their houses were set on fire, and they destroyed therein.

These things being declared to the king, whilst he with his nobles and barons were at dinner, he sent immediately Ranulfe de Glanvile, the lord high steward of England, with diverse other noblemen to accompany him, that they might stay and restrain these so bold enterprises of the Londoners. But all was in vain, for in this so great a tumult none there was that either regarded what the nobility said, or else any whit reverenced their personages, but rather with stern looks and threatening words advised them, and that quickly, to depart. Whereupon they, with good deliberation, thinking it the best so to do, departed; the tumult and insurrection continuing till the next day. At which time also the king, sending certain of his officers into the city, gave them in commandment to apprehend and present some, such as were the chief of the malefactors: of whom three were condemned to be hanged, and so were; the one, for that he had robbed a Christian's house in this tumult; and the other two, for that they fired[4] the houses, to the great danger of the city. After this, the king sent for him who from a Jew was converted to Christianity, and in the presence of those who saw when he was baptized, the king asked him whether he was become a Christian or not? He answering the king said, no, but to the intent that he might escape death, he permitted the Christians to do with him what they listed.[5] Then the king asked the archbishop of Canterbury, other archbishops and bishops being present, what were best to be done with him? Who unadvisedly answering said, "If he will not be a man of God, let him be a man of the Devil," and so revolted[6] he again to Judaism.

Then the king sent his writs to the sheriffs of every county, to inquire for the authors and stirrers of this outrage; of whom three were hanged, diverse were imprisoned. So great was then[7] the hatred of Englishmen against the

---

[4] **fired:** set fire to.   [5] **listed:** wished.   [6] **revolted:** returned.   [7] **then:** at that time.

Jews, that as soon as they began to be repulsed in the court, the Londoners taking example thereof fell upon them, set their houses on fire, and spoiled their goods. The country again, following the example of the Londoners, semblably[8] did the like. And thus the year which the Jews took to be their jubilee,[9] was to them a year of confusion. Insomuch that in the city of York, the Jews obtaining the occupying of a certain castle for their preservation, and afterwards not being willing to restore it to the Christians again, when they saw no other remedy, but by force to be vanquished, first they offered much money for their lives. When that would not be taken, by the counsel of an old Jew amongst them, every one, with a sharp razor, cut another's throat, whereby a thousand and five hundred of them were at that time destroyed. Neither was this plague of theirs undeserved; for every year commonly their custom was to get some Christian man's child from the parents, and on Good Friday to crucify him, in despite of our religion.[10]

[8] semblably: similarly.   [9] jubilee: Leviticus 25 sets out the laws of the יובל (*yovel*) or jubilee, a time of redemption of property and personal freedom, which was to be celebrated every fifty years. September 1189, when these events took place, corresponded with the celebration of the Jewish new year 4950, a jubilee year.   [10] in despite of our religion: to mock our religion.

> ## *From* Fleta   *c. 1290*

Fleta is the title of a Latin textbook of English law that was apparently written around the year 1290; it may take its title from the Fleet prison, where it is conjectured that an imprisoned judge wrote the text. *Fleta* had staying power: Sir Edward Coke cited it as late as 1644 in his *Institutes of the Laws of England* (p. 210). In this excerpt, the author groups arsonists, Christians who have left the faith, sorcerers, sodomites, and traitors with those *"Contrahentes vero cum Iudeis vel Iudeabus,"* that is, those who certainly associate or draw close to Jewish men or Jewish women. Coke understands *contrahentes* to be those marrying Jews. Whatever the original meaning, it is striking to note that even in the seventeenth century, after more than three hundred years in which Jews were not formally permitted to remain in England, the law still called for harsh punishment for "those who have connections with Jews" (p. 304). When Launcelot questions Jessica's status as a Jew in act 3, scene 3, does he undermine the legitimacy of her conversion and marriage to Lorenzo?

Fleta, ed. and trans. H. G. Richardson and G. O. Sayles. Selden Society, vol. 72 (London: Selden Society, 1952), book 1, chapter 35, page 90.

OF ARSON

If anyone in time of peace maliciously burn the house of another, through enmity or for the sake of spoil, and if he be convicted thereof, by appeal or without appeal, he should be punished with a capital sentence.

Apostate Christians, sorcerers and the like should be drawn and burnt. Those who have connection with Jews and Jewesses or are guilty of bestiality or sodomy shall be buried alive in the ground, provided they be taken in the act and convicted by lawful and open testimony. Traitors, who slay their lord or lady, or who lie with their lord's wives or daughters or with the nurses of their lord's children, or forge their lord's seals, or who administer poison secretly to anyone whereof he dies, and are convicted thereof, shall be drawn and hanged, and, if they be women, they shall be burnt.

→ Examination of Roderigo Lopez   *1594*

Roderigo Lopez (d. 1594) was born in Portugal to a family of Jewish origin which had converted to Christianity. He emigrated to England in 1559 where he lived as a denizen or naturalized citizen. He had a very successful career as a doctor; he was a member of the College of Physicians, the doctors' guild, and his first appointment was as house physician at St. Bartholomew's Hospital. He served as physician to a number of powerful noblemen, including Francis Walsingham, secretary to Queen Elizabeth, and the earl of Leicester. In 1586 he became chief physician to Queen Elizabeth; in 1589 she granted him a monopoly over the importation of anise and sumac into England. At the request of the earl of Essex he served as a translator for Don Antonio, a contender to the throne of Portugal whom Essex supported against the rival claims of King Philip of Spain. Philip's agents gave Lopez a valuable piece of jewelry and offered him 50,000 crowns to assist in Antonio's murder. He was also asked to bring about Elizabeth's death for another 50,000 crowns. Lopez ostensibly assented to the plot, though it is not clear that he intended to carry it out. While he did conspire with the enemies of the queen, which constituted a treasonable offence, he may have been doing so at Essex's behest, in order to gain evidence of Spain's violent intentions against England. Lopez was arrested in January 1594 and brought to the tower where he was questioned regarding his involvement in the plot. At the end of February, he was tried before a special commission and found guilty, but the queen delayed signing his death warrant until June, when he was finally executed. Interestingly, the queen granted Lopez's widow the right to most of his property (usually traitors forfeited their goods to

Examination of Roger Lopez and Others, British Library MS Harley 871, 50r–51r, 58v–59v.

A Phisition Iewe.

FIGURE 15 *A Physician Jew, from Nicolas de Nicolay,* The Navigations, Peregrinations, and Voyages Made into Turkey *(1585). In addition to moneylending and trading, Jews frequently served as physicians in the medieval and early modern periods.*

the state); Elizabeth also granted Lopez's son an income to support him while he was in school. The excerpt below is taken from a manuscript record of Lopez's examination in the Tower, just prior to his trial. Manuscripts, or handwritten documents, were and are limited in their accessibility and circulation, in contrast to printed materials. Although this account is a first-hand record of Lopez's interrogation, it is not necessarily an accurate or even unbiased rendering of his answers; notice that he is referred to in the third person and his speech is reported in indirect discourse, not directly quoted. The author seems at pains not only to demonstrate Lopez's guilt in the plot, but also to establish his religious identity as Jewish. By identifying Lopez with Judas, the epitome for Christians of Jewish treachery, the author suggests that the doctor's criminal intent was motivated by his religion.

25 FEBRUARY 1594

The said Doctor Lopez being advised to declare the truth of the practice and speech that passed between him and Steven Ferrera[1] about the poisoning of the Queen's Majesty, he doth confess for the discharge of his conscience that such speech passed between Ferrera and this examinant[2] at his house at London. And he doth likewise confess that after Ferrera had written to the Secretary Ibarra[3] to assure him that the said Doctor Lopez was willing and did offer to poison the queen so he might have the 50,000 crowns, that said Ferrara did acquaint the said Doctor Lopez that he had written in such sort. . . .

But he doth protest that he never meant to do it. But he doth affirm that Ferrera meant verily her Majesty should have been destroyed with poison and the said Doctor said further he told Ferrera that he would minister the poison in a syrup which he said because this examinant knew her Majesty never doth use to take any syrup. . . . He further doth confess that Steven Ferrera told this examinant that if he would offer to the Count of Fuentes[4] this great service to poison her Majesty, he should be sure he should want no money, and hereupon he was contented that the said Ferrera should write to the said Count of Fuentes and the Secretary Ibarra to assure them that the said Doctor should undertake to poison her Majesty. But he sayeth when the money was come he meant to have told her what the King of Spain had sent him to poison her Majesty. And he said further he would have told her Majesty of it long since, but for fear of jealousy her Majesty might have conceived thereby. And the said Doctor Lopez doth further confess he did

---

[1] **Steven Ferrera:** an agent of Philip II, king of Spain.   [2] **this examiniant:** i.e., Lopez, who is being examined or questioned.   [3] **Secretary Ibarra:** King Philip's secretary in the Netherlands.
[4] **Count of Fuentes:** a Spanish administrator in the Netherlands.

often use to say to Steven Ferrera, "When will the money come? If the money were come I was ready to do the service." . . . and all this confession he protested to be true and contain the very truth.

[I]t was not forgotten what the Doctor should do and how he would bestow himself when he had put in practice this wicked purpose. The mention whereof in this narration so often repeated cannot be but very loathsome and grievous, for his intent was to convey himself out of the Realm. . . . And since it is declared he was provided to fly,[5] it is not amiss it should be told whither he meant to retire himself and his family. It is most manifestly discovered he intended to convey himself first to Antwerp, where he was to receive the greatest part of the promised reward for this intended service, and thence to Constantinople. . . , there to live as a Jew amongst the Jews in Turkey where he had nephews and kinfolks and so to end his days. And though both the Portugalls[6] here and those Englishmen that trade thither know him to be come of the basest sort of Jews and amongst them of very mean parents,[7] yet being (as they call them there, New Christians) he here was generally so accompted;[8] and to Ferrera after he had trusted him with this great secret he ordinarily did avow himself to be a very Jew. Besides it is confirmed by letters from Constantinople and out of other places that he is so held and reputed of the Jews and sent ordinarily relief to the Jews and money to be bestowed in Jews' ceremonies. And surely a more Judas-like part since the betraying of Christ our Savior was then never put in use than this treason intended against a most Christian Queen professing Christ and his Gospel.

[5] **provided to fly:** prepared to flee the country.   [6] **Portugalls:** Portuguese.   [7] **the basest sort of Jews and . . . of very mean parents:** alleging that his family was of inferior rank and his parents were of even lower status within that rank.   [8] **generally so accompted:** most people considered him a Christian.

➤ WILLIAM CAMDEN

From The History of Elizabeth, Queen of England
1630
Translated by Robert Norton

William Camden (1551–1623) studied at Oxford, devoting most of his time to antiquarian research. He served as a schoolmaster for several years, enabling him to continue his exploration of the antiquities and topography of England. In 1586 he published the first edition of *Britannia,* a text in Latin that presented

William Camden, *The History of Elizabeth, Queen of England,* trans. Robert Norton (London, 1630), 58–59.

a detailed account of the geography and history of Britain; it was reprinted three times in the four years that followed. Philemon Holland (p. 347) translated the text into an English version that was first published in 1610. Camden also published a number of smaller chronicles and annals that grew out of his research for the *Britannia*. In 1615 he brought out in Latin the first edition of the annals of Queen Elizabeth's reign, which covered the years 1558–88; he completed the compilation in 1617, but did not authorize its publication during his lifetime. After his death it was published in Leyden in 1625 and London in 1627, and was subsequently translated into English, the text of which is excerpted below. Camden situates the Lopez affair in the context of Catholic stratagems to promote the claims of Spain to the throne of England. Paradoxically, New Christians who emigrated to England from Spain and Portugal could be suspected of, and denounced as, being either Catholics or Jews, suggesting an interesting overlap in the way these religious others could be viewed. Camden here asserts that the physician Lopez was a Jew, although Lopez professed Christianity. In Camden's report of the victim's last words, he likens Lopez's purported treason against Elizabeth with the Christian view of the Jewish betrayal of Jesus.

As these learned English fugitives[1] studied to advance by writing, the Infanta[2] of Spain to the Scepter of England, so others of their number secretly attempted the same by the sword, sending privily[3] certain murderers to kill Queen Elizabeth, and some Spaniards by poison.[4] The Spaniards suspecting the fidelity of the English in a matter of so great weight, used the help of Roderigo Lopez of the Jewish sect, the Queen's physician for her household, and of Stephen Ferreira Gama[5] and Emmanuel Loisie,[6] Portugese. (For many Portugese in those days crept into England, as retainers to the exiled Don Antonius[7]) who by means of letters intercepted, being apprehended, were about the end of February arraigned in Guild Hall at London,[8] and charged by their own confessions, to have conspired to make away the Queen by poison. Lopez having been for a long time a man of noted fidelity, was not once suspected (save that outlandish[9] physicians may by bribes and corruption be easily made poisoners and traitors). He confessed that he was drawn by

---

[1] **fugitives:** Camden segues from a discussion of English Catholics (probably William Allen [p. 255] and Robert Parsons [p. 258]) who supported a Spanish claim to the English throne. [2] **Infanta:** daughter of the king of Spain. [3] **privily:** secretly. [4] **and some Spaniards by poison:** and some Spaniards to poison her. [5] **Stephen Ferreira Gama:** agent of Philip II, king of Spain. [6] **Emmanuel Loisie:** probably Manuel Luis Tinoco, a former follower of Don Antonio. [7] **Don Antonius:** Don Antonio was Prior of Crato and the most formidable contender for the Portuguese throne against the claim of King Philip. [8] **Guild Hall at London:** the hall for the corporation of the city of London. [9] **outlandish:** foreign.

FIGURE 16 *Lopez Conspiring to Poison Queen Elizabeth, from George Carleton,* A Thankful Remembrance of God's Mercy *(1624). He is pictured as asking in Latin, "quid dabitis" ("what will you give?"), the same phrase spoken by the Jewish merchant pictured in Figure 10 (see p. 171). The illustration suggests that Jews will do anything for money — even sell their loyalty.*

Andrada,[10] a Portugal[11] to employ his best and secret service for the King of Spain; that he had received from his most inward counselor, Christopher Moro,[12] a rich jewel; that he had diverse times advertised the Spaniard of such things as he could learn; that at length upon a contract for 50,000 ducats, he had promised to poison the Queen. And that this he signified to Count de Fuentez, and Ibara[13] the King's Secretary in the Netherlands.

Stephen Ferreira confessed that Count de Fuentez and Ibara had signified unto him both by letters and word of mouth that there was a plot laid to take away the Queen's life by poison; that he wrote letters by Lopez his dictating,[14] wherein he promised the same, conditionally that 50,000 ducats

[10] **Andrada:** Manuel Andrada, a former courier for Don Antonio, who colluded with the Spanish to betray his master.   [11] **Portugal:** Portuguese.   [12] **Christopher Moro:** Spanish secretary of state.   [13] **Count de Fuentez, and Ibara:** Spanish administrator and Philip's secretary in the Netherlands, respectively.   [14] **by Lopez his dictating:** dictated by Lopez.

should be paid unto him; also that Emmanuel Loisie was secretly sent unto him by Fuentez and Ibara to excite Lopez to dispatch the matter speedily.

Emmanuel confessed that Count Fuentez and Ibara, when he had given them his faith to keep close their counsels, showed him a letter which Andrada had written in Lopez his name about making away with the Queen; and that he himself was likewise sent by Fuentez to deal with Ferreira and Lopez for hastening the Queen's death, and to promise to Lopez himself money, and honors to his children.

At the bar, Lopez spake not much, but cried out, that Ferreira and Emmanuel were wholly composed of fraud and lying; that he intended no hurt against the Queen, but hated the gifts of a tyrant; that he had given that jewel to the Queen which was sent to him from the Spaniard; and that he had no other meaning, but to deceive the Spaniard, and wipe him of his money.

The rest spake nothing for themselves, many times accusing Lopez. They were all of them condemned, and after three months put to death at Tyburn; Lopez affirming that he had loved the Queen as he loved Jesus Christ, which from a man of the Jewish profession was heard not without laughter.

CHAPTER 4

Love and Gender

>|<

Marriage and friendship structure the comic plot of *The Merchant of Venice:* Bassanio simultaneously seeks marriage as a means of repaying his friend Antonio and uses his friendship with Antonio to finance his marriage. Marriage in the early modern period took place between competing motives of love and money, and the competing interests of the lovers and their guardians, both parents and friends. While the law required that husband and wife wed voluntarily, parental approval was a weighty consideration. Popular opinion viewed love as an important ingredient in a good marriage, but counterbalanced this with the need to consider the economics of setting up a household together. The young were expected to heed the marital advice of their elders, whether parents, family friends, or other relations; the suitability of a union was measured in terms of class and property as well as temperament of the couple. While authors on the subject disapproved of people's marrying for financial benefit alone, in practice economic concerns played a central role in the nuptials of the propertied classes. Men and women were expected to obey their parents before marriage; afterward, the wife was expected to obey the husband. The laws and customs governing early modern marriage served the interests of the husband and circumscribed the wife's rights by largely subsuming her legal identity to that of her spouse. Although English women may have experienced a greater degree of personal freedom than those on the continent, most English texts on women and marriage prescribed limitations in female behavior, speech, and public activity. While a wife and her property were subjected to her husband's authority, her desires and actions could never be fully controlled. Anxiety about female chastity ran high in the period, owing to this tension between the ideal of female subordination and the lived experience of marriage. Husbands, however, also had their responsibilities in the formation of marriage and afterward, which included consulting their parents in choosing a spouse, marrying for virtuous reasons, and providing financially for their wives and children.

While marriage was a popular and prevalent institution, same-sex friendship, particularly between men, was sometimes more highly valued. Segregation by gender shaped many areas of social interaction in the period, which facilitated strong bonds of friendship as well as desire between members of the same sex. Although sodomy, usually associated with transgressive behavior perceived to disrupt the social order, was officially condemned, a whole range of homoerotic relations were practiced and implicitly condoned in the period. Both marriage and homosociality, which were not necessarily mutually exclusive, provided a means of establishing important social, economic, and political alliances. However, popular representations of marriage often set it in competition with same-sex friendships, depicting the wife as threatening the friendship or the friend as undermining the marriage.

Women and Marriage

Numerous medieval and early modern authors debated the nature of women and their defining characteristics. For Juan Luis Vives, who expresses Catholicism's preference for celibacy, chastity (defined as abstinence for single women and marital fidelity for wives) was the primary concern for all women (see p. 319). To Vives, women's inferiority necessitates their subordination to fathers and husbands. A wife does not even own her body, according to Vives, and a maid should resist her own desires in choosing a husband and instead be ruled by the advice of her parents. Women are also advised to be circumspect in speech and movement, silent in the presence of others, and away from home only when necessary. Vives does advocate the education of women, though he argues for it on the basis that learning will condition them to more virtuous obedience.

Thomas Becon, representing a Protestant position, expresses views about women's roles similar to those of Vives (p. 327). Wives should stay at home, bear with their husband's bad behaviors, and seek to hide and extenuate his flaws rather than reveal them with chiding. Maids should avoid bad company and refrain from talkativeness. Becon agrees with Vives that daughters should allow parents to influence their marriage choices, but he insists that the child's consent is necessary; he criticizes parents for abusing their authority and arranging marriages for gain. *The Merchant of Venice* engages with and complicates a number of these issues. In act 3, scene 2, when Portia describes herself as an "unlessoned girl" and commits herself to be directed by Bassanio, "as from her lord, her governor, her king" (159, 165), she appears to follow the views of Vives and Becon in taking an appropriately subordinate position to her husband. On the other hand, she also demonstrates a learning superior to all in the play in her court case against Shylock. The secrecy and independence with which she acts to save Antonio, though performed perhaps for Bassanio's benefit, call into question her husband's authority. Here her actions might find more support in the writings of Cornelius Agrippa, who argues for the superiority of women and praises what he views as their preeminence in speech (see p. 332). He even goes so far as to suggest, relying on God's admonition to Abraham to heed Sarah's advice (Genesis 21:12), that men should allow themselves to be led by women. Given how unusual this view was at the time, one cannot rule out the possibility, however, that Agrippa wrote parts if not the whole treatise with tongue in cheek.

As for the issue of filial obedience and parental influence in marriage matters, how does Portia fare in light of the views expressed by Becon and Vives? Is she a completely obedient daughter or does she challenge her father's will? How does her behavior compare to that of Jessica? On the other hand, what kind of concerns do the fathers in the play express about their daughters' marriages? What kind of husband is Portia's father's lottery designed to select? Does it ensure that Portia gets the best husband? What are Shylock's views on his daughter's marriage — does he have her best interests in mind? Does Jessica find a good husband in Lorenzo by following her own desires?

Anxiety about female autonomy and self-assertion was often articulated in terms of women taking on what were defined as masculine roles or behaviors. Philip Stubbes's commentary on women wearing men's clothes tacitly engages the gender contest for superiority (p. 335). He argues that cross-dressing violates biblical law, which was intended to maintain gender distinctions. Stubbes ascribes to women who dress in men's clothes the desire to change their gender; he also claims that clothing has the power somehow to adulterate the gender of the wearer. His vehement castigation of this practice as lascivious and shameful expresses an underlying fear that in donning men's garments, women may be usurping male authority. In *The Merchant of Venice*, while Jessica voices shame about dressing as a male, her disguise nevertheless enables her to act against her father's authority. Portia and Nerissa allude to the transformative and lewd associations of wearing men's clothes, yet are liberated and empowered by them to deploy masculine legal power. They also use their male disguise to get the best of Bassanio and Gratiano, giving the wives "vantage to exclaim on [them]" (3.2.174). Stubbes's association of cross-dressing with lust links him with the authors excerpted in Chapter 1 who write on the transvestite prostitutes of Venice. And in *The Merchant of Venice*, while Portia is not promiscuous, like the simultaneously threatening and alluring Venetian prostitutes, she is sought by men the world over for her beauty; she is witty and articulate; she circulates freely in public; she disguises herself as a man; and she is accused of sexual impropriety. The play ends on a bawdy note, with Gratiano promising to "keep safe Nerissa's ring," or genitals, perhaps by keeping it occupied himself. How is heterosexual desire represented in the play in terms of power relations and marital harmony?

Early modern writers also offered advice to prospective husbands about choosing and managing their wives. Alexander Niccholes criticizes men who select wives based solely on wealth or beauty, when offspring should be their main concern (see p. 336). He characterizes marriage as a voyage, in which a husband risks "his peace, his freedom, his liberty, his body; yea, and sometimes his soul too" (p. 338). Bassanio clearly views his marriage in terms of an adventure, but how are we to weigh his motives in selecting Portia for his spouse: in describing her, he mentions that she is rich, fair, and virtuous (1.1.160–62). Does the order of these qualities matter to him? In what context does he present his marriage plans in his discussion with Antonio? Does he "give and hazard all he hath" (2.7.16), as the lead casket warns, in wooing Portia? How should we evaluate his marriage, and that of Nerissa and Gratiano, at the play's conclusion? What information does the play provide about Lorenzo's motives in marrying Jessica and about the subsequent success of their union? Niccholes's invocation of magic rings to insure a woman's fidelity resonates with the association of rings and adultery at the play's conclusion. He argues that a husband who exercises his authority and refrains from indulging his wife reduces his chances of being cuckolded. However, the husband needs to attend to his own behavior as well in this endeavor: he needs to set a good example for the wife, prevent his desire for gain from compromising her position, insure that he provides her with a sufficient income, and love her. While neither Portia nor Nerissa is unfaithful to her husband, how are we supposed to understand the threat of their infidelity?

The centrality of finance in the play is evident in its concern with the issue of marital property. Thomas Smith summarizes the financial and legal alterations that marriage worked on women (p. 339). The wife comes under the power of her husband upon marriage; her movable property held before the union becomes his, as does anything she acquires afterward. She loses her property rights, her surname, and with them certain rights in the common law to litigate independently. Smith prescribes a strict separation of the wife's private sphere from the husband's public one, but he argues that the Englishwoman's lot compares favorably to that of her European sisters. Although English widows were entitled to only one third of their husband's chattels or goods — and sometimes received even less — they were frequently named as executrixes of the estate over which they exercised a great deal of control. Smith provides one more intriguing comment on the property rights of wives, which speaks specifically to Portia's circumstances: "If there be any private pacts, covenants, and contracts made before the marriage betwixt the husband and the wife, . . . those have force and [must] be kept according to the firmity and strength in which they are made" (pp. 341–42). When Portia gives herself and all of her property in presenting her ring to Bassanio in act 3, she does so on the condition that he never part from the ring. When he gives it away, even though he is only giving it back to her, what happens to Bassanio's possession of his wife's property? Even before he breaks his promise, do we see Bassanio take control of Portia's property? Note the ways in which she refers to her money and household for the remainder of the play. What do her words reveal about her financial and legal status?

Friendship and Homosociality

Marriage secures property and legal rights for the husband, and most writers on the subject express concerns that such financial considerations might outweigh sentimental and godly motives for marriage. In contrast, the rhetoric of friendship, particularly same-sex friendship, suggests that this relationship depends on virtue and selflessness rather than on personal gain and control. For Sir Thomas Elyot, who draws heavily from Cicero's treatise on the subject, friends act without greed or foolhardiness and are characterized by liberality, constancy, and selflessness; he goes so far as to say that this bond even secures the social order (see p. 342). How does this model of friendship help us understand Antonio's relationship with Bassanio? Are they equally good friends to each other? Are both selfless in their actions toward the other? Friendship also establishes a marriage of minds and goods through unity of person; Elyot represents the friends Titus and Gisippus as so totally bound up with each other "that they seemed to be one in form and

FIGURE 17 *Two Young Friends Hunting, from Primer of Salisbury (1542). Here is depicted the camaraderie of male friendship.*

FIGURE 18 *Man and Woman Wooing, from Primer of Salisbury (1542). The other woman is an attendant.*

FIGURE 20 *Wedding Ceremony, from Primer of Salisbury (1542).*

FIGURE 19 *Men Riding with Woman, from Primer of Salisbury (1542). The men seem to be looking at each other, ignoring the woman placed between them. Above, the sign of Gemini is depicted as two figures, possibly men, embracing.*

personage, . . . [and] nature wrought in their hearts such a mutual affection, that their wills and appetites daily more and more . . . confederated themselves" (p. 344). Their friendship discourages Gisippus from even contemplating marriage: "But the young man, having his heart already wedded to his friend Titus and his mind fixed to the study of philosophy, fearing that marriage should be the occasion to sever him both from the one and the other, refused of long time to be persuaded" (p. 345). His fear that marriage will impede the friendship implies that the two draw on the same emotional source. Bassanio's friendship with Antonio appears to facilitate his marriage to Portia, but are these relationships in concord or conflict with each other? If Portia and Antonio are in some sense competing for Bassanio, what do they want from him? Who wins? The tension between marriage and friendship that appears in a number of comedies in the period reflects contemporary social conditions: "Young men of a certain age in Renaissance England had, then, to reconcile between two conflicting demands: the emotional intensity of male bonds as they were fostered by Renaissance patriarchy and the necessity of marrying to acquire full status within that patriarchy" (Smith 65). Plutarch gives voice to this conflict in his debate on love; in translating it for an early modern English audience, Philemon Holland vehemently condemns homoerotic desire and expresses anxiety about its favorable presentation in Plutarch's story (p. 347). In so doing he nevertheless attributes a powerful attraction to homoerotic love. Although marriage ultimately wins out in this story, a strong case is made for the superiority of homoerotic relations. Protogenes takes the argument a step further, characterizing heterosexual relations as lust and arguing that homoerotic relations constitute a superior and more virtuous form of sexuality and true love. While early modern discourse privileged male friendship and castigated sodomy, many important social, political, and economic relations were homosocial. Friendship implied a physical as well as emotional intimacy, which included the common practices of sharing beds and embraces. Such actions do not necessarily indicate the existence of a sexual relationship, but they do not preclude it either; as Alan Bray argues, "[the] shadow [of sodomy] was never far from the flower-strewn world of Elizabethan friendship and it could never wholly be distinguished from it" (Bray 4–5, 8). What evidence does the play offer about the nature of the relationship between Antonio and Bassanio? What kind of statements do you think the play makes about same-sex relationships and heterosexual ones? How do ideas of love and gender relate in the play?

➤ JUAN LUIS VIVES

From The Instruction of a Christian Woman *1529*

Translated by Richard Hyde

Juan Luis Vives (1492–1540) was born in Valencia and was educated there and in Paris. In 1520 he began teaching at the University of Louvain; shortly afterward, he agreed to a request from Erasmus (who was preparing a new edition of the works of St. Augustine) to write a commentary on Augustine's *The City of God.* Vives lived much of his life in Bruges (a city in what is now Belgium), where he met the visiting English rulers, Henry VIII and Katherine of Aragon. Vives had already received a pension from his countrywoman the queen, and he dedicated *The City of God* to the king. In 1523 he journeyed to England to seek out royal patronage; he resided and apparently lectured at Oxford. He was inevitably drawn into the controversy of Henry's divorce from Katherine, and managed to anger both the king and the queen, siding with the queen, but refusing, after an imprisonment, to serve as one of her defenders. He spent the rest of his life writing, authoring a number of works of theology, grammar, philosophy, law, and history. The *Instruction of a Christian Woman,* dedicated to Katherine of Aragon, articulates a conservative and restrictive view of female behavior, and probably reflects Continental and Mediterranean attitudes more than English attitudes. Nevertheless, it was an extremely popular book and saw four subsequent editions in the fifty years after the first publication of the English translation. The excerpt that follows presents Vives's somewhat progressive advocacy of female education, though he argues against their freedom of speech and movement. He presents a skeptical view of romantic love and emphasizes the necessity of the parents' involvement in selecting a spouse. Perhaps most radical is Vives's formulation of wifely chastity — going so far as to argue that her husband's power over her and her body extends even to controlling her sexual continence.

OF THE LEARNING OF MAIDS

Of maids, some be but little meet for learning: likewise as some men be unapt, again some to be even born unto it, or at least not unfit for it. Therefore they that be dull are not to be discouraged, and those that be apt, should be heart[en]ed and encouraged. I perceive that learned women be suspected of many: as who sayeth, the subtlety of learning should be nourishment for the maliciousness of their nature. Verily, I do not allow in a subtle and crafty

Juan Luis Vives, *A Very Fruitful and Pleasant Book Called the Instruction of a Christian Woman* (1523), trans. Richard Hyde (London, 1529), C1r–C2v, C5v–C7r, I2r–I3v, K8v, M1v–M3v, O1v–O3v, O7r, P4r–P8v.

woman such learning as should teach her deceit and teach her no good manners and virtues. Notwithstanding, the precepts of living and the examples of those that have lived well and had knowledge together of holiness be the keepers of chastity and pureness, and the copies of virtues, and pricks to prick and to move folks to continue in them. . . . But you shall not lightly[1] find an ill woman, except it be such a one, as either knoweth not, or at the least way considereth not what chastity and honesty is worth. . . . nor pondreth[2] what bodily pleasure is, how vain and foolish a thing, which is not worth the turning of a hand, in respect that she should cast away that which is the goodliest treasure that a woman can have. And she that hath learned in books to cast this[3] and such other things, and hath furnished and fenced her mind with holy counsels, shall never find to do any villainy. For if she can find in her heart to do naughtily, having so many precepts of virtue to keep her, what should we suppose she[4] should do, having no knowledge of goodness at all? And truly, if we would call the old world to remembrance, and rehearse their time, we shall find no learned woman that ever was ill.

But here, peradventure, a man would ask, what learning a woman should be set unto, and what shall she study? I have told you, the study of wisdom, which doth instruct their manners and inform their living and teacheth them the way of good and holy life. As for eloquence, I have no great care, nor a woman needeth it not, but she needeth goodness and wisdom. Nor it is no shame for a woman to hold her peace, but it is a shame for her and abominable to lack discretion and to live ill. . . . When she shall be taught to read, let those books be taken in hand that may teach good manners. And when she shall learn to write, let not her example be void verses nor wanton or trifling songs, but some sad sentences prudent and chaste, taken out of the Scripture, or the sayings of philosophers, which by often writing she may fasten better in her memory. And in learning, as I point[5] none end to the man, no more I do to the woman: saving it is meet that the man have knowledge of many and diverse things that may both profit himself and the commonwealth, both with the use and increasing of learning. But I would the woman should be altogether in that part of philosophy that taketh upon him[6] to inform and teach, and amend the conditions.

Finally, let her learn for herself alone and her young children or her sisters in our Lord. For it neither becometh a woman to rule a school, nor to live amongst men, or speak abroad, and shake off her demureness and honesty, either all together, or else a great part; which if she be good, it were better be at home within and unknown to other folks, and in company to hold

---

[1] **lightly:** easily.   [2] **pondreth:** ponders.   [3] **this:** bodily pleasure.   [4] **she:** text reads "we" here.
[5] **point:** appoint.   [6] **him:** it.

her tongue demurely, and let few see her, and none at all hear her. The apostle Paul, the vessel of election, informing and teaching the Church of the Corinths with holy precepts, sayeth: "Let your women hold their tongues in congregations. For they be not allowed to speak but to be subject as the law biddeth. If they would learn anything, let them ask their husbands at home."[7] And to his disciple Timothy, he writeth on this wise: "Let a woman learn in silence with all subjection."[8] But I give no license to a woman to be a teacher, nor to have authority of the man, but to be in silence. For Adam was the first made, and after, Eve; and Adam was not betrayed; the woman was betrayed into the breach of the commandment.[9] Therefore, because a woman is a frail thing and of weak discretion, and that may lightly be deceived, which thing our first mother Eve showeth, whom the Devil caught with a light argument; therefore a woman should not teach, lest when she hath taken a false opinion and belief of any thing, she spread it into the hearers by the authority of mastership, and lightly bring other[10] into the same error, for the learners commonly do after the teacher with good will.

How the Maid Shall Behave Herself Forth Abroad

Forth she must needs go sometimes, but I would it should be as seldom as may be, for many causes. Principally because as oft as a maid goeth forth among people, so often she cometh in judgment and extreme peril of her beauty, honesty, demureness, with shamefastness, and virtue. For nothing is more tender than is the fame and estimation[11] of women, nor nothing more in danger of wrong; insomuch that it hath been said and not without cause to hang by a cobweb, because those things that I have rehearsed be required perfect in a woman, and folk's judgments be dangerous to please and suspicious. . . .

But afore she go forth at door, let her prepare her mind and stomach none otherwise than if she went to fight. Let her remember what she shall hear, what she shall see, and what herself shall say. Let her consider with herself that something shall chance on every side that shall move her chastity and good mind. Against these darts of the devil flying on every side, let her take the buckler of stomach[12] defended with good examples and precepts, and a firm purpose of chastity, and a mind ever bent toward Christ.

---

[7] 1 Corinthians 14:34–35. [8] 1 Timothy 2:10–14. [9] Genesis 3:1–6; the snake "betrays" Eve into breaking the commandment that forbids eating the fruit of the tree of knowledge of good and evil. [10] **other**: others. [11] **fame and estimation**: reputation. [12] **buckler of stomach**: a shield of valor.

OF LOVING

Love is bred by reason of company and communication with men; for among pleasures, feasts, laughing, dancing, and volupties[13] is the kingdom of Venus and Cupid. And with these things folk's minds be enticed and snared, and specially the women's, on whom pleasure hath sorest dominion. O miserable young woman, careful[14] mayest thou be if thou depart out of that company entangled already; how much better had it been for thee to have bidden at home and rather to have broken a leg of thy body than a leg of thy mind? Howbeit, yet I will go about to find a remedy to save thee from taking if thee be untaken;[15] and if thou be taken, that thou mayest [escape] out again. . . . Love of the beauty is a forgetting of reason and the next thing unto frenzy, a foul vice, and an unmannerly for an whole mind. It troubleth all the wits, it breaketh and abateth high and noble stomachs, and draweth them down from the study and thinking of high and excellent things unto low and vile, and causeth them to be full of groaning and complaining, to be angry, hasty, foolhardy, strait in ruling, full of vile and servile flattering, unmeet[16] for everything, and at the last unmeet for love itself. . . . Give none ear unto the lover, no more than thou wouldest do unto an enchanter or a sorcerer; for he cometh pleasantly and flattering, first praising the maiden showing her how he is taken with the love of her beauty, and that he must be dead for her love. For these lovers know well enough the vainglorious minds of many which have a great delight in their own praises wherewith they be caught like as the birder beguileth the birds.[17] He calleth thee fair, proper, witty, well-spoken, and of gentle blood, whereof peradventure thou art nothing at all, and thou, like a fool, art glad to hear those lies and weenest[18] that thou dost seem so indeed when thou art never a whit so. . . .

He sayeth he shall die for thee, yea, and that he dieth even straightaway. Believest thou that? A fool; let him show thee how many have died for love among so many thousands as have been lovers. Love doth pain sometimes, but it never slayeth. Or though he did die for thee, yet it were better for thee to let him perish than be perished thyself, and that one should perish rather than twain.

HOW THE MAID SHALL SEEK A HUSBAND

The wise poet Virgil signifyeth that it becometh not a maid to talk where her father and mother be in communication about her marriage, but to leave all that care and charge wholly unto them which love her as well as her self

---

[13] **volupties:** delights. [14] **careful:** full of grief. [15] **untaken:** free from love. [16] **unmeet:** unfit.
[17] **the birder beguileth the birds:** the bird-catcher traps birds. [18] **weenest:** thinks.

doth. And let her think that her father and mother will provide no less diligently for her than she would for herself, but much better, by the reason they have more experience and wisdom. Moreover, it is not comely for a maid to desire marriage, and much less to show herself to long therefore. . . . Therefore, when the father and the mother be busy about their daughter's marriage, let her help the matter forward with good prayer and desire of Christ with pure affection that she may have such a husband which shall not let nor hinder her from virtuous living, but rather provoke, exhort, and help her unto it. . . .

It is a great charge for a man to seek a husband for his daughter, neither it ought not to be gone about negligently. It is a knot that cannot be lightly loosed; only death undoeth it. Wherefore the fathers and mothers procure unto their daughters either perpetual felicity if they marry them to good men or perpetual misery, marrying them unto ill men. Here is much to be studied and great deliberation to be taken with good advisement and counsel afore a man determine ought. For there is much weariness in marriage and many pains must be suffered. There is nothing but one[19] that shall cause marriage to be easy unto a woman, that is, if she chance on a good and wise husband. O foolish friends, and maids also, that set more by them that be fair, or rich, or of noble birth than them that be good, and cast yourselves into perpetual care. For if thou be married to a fair one, he will be proud of his person; and if thou marry to a rich one, his substance maketh him stately; and if thou be married to one of great birth, his kindred exalteth his stomach[20] . . . and in very deed it were better to be married unto an image or a picture or unto a painted table than to be married to a vicious or a foolish or a brainless man.

But they that would keep the nature of things whole and pure, neither corrupt them with wrong understanding, should reckon that wedlock is a band and coupling of love, benevolence, friendship, and charity; comprehending within it all names of goodness, sweetness, and amity. Therefore let the maid neither catch and deceive by subtlety him that should be her inseparable fellow, nor pull and draw by plain violence, but take and be taken by honest, simple, plain, and good manner, that neither of them complain with both their harms, or say they were deceived or compelled.

Of Two [of] the Greatest Points in a Married Woman

Among all other virtues of a married woman, two there ought to be most special and greatest, the which only if she have them may cause marriage to

[19] **one:** one thing.   [20] **his kindred exalteth his stomach:** his family increases his pride.

be sure, stable, durable, easy, light, sweet, and happy; and again, if one be lacked, it shall be unsure, painful, unpleasant, and intolerable, yea, and full of misery and wretchedness. These two virtues that I mean be chastity and great love toward her husband. The first she must bring with her forth of her father's house. The second she must take after she is once entered in at her husband's door; and both father and mother, kinfolks, and all her friends left, she shall reckon to find all these in only her husband. And in both these virtues he shall represent the image of the holy church, which is both most chaste and most faithfully doth keep truth and promise unto her spouse, Christ. . . .

A married woman ought to be of greater chastity than an unmarried. For if that thou then pollute and defile thy chastity, as God forbid thou shouldest, hark, I pray thee, how many thou shalt offend and displease at once with one wicked deed. How many revengers thou shalt provoke against thee. They be so many and so heinous that among some a man can make no difference, but I shall gather them without any order and set them before their eyes. First thou offendest two, which ought to be unto thee both most in price and most dear and best, that is to say, almighty God, by whose means ye were coupled together and by whose power thou hast made oath to keep the pureness of the body. And next unto God, thou offendest thine husband, unto whom only thou hast given thyself, in whom thou breakest all loves and charities if thou once be defiled. For thou art unto him as Eve was unto Adam, that is to say, his daughter, his sister, his companion, and his wife, and as I might say another himself.

Wherefore, thou desperate woman that hast abused thyself so, thou farest in like manner as though thou haddest strangled, destroyed or murdered thyself. Thou hast broken the greatest band that can be in the world. Thou has broken, thou false woman, the most holy band of temporal law, that is to say, thy faith and thy truth, which once given, one enemy in the field will keep to another though he should stand in danger of death, and thou like a false wretch doth not keep it to thine husband, which ought to be more dear unto thee by right than thyself. Thou defilest the most pure church, which helped to couple thee; thou breakest worldly company; thou breakest the laws; thou offendest thy country; thou beatest thy father with a bitter scourge; thou beatest thy sorrowful mother, thy sisters, thy brethren, thy kinsfolk, alliances, and all thy friends; thou givest unto the company once an example of mischief and castest an everlasting blot and shame upon thy kin; thou, like a cruel mother, castest they children into such a necessity that they can never hear speak of their mother without shame nor of their father without doubting. What greater offense can they do; or what great wickedness can they infect themselves withal that destroy their country and perish all laws and justice, and murder their fathers and mothers, and finally defile and mar all things both spiritual and temporal? What good man or God, thinkest thou, can favor thee that dost so? All thy country folks,[21] all rights and laws, thy country itself, thy parents, all thy kinsfolk and thine husband himself shall damn and punish thee. Almighty God will avenge most rigorously his majesty so displeased and offended of thee.

And know thou this, woman, that the chastity and honesty which thou hast is not thine, but committed and betaken unto thy keeping by thine husband. Wherefore thou dost the more wrong to give away that thing which is another body's, without the owner's license. And therefore the married woman of Lacedemon,[22] when a young man desired of her that unhonest thing,[23] answered him, I would grant thee thine asking, young man, if it were mine own to give that thou askest, but that thing which thou wouldest have while I was unmarried was my father's and now is my husband's. She made him a merry and wise answer. But St. Paul speaketh full wisely for the [ad]monition of good women where he teacheth the church of God, saying: "A woman hath no power of her own body, but her husband."[24] Which saying ought so much to keep a woman, except[25] she be too ungracious, from all filthy acts, that St. Augustine doth not allow perpetual chastity[26] in a married woman, without[27] her husband be content with the same. . . . For a woman hath no power of her own body, no not unto the goodness of continence.[28]

How She Shall Behave Herself unto Her Husband

[If] it be true that men do say that friendship maketh one heart of two, much more truly and effectually ought wedlock to do the same, which far passeth all manner both friendship and kindred. Therefore, it is not said that wedlock doth make one man, or one mind, or one body of two, but clearly one person. Wherefore the words that the man spake of the woman, saying for her sake a man should leave both father and mother and abide with his wife, the same words the woman ought both to say and think with more reason. For although there be one made of two, yet the woman is as daughter unto her husband, and of nature more weaker. Wherefore she needeth his

---

[21] **country folks:** fellow country men and women.   [22] **Lacedemon:** from Sparta, a military city-state in ancient Greece.   [23] **unhonest thing:** adultery.   [24] 1 Corinthians 7:4; St. Paul says here that husbands have control over their wives' bodies and wives have control over their husbands'. Vives omits the second part of this statement.   [25] **except:** unless.   [26] **perpetual chastity:** celibacy.   [27] **without:** unless.   [28] **For a woman . . . continence:** because her body belongs to her husband, a wife cannot independently choose to remain celibate.

aid and succor. Wherefore if she be destitute of her husband, deserted and left alone, she may soon take hurt and wrong. Therefore if she be with her husband, where he is, there hath she both her country, her house, her father, her mother, her friends, and all her treasure.

Neither I would that she should love her husband as one loveth his friend or his brother, that is to say, I will that she shall give him great worship, reverence, great obedience, and service also; which thing not only the example of the old world teacheth us, but also all laws, both spiritual and temporal, and Nature herself cryeth and commandeth that the woman shall be subject and obedient to the man. And in all kinds of beasts the females obey the males, and wait upon them, and fawn upon them, and suffer themselves to be corrected of them. Which thing Nature showeth must be and is convenient to be done. Which, as Aristotle in his book of beasts showeth, hath given less strength and power unto the females of all kinds of beasts than to the males and more soft flesh and tender hair. Moreover, these parts which nature hath given for weapons of defense unto beasts, as teeth, horns, spurs, and such other, the most part of females lack, which their males have, as harts[29] and boars. And if any females have any of these, yet be they more stronger in the males, as horns of bulls be more stronger than of kine.[30] In all the which things Nature showeth that the male's duty is to succor and defend, and the female's to follow and to wait upon the male and to creep under his aid and obey him, that she may live the better.

But let us leave the examples of beasts which make us ashamed of ourselves without[31] we pass them in virtue, and let us ascend up unto man's reason. . . . For in wedlock the man resembleth the reason and the woman the body. Now reason ought to rule and the body to obey if a man will live. Also St. Paul sayeth the head of the woman is the man.[32] Here now I enter into the divine commandments, which in stomachs of reasonable people ought of reason to bear more rule and value than laws, more than all man's reasons, and more than the voice of Nature herself. God the maker of the whole world in the beginning, when the world was yet but rude[33] and new, giving laws unto mankind, he gave this charge unto the woman. Thou shalt be under thine husband's rule, and he shall have dominion over thee.[34] . . . But foolish women do not see how sore they dishonest themselves that take the sovereignty of their husbands,[35] of whom all their honor must come. And so in seeking for honor, they lose it. For if the husband lack honor, the wife must needs go without it. Neither kindred, riches, nor wealth can avail her. For

---

[29] harts: male deer.   [30] kine: cows.   [31] without: unless.   [32] head of the woman . . . man: 1 Corinthians 11:3.   [33] rude: uncultivated.   [34] Thou shalt . . . dominion over thee: Genesis 3:16.
[35] how sore . . . husbands: how seriously they dishonor themselves by ruling over their husbands.

who will give any honor to that man whom he seeth mastered by a woman. And again, if thy husband be honorable, be thou never so low of birth, never so poor, never so uncomely of face, yet canst thou not lack honor. . . .

Nor let [women] not love goodly men for their beauty, nor rich men for their money, nor men of great authority for their honor; for if they do so, then shall they hate the sickly, the poor, and those that bear no rule. If thou have a learned husband, learn good holy lessons of[36] him; if he be virtuous, do after him. . . . [I]f she chance upon an infortunate husband, neither hate nor despise him therefore, but rather contrary. She ought, if he be poor, to comfort him, and advertise[37] him to call into remembrance that virtue is the chief riches. . . . But beware thou fall not into such a wicked mind to will him for lucre of money[38] to occupy any unhonest crafts or to do any unhappy deeds that thou mayest live more delicately, or more wealthily, or go more gaily and gorgeously arrayed, or dwell in more goodly housing; and at few words, compel not him to use any filthy occupation or drudgery for thy welfare, nor to sweat and to toil that thou mayst lie at ease. For it were better for thee to eat brown bread and drink clay and mirey[39] water than cause thy husband to fall unto any slubbery[40] work or stinking occupation and exceeding labor for to escape thy scolding and chiding at home. For the husband is his own ruler and his wife's lord, and not her subject; neither the wife ought to crave any more of her husband than she seeth she may obtain with his heart and good will, wherein many women do amiss which with their ungoodly crying and unreasonable calling, craving and bullying upon them, driveth them to seek unlawful means of living and to do ungracious deeds, to bear out with all their[41] gluttony and vain pride. . . . but thou, good daughter that wilt do well, shalt not withdraw thine husband from goodness, but rather exhort him unto virtue though thou shouldest be sure to lose all thy goods.

---

[36] of: from.   [37] advertise: counsel.   [38] lucre of money: financial gain.   [39] mirey: muddy.
[40] slubbery: messy, dirty.   [41] their: the wives.

> **THOMAS BECON**

From The Catechism   c. *1550*

Thomas Becon (1512–1567) was educated at Cambridge; he became a minister in 1538 and was appointed to a small parish. His controversial writings attacking Catholicism landed him in trouble in 1543 when he was forced to recant his

Thomas Becon, *The Catechism* (c. 1550) Parker Society vol. 3, ed. John Ayre (Cambridge: Cambridge University Press, 1844), 343–44, 368, 369, 371–72.

doctrine. He subsequently made a living as a teacher until the accession of the Protestant Edward VI, when he was awarded a number of prestigious posts, including that of chaplain to the Protector Somerset, the *de facto* ruler during the minority of Edward VI. After Edward's death Becon was imprisoned and on his release fled to Strasbourg, where he continued to publish polemics against Catholicism. He returned to England on Elizabeth's accession and was given several ecclesiastical posts, which he filled until his death. His *Catechism,* presented as a dialogue between a father and son, prescribes the proper godly behavior of members of a Protestant family. Although doctrinally opposed to the Catholic Juan Luis Vives (p. 319), Becon advances similar views about the behavior of women. His discussion on the behavior of daughters raises questions about the behaviors of both Portia and Jessica; would an early modern audience expect these characters to be more wary of their lovers and obedient to their fathers? Could conventions of comic drama for the period, which frequently represent and condone children disobeying their parents in selecting a spouse, license their behavior? Alternatively, as the Son in Becon's *Catechism* points out, children must also have a say in the selection of their spouses and parents should refrain from viewing marriage solely in terms of economics. How would this view apply to the lottery devised by Portia's father, or Shylock's mercenary attitude toward his daughter? Does the lottery insure a better husband for Portia than the one Jessica selects on her own? What are the economics of each marriage? Are they both happy marriages?

SON: The third point of a virtuous matron is to look unto her house; to provide that nothing perish, decay, or be lost, through her negligence; to see that whatsoever be brought into the house by the industry, labor, and provision of her husband, be safely kept and warely[1] bestowed; not only to command other to do things, but also to set hand to the business herself; never to be idle, but always to be well occupied; to be an example of all godliness and honesty to her household; to reprove vice sharply in her servants, and to commend virtue; not to meddle with other folks' business abroad, but diligently to look upon her own at home; not to go unto her neighbors' houses, to tattle and prattle after the manner of light housewives; not to be tavern-hunters; not idly and wantonly to gad[2] abroad, seeking new customers;[3] not to resort unto places where common plays, interludes, and pastimes be used; not to accompany herself with any light persons, but only with such as be sober, modest, grave, honest, godly, virtuous, housewifely, thrifty, of good name, well reported, etc.; and

---

[1] **warely**: prudently.   [2] **gad**: wander.   [3] **customers**: familiar associates.

in fine,[4] continually to remain at home in their house diligently and virtuously occupied, except urgent, weighty, and necessary causes compel her to go forth, as to go unto the church, to pray or to hear the word of God, to visit her sick neighbors or to help them, to go to the market to buy things necessary for her household, etc. . . .

SON: The fourth point of an honest and godly matron is patiently and quietly to bear the incommodities[5] of her husband; to dissemble, cloak, hide, and cover the faults and vices of her husband; not to upbraid nor cast them in his teeth;[6] not to exasperate or sharpen her husband's mind through her churlishness, but rather with her soft, gentle, and sober behavior to quiet him, to pacify his anger, to mitigate his fury, and, as they use to say, to make him of a lion a lamb. So sayeth St. Peter: "Ye wives, be in subjection to your husbands, that even they, which obey not the word, may without the word be won of the conversation of the wives, while they behold your chaste conversation coupled with fear" [1 Peter 3:1–2]. . . .

SON: If all women address themselves unto the practice of this godly act . . . and use the like gentleness and sober behavior toward their husbands, more love, peace, amity, quietness, and concord should be found among married folk than is at this present day. But some women are more like the furies of hell. . . . For their whole delight and pleasure is to scold, to brawl, to chide, and to be out of quiet with their husbands; so far is it off[7] that with their godly conversation and gentle behavior they go about to maintain amity and concord in their houses. And when they are reproved for their misdemeanor toward their husbands, they shame not to answer: "A woman hath none other weapon but her tongue, which she must needs put in practice. They have been made dolts and fools long enough: it is now high time to take hart of grease[8] unto them. There is no worm so vile, but if it be trodden upon it will turn again, etc." . . .

SON: . . . Nothing doth so greatly hinder the good name and fame of maids, as keeping company with naughty packs, and persons of a dissolute and wanton life (for every man proves such as he is with whom he is conversant); and contrariwise, nothing doth so much commend, advance, and set forth their good name and fame, as resorting unto such as are well reported, and of an honest disposition;

[4] **in fine:** in conclusion.   [5] **incommodities:** shortcomings.   [6] **cast them in his teeth:** throw them in his face.   [7] **so far is it off:** rather than.   [8] **hart of grease:** a fat, male deer. Ayre's gloss suggests that this fatness might indicate "ease of spirit" (note 3, 345). I take the sense to mean that wives, albeit wrongly in the author's eyes, feel that they should gain relief from the criticism of their husbands by answering back.

therefore shall it be requisite that all godly maids to refrain themselves from keeping company with light, vain, and wanton persons, whose delight is in fleshly and filthy pastimes, as singing, dancing, leaping, skipping, playing, kissing, whoring, etc. All such must they avoid, if they tender[9] their good name; which once lost, they are no more of estimation, but contemned and despised of all good and godly persons. . . .

SON: . . . this also must honest maids provide, that they be not full of tongue, and of much babbling, nor use many words, but as few as they may, yea, and those wisely and discreetly, soberly and modestly spoken, ever remembering this common proverb: "A maid should be seen, and not heard." Except the gravity of some matter do require that she should speak, or else an answer is to be made to such things as are demanded of her, let her keep silence. For there is nothing that doth so much commend, advance, set forth, adorn, deck, trim, and garnish a maid, as silence. And this noble virtue may the virgins learn of that most holy, pure, and glorious virgin Mary, which, when she either heard or saw any worthy and notable thing, blabbed it not out straightways to her gossips,[10] as the manner of women is at this present day; but, being silent, she "kept all those sayings" secret, and "pondered them in her heart," saith blessed Luke [Luke 2:19]. . . .

SON: Finally, when the time cometh that they feel themselves apt unto marriage, and are desirous to contract matrimony, to the end that they may avoid all uncleanness (and bring forth fruit according to God's ordinance, as their parents have done before them) they must diligently take heed, that they presume not to take in hand so grave, weighty, and earnest matter, nor entangle themselves with the love of any person, before they have made their parents, tutors, friends, or such as have the governance of them, privy of their intent; yea, and also require their both counsel and consent in the matter, and by no means to establish or appoint anything in this behalf without the determination of their rulers. For this is part of the honor that the children owe to their parents and tutors by the commandment of God, even to be bestowed in marriage as it pleaseth the godly, prudent, and honest parents or tutors to appoint; with this persuasion, that they,[11] for their age, wisdom, and experience, yea, and also for the tender love, singular benevolence, and hearty goodwill that they

---

[9] tender: value. [10] gossips: friends. [11] they: parents.

bear toward them, both know and will better provide for them than they be able to provide for themselves. . . .

SON: Let all godly maids take heed therefore that they snarl not themselves with the love of any other, nor marry with any person before they have the good will of their parents. Let them receive no tokens of any man, nor be too much familiar with any person in the way of marriage; but if any be suitors unto them for to marry with them, and they could well be contented for their godly qualities and honesty of life to take them unto their husbands in the fear of God, let them first of all open the matter to their parents or tutors, being contented to be ruled by them, submitting their own judgment to the judgment or wisdom of their superiors, and praying also unto God that he may rule and govern the hearts of their parents on such sort, that they may appoint that thing which he knoweth to make most unto his glory, and unto the salvation of their souls. If their parents do agree unto their desire, let them thank God for it, and know that God is the author of this their marriage, and that, as he hath brought them together, so will he bless both them and their marriage, and give good success unto them in all their godly and honest attempts, so that they shall lack no good thing. But if their parents do not consent, let the children be content, and think that it is for the best, and that they[12] see more in the matter than they[13] themselves can perceive, being far inferior to them in age, wisdom, discretion, knowledge, reason, experience, etc. And let them think that this also cometh of God, which ruleth the hearts of their parents, and turneth them which way it pleaseth him, and all for their commodity and profit, for their health and wealth. Yea, let them think that God is their father, and they his children; again, that he is their creator, and they his creatures; and therefore that he will not neglect them, but so provide for them as a father for his children, and never forsake them so long as they live in the obedience of his holy word. . . .

FATHER: Ought not the consent of the children also to be considered in this behalf no less than the authority of the parents?

SON: God forbid else! For we read, that when Rebecca was promised that she should go with Abraham's servant to be married unto Isaac, they said: "We will call the damsel, and inquire at her mouth. And they called forth Rebecca, and said unto her, Wilt thou go with this man? And she answered, I will go," etc. [Genesis 24:57–58]. Here see

---

[12] they: the parents.   [13] they: the children.

we, that though the authority of the parents be great over their children, yet in the matter of marriage the consent of the children may
not be neglected. For parents must so use their authority, that they do
not abuse it. They abuse it, when it turns unto the hindrance, incommodity,[14] and destruction of their children. The parents therefore
must so place their children in marriage, as may profit, and not hinder them, yea, and that with the good will and consent of the children, to whom the matter chiefly pertain; that the authority of the
parents and the consent of the children may go together, and make
perfect an holy and blessed marriage.

FATHER: This is commendable. But some parents greatly abuse their
authority, while they sell their children to others for to be married for
worldly gain and lucre, even as the grazier selleth his oxen to the
butcher to be slain, having no respect to the person, whether he be
godly or ungodly, honest or unhonest, wise or foolish, etc.

---

[14] incommodity: disadvantage.

→ CORNELIUS AGRIPPA

From Of the Nobility and Excellency of Womankind
1542
Translated by David Clapham

Cornelius Agrippa (1486–1535) was a German humanist who studied law, medicine, and theology at the University of Cologne. He practiced multiple professions in his lifetime, serving variously as a soldier, lawyer, court historian, and
author. He had a lifelong interest in the occult, which he studied in the form of
Kabbalah (Jewish mystical teachings), and which he wrote about in his *De
occulta philosophia*, a study of magic and esoteric knowledge. Agrippa wrote the
De nobilitate et praecellentia foeminei sexus declamatio (Of the Nobility and Excellency of Womankind) in 1509 in the hopes of gaining the patronage of Margaret
of Austria, Regent of the Netherlands and the Franche-Comté (a section of
France bordering on Switzerland). However, he was forced to leave the area
after he was denounced as a "Judaizing heretic" and the work was not published
until 1529 (Bietenholz 17). The following excerpt demonstrates that it was
possible in the period to criticize the cultural assumption that women were

---

Cornelius Agrippa, *Of the Nobility and Excellence of Womankind* (1529), trans. David Clapham
(1542), A2r–A3v, C3v–C4, F8r–G2v.

subordinate to men and that learned reasoning could support the claim of female superiority just as easily as it could support the male claim. Agrippa makes a very modern, albeit possibly ironic, argument for the culturally constructed inferiority of women, who "subdued as it were by force of arms, are constrained to give place to men and to obey their subduers, not by no natural, no divine necessity or reason, but by custom, education, fortune, and a certain tyrannical occasion" (p. 334). While Juan Luis Vives (p. 319) and Thomas Becon (p. 327) suggest that a restricted role for women's speech and behavior constitutes, or should constitute, the norm, Agrippa argues for a much more active and powerful position for women. How might Portia's autonomous and authoritative actions in *The Merchant of Venice* be viewed from this perspective?

Almighty God, the maker and nourisher of all things, the father and goodness of both male and female, of his great bountifulness has created mankind like unto himself, he made them man and woman. The diversity of which two kinds stand only in the sundry situation of the bodily parts, in which the use of generation requireth a necessary difference. He has given but one similitude and likeness of the soul to both male and female, between whose souls there is no manner difference of kind. The woman hath that same mind that a man hath, that same reason and speech, she goeth to the same end of blissfulness, where shall be no exception of kind. . . . And thus between man and woman by substance of the soul, one hath no higher preeminence of nobility above the other, but both of them naturally have equal liberty of dignity and worthiness. But all other things the which be in man besides the divine substance of the soul, in those things the excellency and noble womanhood in a manner infinitely doth excel the rude gross kind of men. . . .

Now let us speak of speech and language, which is the gift of God, and by which one thing we pass and are better than all other brute beasts: Trismegistus Mercurius[1] judgeth it to be of as great a price, as much worth, and as good a thing, as immortality. And Hesiod[2] nameth it the chiefest treasure of mankind. And is not a woman better spoken, more eloquent, more copious and plentiful of words than a man? Do not all we, that be men, learn first to speak of[3] our mothers, or of our nurses? Truly nature herself, the former of things, sagely providing for mankind, gave this gift to womankind, that scarce in any place ye shall find a dumb[4] woman. Is it not right fair and commendable, that women should excel men in that thing, in which men chiefly pass all other beasts? . . .

---

[1] **Mercurius**: Trismegistus Mercurius or Hermes is the name ascribed to the author of various Neoplatonic texts dating from the third century, c.e. (*Oxford Companion to English Literature*, s.v. Hermes Trismegistus).   [2] **Hesiod**: Greek poet.   [3] **of**: from.   [4] **dumb**: silent, mute.

Thou will say, that[5] is now forbidden by laws, abolished by custom, extinct by education. For anon as[6] a woman is born even from her infancy, she is kept at home in idleness, and as though she were unmeet[7] for any higher business, she is permitted to know no farther than her needle and thread. And then when she cometh to age, able to be married, she is delivered to the rule and governance of a jealous husband, or else she is perpetually shut up in a close nunnery. And all offices belonging to the commonweal be forbidden them by the laws. Nor it is not permitted to a woman, though she be very wise and prudent, to plead a cause before a judge. Furthermore, they be repelled in jurisdiction, in arbiterment, in adoption, in intercession, in procurement, or to be guardians or tutors, in causes testamentary and criminal.[8] Also they be repelled from preaching of God's word, against express and plain scripture, in which the holy ghost promised unto them by Joel the prophet, saying: And your daughters shall prophesy and preach [Joel 8]. . . . But the unworthy dealing of the later law makers is so great that, breaking God's commandment to establish their own traditions, they have pronounced openly that women, otherwise in excellency of nature, dignity, and honor most noble, be in condition more vile than all men. And thus by these laws the women, being subdued as it were by force of arms, are constrained to give place to men and to obey their subduers, not by no natural, no divine necessity or reason, but by custom, education, fortune, and a certain tyrannical occasion. . . .

For this is the order in the church, that men in ministration shall be preferred before women: like as the Jews in promission[9] are before the Greeks: yet nevertheless God is no exceptor of persons [Romans 2 and Acts 10]. For in Christ neither male nor female is of value, but a new creature. And many things were permitted unto men for the hardness and cruelty of their hearts against women (as in times past divorces were granted unto the Jews) which for all that nothing hurteth the dignity of women. But when men commit offence and err, the women have power of judgment over them, to the great shame and rebuke of men. And that Queen Saba shall judge the men of Jerusalem. Therefore they, which being justified by faith are become the sons of Abraham, the children, I say, of promission, be subdued to a woman and bound by the commandment of God, saying to Abraham: what soever Sara saith unto thee, follow it [Genesis 21].

[5] that: Agrippa has just completed a catalogue of women's ancient legal rights.   [6] anon as: as soon as.   [7] unmeet: unfit.   [8] Furthermore . . . criminal: referring to women's legal disabilities in the civil law.   [9] promission: promise; Agrippa applies a theological idea to the relative status of men and women. As the Jews had a prior relation to God before the Gentiles, so men might argue their priority over women. However, Agrippa rejects this argument using a number of biblical passages, ultimately concluding that men are bound by God to follow women.

→ PHILIP STUBBES

From The Anatomy of Abuses

1583

Philip Stubbes (1555?–1610?) studied at both Cambridge and Oxford. In the early 1580s, he began his career as an author and Puritan pamphleteer with the publication of several ballads on religious topics. *The Anatomy of Abuses* (1583) presents a Puritan critique of contemporary English mores; it was extremely popular, precipitating a second edition in the same year, and two more editions by 1595. In it, Stubbes condemns the fashion of women's wearing masculine attire; he sees cross-dressing as representing a desire to change genders. If such a transformation were possible, it would threaten the superior position of men in the gender hierarchy, since it would dissolve the distinction between women and men. How could Stubbes's disapproval of female transvestism inform our response to Jessica's shame at dressing as a page? Does she express similar concern about other transgressive elements of her behavior? How does Portia's attitude toward cross-dressing differ from Jessica's? Does her behavior change with her clothes in the way that Stubbes fears?

A Particular Description of the Abuses of Women's Apparel in Ailgna.[1]

Thus having given thee a superficial view, or small taste (but not discovered the hundredth part) of the guises of Ailgna in men's apparel, and of the abuses contained in the same, now will with like celerity of matter impart unto thee, the guise and several abuses of the apparel of women there used also: wherefore give attentive ear. . . .

The women also there have doublets and jerkins[2] as men have here, buttoned up the breast, and made with wings,[3] welts[4] and pinions[5] on the shoulder points, as man's apparel is, for all the world. And though this be a kind of attire appropriate only to man, yet they blush not to wear it, and if they could as well change their sex, and put on the kind[6] of man, as they can wear apparel assigned only to man, I think they would as verily become men indeed as now they degenerate from godly sober women, in wearing this wanton lewd kind of attire, proper only to man.

It is written in the 22nd chapter of Deuteronomy, that what man so ever weareth woman's apparel is accursed, and what woman weareth man's

---

[1] Ailgna: Anglia, or England, reversed.   [2] jerkins: male upper body garments, jackets.   [3] wings: projecting pieces of garment on or near the shoulder.   [4] welts: trimming.   [5] pinions: shoulder embellishments.   [6] kind: gender.

Philip Stubbes, *The Anatomy of Abuses: Containing a Discovery, or Brief Summary of Such Notable Vices and Imperfections as Now Reign in Many Countries of the World; but (Especially) in a Very Famous Island Called Ailgna* (London: Richard Jones, 1583), E7v, F5r.

apparel is accursed also. Now, whether they be within the bands and limits of that curse, let them see to it themselves. Our apparel was given us as a sign distinctive to discern betwixt sex and sex, and therefore, one to wear the apparel of another sex, is to participate with the same, and to adulterate the verity of his own kind.

Wherefore these women may not improperly be called Hermaphrodite, that is, monsters of both kinds, half women, half men. . . .

I never read nor heard of any people, except[7] drunken with Circe's[8] cups, or poisoned with the exorcisms of Medea[9] that famous and renowned sorceress, that ever would wear such kind of attire; as is not only stinking before the face of God, offensive to man, but also painteth out to the whole world, the venereous[10] inclination of their corrupt conversation.

---

[7] **except:** unless.   [8] **Circe:** a sorceress in Homer's *Odyssey* whose potions transform men into swine.   [9] **Medea:** a sorceress of Greek myth associated with poisoning.   [10] **venereous:** lecherous.

→ ALEXANDER NICCHOLES

From A Discourse of Marriage and Wiving   *1615*

Alexander Niccholes' *Discourse of Marriage and Wiving* looks at a marriage from the perspective of the prospective husband. Niccholes presents marriage as a kind of adventure that entails numerous risks, although he suggests these can be mitigated by exercising wisdom and common sense in choosing a spouse. In *The Merchant of Venice,* Bassanio voices this view of marriage, representing his — and others' — wooing of Portia as a heroic enterprise, such as that performed by the mythic Jason (1.1.167–71). However, Niccholes counsels against the popular custom of choosing a wife by her wealth or beauty, the two qualities Bassanio mentions first when describing Portia. Do we see his desire for her as a romantic quest or as an economic venture? Does the conclusion of the play raise questions about the durability of a marriage based on physical attraction? In early modern British culture, a woman's beauty, which makes her sought after when single, poses problems after she is married. Her husband then faces the anxiety of being rendered a cuckold by an adulterous wife. Niccholes advises against reacting to this threat with magic rings, coercion, and jealousy; while he does recommend some manipulative measures, the main thrust of his advice is that the husband should behave himself, set a good example, provide for the economic stability of the household, and, finally, truly love his wife. What judgment does the conclusion of the play render about the various marriages contracted in it?

Alexander Niccholes, *A Discourse of Marriage and Wiving* (London, 1615), 7–10, 33, 35.

CHAPTER 3

Worldly choice what it is, or how, for the most part men choose their wives.

It is a fashion much in use in these times to choose wives as chapmen[1] sell their wares, with *Quantum dabitis*? What is the most you will give? And if their parents, or guardians shall reply their virtues are their portions, and others have they none, let them be as dutiful as Sara,[2] as virtuous as Anna,[3] as obedient as the Virgin Mary, these (to the wise man, every one a rich portion, and more precious than the gold of Ophire[4]) shall be nothing valued, or make up where wealth is wanting; these may be adjuncts or good additions, but money must be the principal, of all that marry, and (that scope is large) there are but few that undergo it for the right end and use, whereby it comes to pass that many attain not to the blessedness therein. Some undergo this curse instead of blessing, merely for lust choosing their wives most unfitly, as adulterers, and such are said to marry by the eye, looking no further than a carnal beauty is distinguished, which consists in the outward shape and lineaments of the body, as in gait, gesture, countenance, behavior, etc. And for such a one so she be fair, and can kiss, she has portions enough for such a pirate: but when this flower withers, as it is of no continuance,[5] for diseases blast it, age devours it, discontent does wither it (only virtue is not foiled by these adversaries) what shall continue love as then to the end; their winter sure shall be full of want, full of discontent, that thus grasshopper-like respected their summer.[6] There are others that marry to join wealth to wealth, and those are said to marry by the fingers' ends. Some others there are that take their wives from the report or good liking of others, and those are said to take their wives upon trust, and such I hope are not seldom deceived in their venture. There are some that marry for continuance of posterity, and those come nearest to the true intent, for the end of marriage is *proles*, issue. It was the primal blessing, "increase and multiply" (Genesis 1:28): God has given and bequeathed many precepts and commandments to mankind, yet of all that ever he delivered, never was there any better observed (for the letter) than this.

---

[1] chapmen: merchants, dealers.   [2] Sara: the wife of Abraham in Genesis.   [3] Anna: could refer either to Hannah in 1 Samuel 1, Anna the prophetess in Luke 2:36–38, or Anne the mother of the Virgin Mary.   [4] Ophire: a location in which fine gold is found, as mentioned in the Hebrew Bible, for example 1 Kings 9:28.   [5] it is of no continuance: it is not eternal.
[6] grasshopper-like . . . summer: refers to the tale of the grasshopper and the ant; while the latter worked hard all summer long, the former was idle, and went hungry when winter came. Here, "summer" is youth and "winter" is old age.

CHAPTER 4

How to choose a good wife from a bad.

The first aim that I would give to him, that would adventure this voyage (for marriage is an adventure, for whosoever marries adventures, he adventures his peace, his freedom, his liberty, his body; yea, and sometimes his soul too) is, that the . . . choice of his wife [be] . . . grounded upon some of these promising likelihoods, videlicet,[7] that she be of a sober and mild aspect, courteous behavior, decent carriage, of a fixed eye, constant look, and unaffected gait, the contrary being oftentimes signs of ill portent and consequence. For as the common saying is, an honest woman dwells at the sign of an honest countenance, and wild looks (for the most part) accompany wild conditions; a rolling eye is not fixed, but would fix upon objects it likes, it looks for, and affected nicety is ever a figure of lascivious petulancy.

Next regard, according as thine estate and condition shall best instruct thee, the education, and quality, of her thou has so elected; her personage not being unrespected, for love looks sometimes as well with the eye of the body, as with the mind, and beauty in some begets affection, and affection augmenteth love, whereas the contrary would decrease, and diminish it, and so bring thee to a loathed bed, which must be utterly taken heed of, for the dangerous consequences that follow. Therefore, let thy wisdom so govern thy affection, that as it seize not up deformity to thine own proper use, for some sinister respect to be shortly after repented of; so likewise (for the mean[8] is everbest) that it level[9] not at so high and absolute endowment and perfection, that every carnal eye shall bethink thee injury, that every goatish disposition shall level to throw open thy enclosures.[10]

CHAPTER II

The best way to keep a woman chaste.

Is not the magician's ring, nor the Italian's lock,[11] nor a continual jealousy, ever watching over her, nor to humor her will in idle fancies, adorn her with new fangles (as the well-appayed[12] folly of the world in this kind can witness). But for him that would not be basely mad with the multitude, would not bespeak[13] folly to crown him, would not set that to sale that he would not have sold . . . [he should] adorn her decently, not dotingly; thriftily, not

---

[7] **videlicet:** namely.   [8] **mean:** the middle.   [9] **level:** aim.   [10] **goatish:** lustful; **throw open thy enclosures:** (metaphorically) to seduce your wife. In short, too beautiful a wife might excite the interest of others.   [11] **Italian's lock:** chastity belt.   [12] **well-appayed:** requited.   [13] **bespeak:** ask.

lasciviously; to love her seriously, not ceremoniously; to walk before her in good example (for otherwise how can thou require that of thy wife that thou are not, wilt not be thyself? . . . Would thou expect thy wife a conqueror when thou thyself lies foiled at the same weapon?)[14] to acquaint her with, and place about her, good and chaste society, to busy and apply her mind and body in some domestic, convenient and profitable exercises, according to her education and calling, for example to the frailty of that whole sex hath a powerful hand,[15] as it shall induce either to good or evil. . . .

Therefore whoever thou art that would not wink at such a shame that, so profit doth succeed, would not regard whether hand brought it in,[16] use a good endeavor, such foresight and wariness as may provide for competency, prevent indigence and want, two great allayers of affection, and a main inciter of impatient bearers to this folly and abuse, and above all seek to plant in her religion, for so she cannot love God but withal she must honor thee; increase her knowledge in good things, and give her certain assurance and testimony of thy love, that she may with hers again the more reciprocally equal thy affection. For true love has no power to think much less act amiss: and these, discreetly put in practice, shall more preserve at all times and temptations, than spies or eyes, jealousy, or any restraint, for these[17] sometimes may be deluded, or overwatched,[18] or prevented by opportunity, but this[19] never.

---

[14] **Would thou . . . weapon:** would you expect your wife to be able to conquer temptation when you yourself cannot?   [15] **example . . . hath a powerful hand:** example exercises a strong influence over feminine frailty.   [16] **whoever thou . . . brought it in:** one who ignores shame, and so long as he makes a profit, doesn't care how it is made. Niccholes is referring to a husband who prostitutes his wife.   [17] **these:** the former.   [18] **overwatched:** fatigued.   [19] **this:** i.e., true love.

SIR THOMAS SMITH

From De Republica Anglorum   *1583*

Sir Thomas Smith (1513–1577) received bachelor's and master's degrees at Cambridge, and subsequently taught natural philosophy and Greek there. He also studied abroad, receiving a doctorate in civil law in Padua, and a doctorate of law at Cambridge in 1542; he became Regius Professor of civil law at Cambridge two years later. His Protestant views helped advance his career during the reign of Edward VI; he served as clerk of the Privy Council and Master of the Court of Requests (which handled the claims of poor litigants), among other posts. After

---

Thomas Smith, *De Republica Anglorum: The Manner of Government or Policy of the Realm of England* (London, 1583), 101–05.

Mary's succession, Smith was protected by a powerful ally and served in Parliament, but became more active in public life after the accession of Elizabeth, who appointed him ambassador to France, and subsequently, member of the Privy Council and Secretary of State. In addition to being a respected statesman, he was also an author and classical scholar. His principal work is *De Republica Anglorum* (On the Republic of England), considered "the most important description of the constitution and government of England written in the Tudor age" (*DNB*, s.v. Thomas Smith). In this excerpt, he considers the legal status and property rights of wives, noting that they in effect lose both to their husbands upon marriage. Smith does note that wives who know how to handle their husbands may better their economic position, and points out that contracts made before marriage with regard to the woman's property must be observed after marriage. In *The Merchant of Venice*, what happens to Portia's property when she agrees to marry Bassanio? Does she impose any conditions or contractual exceptions regarding her wealth before marriage? Who ultimately controls this property at the end of the play?

Of Wives and Marriages

The wives in England be as I said in *potestate maritorum*,[1] not that the husband has *vitae ac necis potestatem*,[2] as the Romans had in the old time of their children, for that is only in the power of the prince, and his laws, as I have said before. But that whatsoever they have before marriage, as soon as marriage is solemnized, is their husbands; I mean of money, plate, jewels, cattle,[3] and generally all movables. For as for land and heritage followeth the succession, and is ordered by the law . . . as I shall say hereafter: and whatsoever they get after marriage they get to[4] their husbands. They neither can give nor sell anything either of their husbands or their own. Hers no movable thing is by the law of England *constanti matrimonio*, but as *peculium servi aut filii familias*;[5] and yet in movables at the death of her husband she can claim nothing, but according as he shall will by his Testament, no more than his son can, all the rest is in the disposition of the executors if he die testate.[6] Yet in London and other great cities they have that law and custom, that when a man dieth, his goods be divided into three parts. One third is employed upon the burial and the bequests which the testator maketh in his testament. Another third part the wife hath as her right, and the third part is the due and right of the children, equally to be divided among them. . . .

---

[1] *potestate maritorum:* in the husband's power.   [2] *vitae ac necis potestatem:* power over life and death.   [3] **cattle:** chattel, or personal property.   [4] **get to:** belongs to.   [5] *constanti matrimonio, . . . aut filii familias:* for the duration of the marriage, but as household property of the servant or minor.   [6] **testate:** having made a will.

But to turn to the matter which we now have in hand, the wife is so much in the power of her husband, that not only her goods by marriage are straight made her husband's, and she loseth all her administration which she had of them: but also where all English men have name and surname, as the Romans had, Marcus Tullius, Caius Pompeius, Caius Julius, whereof the name is given to us at the font,[7] the surname is the name of the gentility and stock the which the son doth take of the father always, as the old Romans did, our daughters so soon as they be married lose the surname of their father and of the family and stock whereof they do come, and take the surname of their husbands, as transplanted from their family into another. . . .

Although the wife be (as I have written before) *in manu and potestate mariti*,[8] by our law yet they be not kept so straight as in mew[9] and with a guard as they be in Italy and Spain, but have almost as much as liberty as in France, and they have for the most part all the charge of the house and household (as it may appear by Aristotle and Plato the wives of Greece had in their time) which is indeed the natural occupation, exercise, office and part of a wife. The husband to meddle with[10] the defense either by law or force, and with all foreign matters which is the natural part and office of the man, as I have written before. And although our law may seem somewhat rigorous toward the wives, yet for the most part they can handle their husbands so well and so doulcely,[11] and specially when their husbands be sick: that where the law giveth them nothing, their husbands at their death of their good will give them all. And few there be that be not made at the death of their husbands either sole or chief executrixes of his last will and testament, and have for the most part the government of the children and the portions; except it be in London, where a peculiar order is taken by the city much after the fashion of the civil law.

All this while I have talked only of movable goods: if the wife be an inheritrix and bring land with her to the marriage, that land descendeth to her eldest son, or is divided among her daughters. Also the manner is that the land which the wife bringeth to the marriage or purchases afterwards, the husband cannot sell nor alienate the same, no not with her consent, nor she herself during the marriage, except that she be sole examined by a judge at the common law. . . .

This which I have written touching marriage and the right in movables and unmovables which come thereby, is to be understood by the common law when no private contract is more particularly made. If there be any

---

[7] **at the font:** at baptism.   [8] *in manu and potestate mariti:* in the hand and power of the husband.   [9] **mew:** a confined space.   [10] **to meddle with:** to concern himself with.   [11] **doulcely:** sweetly.

private pacts, covenants, and contracts made before the marriage betwixt the husband and the wife, by themselves, by their parents, or their friends, those have force and be kept according to the firmity and strength in which they are made. And this is enough of wives and marriage.

→ **SIR THOMAS ELYOT**

From The Book Named the Governor   *1531*

Sir Thomas Elyot (1490?–1546) may have studied at a university; as a well-to-do country gentleman, he spent much of the early part of his career in various clerkships and local government positions. His first book, *The Book Named the Governor* (1531), was dedicated to Henry VIII; its purpose was to educate men in the qualities necessary for public service. The treatise was very popular at court, and probably brought about Elyot's appointment as ambassador to the court of Emperor Charles V. He continued in various diplomatic and local roles, but writing was the main work of his life. He translated a number of Greek and Latin texts, in addition to compiling a Latin-English dictionary and producing a number of prose dialogues; he aided the cause of vernacular prose by writing all of his works in English. The excerpt that follows from *The Book Named the Governor* offers a description of sincere friendship. In what ways does Elyot's ideal friend resemble Antonio of *The Merchant of Venice*? Does Bassanio fit this depiction as well? The illustrative story that follows the general definition presents the love between two friends, Titus and Gisippus, who, in spite of the conflict they have over a young woman, are primarily bonded to each other. When Titus falls in love with his friend's fiancée, Gisippus renounces her in order to preserve the friendship. The woman is never named or consulted about the arrangement; her exchange from one friend to the other suggests that the devotion of the two men is greater than that between each man and the woman. In *The Merchant of Venice,* what tensions does Bassanio experience in attempting to be a good friend to Antonio and a devoted lover and husband to Portia? To whom does he owe the greater allegiance? Who has the stronger hold on him at the play's end and why?

Sir Thomas Elyot, *The Book Named the Governor* (London, 1531), S2v–S3r, S4r–S5r, S6r–S7v, S9r–S13r.

The true description of amity or friendship.

But now let us ensearch what friendship or amity is. Aristotle sayeth that friendship is a virtue, or joineth with virtue; which is affirmed by Tully[1] saying, that friendship cannot be without virtue, ne[2] but in good men only. Who be good men, he after declareth to be those persons, which so do bear themselves and in such wise do live, that their faith, surety, equality, and liberality be sufficiently proved. Ne that there is in them any covetise,[3] willfulness, or foolhardiness, and that in them is great stability or constancy; then suppose I (as they be taken) to be called good men, which do follow (as much as men may) nature, the chief captain or guide of man's life. Moreover the same Tully defineth friendship in this manner, saying that it is none other thing but a perfect consent of all things appertaining as well to God as to man, with benevolence and charity; and that he knoweth nothing given of God (except sapience[4]) to man more commodious. Which definition is excellent and very true. For in God, and all things that cometh of God, nothing is of more great estimation than love, called in Latin *amor,* whereof *amicitia* cometh, named in English friendship or amity; the which taken away from the life of man, no house shall abide standing, no field shall be in culture. And that is lightly[5] perceived, if a man do remember what cometh of dissension and discord. Finally he seemeth to take the sun from the world, that taketh friendship from man's life. . . .

But now to resort to speak of them in whom friendship is most frequent, and they also thereto be most aptly disposed. Undoubtedly it be specially they which be wise and of nature inclined to beneficence, liberality, and constancy. For by wisdom is marked[6] and substantially discerned the words, acts, and demeanor of all men between whom happeneth to be any intercourse or familiarity, whereby is engendered a favor or disposition of love. Beneficence, that is to say, mutually putting to their study and help in necessary affairs, induceth love. They that be liberal do withhold or hide nothing from them whom they love, whereby love increaseth. And in them that be constant is never mistrust or suspicion, nor any surmise or evil report can withdraw them from their affection, and hereby friendship is made perpetual and stable. . . .

Now let us try out what is that friendship that we suppose to be in good men. Verily it is a blessed and stable connection of sundry wills, making of

---

[1] Tully: Marcus Tullius Cicero (106–43, B.C.E.), Roman statesman and orator who wrote a treatise on friendship, *De amicitia.*   [2] ne: nor.   [3] covetise: greed.   [4] sapience: wisdom.   [5] lightly: easily.   [6] marked: noticed.

two persons one in having and suffering. And therefore a friend is properly named of philosophers "the other I." For that in them is but one mind and one possession, and that which more is, a man more rejoiceth at his friend's good fortune than at his own. . . .

Undoubtedly, that friendship which doth depend either on profit or else in pleasure, if the ability of the person which might be profitable, do fail or diminish, or the disposition of the person which should be pleasant, do change or appair,[7] the ferventness of love ceaseth, and then is there no friendship.

The wonderful history of Titus and Gisippus, and whereby is fully declared the figure of perfect amity.

But now in the midst of my labor, as it were to pause and take breath, and also to recreate the readers, which, fatigued with long precepts, desire variety of matter, or some new pleasant fable or history, I will rehearse a right goodly example of friendship. Which example, studiously read, shall minister to the readers singular pleasure and also incredible comfort to practice amity.

There was in the city of Rome a noble senator named Fulvius, who sent his son called Titus, being a child, to the city of Athens in Greece (which was the fountain of all manner of doctrine), there to learn good letters, and caused him to be hosted with a worshipful man of that city called Chremes. This Chremes happened to have also a son called Gisippus, who not only was equal to the said young Titus in years, but also in stature, proportion of body, favor, and color of visage, countenance, and speech. The two children were so like that without much difficulty it could not be discerned of their proper[8] parents which was Titus from Gisippus, or Gisippus from Titus. These two young gentlemen, as they seemed to be one in form and personage, so, shortly after acquaintance, the same nature wrought in their hearts such a mutual affection, that their wills and appetites daily more and more so confederated themselves, that it seemed none other, when their names were declared, but that they had only changed their places, issuing (as I might say) out of the one body, and entering into the other. They together and at one time went to their learning and study, at one time to their meals and refection;[9] they delighted both in one doctrine, and profited equally therein; finally they together so increased in doctrine[10] that within a few years, few within Athens might be compared unto them. At the last died Chremes, which was not only to his son, but also to Titus, cause of much sorrow and heaviness. Gisippus, by the goods of his father, was known to be

---

[7] **appair:** deteriorate.   [8] **proper:** own.   [9] **refection:** refreshment.   [10] **doctrine:** learning.

a man of great substance, wherefore there offered to him great and rich marriages. And he then being of ripe years and of an able and goodly personage, his friends, kin, and allies exhorted him busily to take a wife, to the intent he might increase his lineage and progeny. But the young man, having his heart already wedded to his friend Titus and his mind fixed to the study of philosophy, fearing that marriage should be the occasion to sever him both from the one and the other, refused of long time to be persuaded; until at last, partly by the importunate calling on of his kinsmen, partly by the consent and advice of his dear friend, Titus, thereto by other desired, he assented to marry such one as should like him. . . .

[Gisippus becomes engaged to a young women and introduces her to Titus, who falls deeply in love with her. Titus falls ill with love and shame; Gisippus presses him to reveal what ails him.]

With which words the mortal sighs renewed in Titus, and the salt tears burst out of his eyes in such abundance, as it had been a land flood running down of a mountain after a storm. That beholding Gisippus, and being also resolved into tears, most heartily desired him and (as I might say) conjured him that for the fervent and entire love that had been, and yet was, between them, he would no longer hide from him his grief, and that there was nothing to him so dear or precious (although it were his own life) that might restore Titus to health, but that he should gladly and without grudging employ it. With which words, obtestations,[11] and tears of Gisippus, Titus, constrained, all blushing and ashamed, holding down his head, brought forth with great difficulty his words in this wise.[12] "My dear and most loving friend, withdraw your friendly offers, cease of your courtesy, refrain your tears and regrettings, take rather your knife and slay me here where I lie, or otherwise take vengeance on me, most miserable and false traitor unto you, and of all other most worthy to suffer most shameful death. For whereas God of nature, like as he hath given to us similitude in all the parts of our body, so had he conjoined our wills, studies, and appetites together in one, so that between two men was never like concord and love, as I suppose. And now notwithstanding, only with the look of a woman, those bonds of love be dissolved, reason oppressed, friendship is excluded; there avails no wisdom, no doctrine, no fidelity or trust; yea, your trust is the cause that I have conspired against you this treason. Alas, Gisippus, what envious spirit moved you to bring me with you to her whom ye have chosen to be your wife, where I received this poison? I say, Gisippus, where was then your wisdom, that ye remembered not the fragility of our common nature? What needed

---

[11] **obtestations:** supplications.   [12] **in this wise:** in this way.

you to call me for a witness of your private delights? Why would ye have me see that, which you yourself could not behold without ravishing of mind and carnal appetite? Alas, why forget ye that our minds and appetites were ever one? And that also what so ye liked was ever to me in like degree pleasant? What will ye more? Gisippus, I say your trust is the cause that I am entrapped; the rays or beams issuing from the eyes of her whom ye have chosen, with the remembrance of her incomparable virtues, hath thrilled throughout the midst of my heart, and in such wise burneth it, that above all things I desire to be out of this wretched and most unkind life, which is not worthy the company of so noble and loving a friend as ye be." And therewith Titus concluded his confession with so profound and bitter a sigh, received with tears, that it seemed that all his body should be dissolved and relented into teardrops.

But Gisippus, as he were therewith nothing astonished or discontented, with an assured countenance and merry regard, embracing Titus and kissing him, answered in this wise, "Why, Titus, is this your only sickness and grief that ye so uncourteously have so long concealed, and with much more unkindness kept it from me then ye have conceived it? I acknowledge my folly, wherewith ye have with good right upbraided me, that, in showing to you her whom I loved, I remembered not the common estate of our nature, nor the agreeableness, or (as I might say) the unity of our two appetites; surely that default can be by no reason excused. Wherefore it is only I that have offended. For who may by right prove that ye have trespassed, that by the inevitable stroke of Cupid's dart are thus bitterly wounded? . . .

I confess to you, Titus, I love that maiden as much as any wise man might possible, and take in her company more delight and pleasure than of all the treasure and lands that my father left to me, which ye know was right abundant. But now I perceive that the affection of love toward her surmounteth in you above measure. What, shall I think it of a wanton lust or sudden appetite in you, whom I have ever known of grave and sad disposition, inclined always to honest doctrine, fleeing all vain dalliance and dishonest pastime? Shall I imagine to be in you any malice or fraud, since from the tender time of our childhood I have always found in you, my sweet friend Titus, such a conformity with all my manners, appetites, and desires, that never was seen between us any manner of contention? Nay, God forbid that in the friendship of Gisippus and Titus should happen any suspicion, or that any fantasy should pierce my head, whereby that honorable love between us should be the mountenance of a crumb perished.[13] Nay, nay,

---

[13] **any suspicion . . . the mountenance of a crumb perished:** no suspicion or fantasy should diminish the honorable love between us even in the smallest amount.

Titus, it is (as I have said) the only providence of God. She was by him from the beginning prepared to be your lady and wife. For such fervent love entereth not into the heart of a wise man and virtuous, but by a divine disposition; whereat if I should be discontented or grudge, I should not only be unjust to you, withholding that from you which is undoubtedly yours, but also obstinate and repugnant against the determination of God; which shall never be found in Gisippus. Therefore, gentle friend Titus, dismay you not at the chance of love, but receive it joyously with me, that am with you nothing discontented, but marvelous glad, since it is my hap to find for you such a lady, with whom ye shall live in felicity, and receive fruit to the honor and comfort of all your lineage. Here I renounce to you clearly all my title and interest that I now have or might have in that fair maiden. . . ."

With these words, Titus began to move, as it were, out of a dream, and doubting whether he heard Gisippus speak, or else saw but a vision, lay still as a man abashed. But when he beheld the tears trickling down by the face of Gisippus, he then recomforted him, and thanking him for his incomparable kindness, refused the benefit that he offered, saying that it were better that a hundred such unkind wretches, as he was, should perish, than so noble a man as was Gisippus should sustain reproach or damage. But Gisippus eftstoons[14] comforted Titus, and therewith swore and protested that with free and glad will he would that this thing should be in form aforesaid accomplished, and therewith embraced and sweetly kissed Titus.

[14] **eftstoons:** soon after.

PHILEMON HOLLAND

From Plutarch's Morals   *1603*

Philemon Holland (1552–1637) received degrees from both Cambridge and Oxford, and acquired his medical degree by 1595. He had a small medical practice, but spent most of his time translating a number of lengthy works from Latin and Greek into English, including Plutarch's *Morals* excerpted here. Plutarch sets out a debate still crucial in the early modern era over the relative superiority of heterosexual and homoerotic relations. While Holland's summary articulates his clear disapproval of the homoerotic, the general misogyny of the early modern period would have supported Protogenes' arguments in favor of

---

Philemon Holland, *The Philosophy Commonly Called the Morals, Written by the Learned Philosopher Plutarch of Chaeronea, Translated out of Greek into English, and Conferred with Latin and French* (London, 1603), 1130, 1132–33.

the primacy of relations between men: since women were considered inferior to men, so were relationships with them. How does *The Merchant of Venice* represent the relationship between Antonio and Bassanio? Is the intensity of affection for the other the same in both friends? How do Bassanio's relationships with Portia and Antonio play out some of the issues raised in the debate that follows?

OF LOVE

The Summary

This dialogue is more dangerous to be read by young men than any other treatise of Plutarch, for that there be certain glances here and there against honest marriage, to uphold indirectly and underhand the cursed and detestable filthiness covertly couched under the name of love of young boys. But minds guarded and armed with true chastity and the fear of God, may see evidently in this discourse the miserable estate of the world, in that there be found patrons and advocates of so detestable a cause; such I mean as in this book are brought in under[1] the persons of Protogenes and Pisias. Meanwhile they may perceive likewise in the combat of matrimonial love against unnatural pederasty[2] not to be named, that honesty hath always means sufficient to defend itself for[3] being vanquished, yea and in the end to go away with the victory. Now this treatise may be comprised of four principal points: of which, the first . . . containeth the history of Ismenodora, enamored upon a young man named Bacchon; whereupon arose some difference and dispute, of which Plutarch and those of his company were chosen arbitrators. Thereupon Protogenes seconded by Pisias (and this is the second point), setting himself against Ismenodora, disgraceth and discrediteth the whole sex of womankind, and praises openly enough the love of males. But Daphnaeus answers them so fully home and pertinently to the purpose, that he discovereth and detecteth all their filthiness, and confuseth them as behooveful[4] it was, showing the commodities and true pleasure of conjugal love.[5]

[A marriage is proposed between Bacchon and the widow Ismenodora; when the former cannot decide whether or not to marry, he consults his friends.]

Howbeit in the end, shaking off all others, [Bacchon] referred himself to Anthemion and Pisias, for to tell him their minds upon the point, and to advise him for his best. Now was Anthemion his cousin german,[6] one of

---

[1] **brought in under:** represented in. [2] **pederasty:** sexual love of boys. [3] **for:** from.
[4] **behooveful:** necessary. [5] **conjugal love:** marital, or heterosexual love. [6] **cousin german:** cousin or close relative.

good years, and elder than himself far. And Pisias, of all those that made love unto him,[7] most austere: and therefore he both withstood the marriage, and also checked Anthemion, as one who abandoned and betrayed the young man unto Ismenodora. Contrariwise, Anthemion charged Pisias, and said he did not well, who, being otherwise an honest man, yet herein imitated lewd lovers, for that he went about to put his friend beside a good bargain[8] (who now might be sped[9] with so great a marriage, out of so worshipful an house, and wealthy besides), to the end that he might have the pleasure to see him a long time stripped naked in the wrestling place, fresh, still, and smooth, and not having touched a woman. But because they should not by arguing thus one against another, grow by little and little into heat of choler,[10] they chose for umpires and judges of this their controversy, [Plutarch] and those who were of his company, and thither they came. Assistant also there were unto them other of their friends, Daphnaeus to the one, and Protogenes to the other, as if they had been provided of set purpose to plead a cause. As for Protogenes who sided with Pisias, he inveighed verily with open mouth against dame Ismenodora. Whereupon Daphnaeus: "O Hercules" (quoth he) "what are we not to expect, and what thing in the world may not happen? In case it be so[11] that Protogenes is ready here to give defiance and make war against love, who all his life both in earnest and in game, hath been wholly in love, and all for love, which hath caused him to forget his book, and to forget his natural country . . . your Cupid, Protogenes. 'With his light wings displayed and spread, / Hath over seafull swiftly fled' . . . From out of Sicily to Athens, to see fair boys, and to converse and go up and down with them" (for to say a truth, the chief cause why Protogenes made a voyage out of his own country, and became a traveler, was at the first this and no other). Hereat the company took up a laughter, and Protogenes: "Think you" (quoth he) "that I war not against love, and not rather stand in the defense of love against lascivious wantonness, and violent intemperance, which by most shameful acts and filthy passions, would perforce challenge and break into the fairest, most honest, and venerable names that be?" "Why" (quoth Daphnaeus then) "do you term marriage and the secret of marriage (to wit, the lawful conjunction of man and wife) most vile and dishonest actions, than which there can be no knot nor link in the world more sacred and holy?" "This bond, in truth, of wedlock" (quoth Protogenes) "as it is necessary for generation, is by good right praised by politicians and law-givers, who recommend the same highly unto the people and

---

[7] **of all those that made love unto him:** of all Bacchon's friends who love him.   [8] **to put his friend beside a good bargain:** i.e., to talk him out of a good marriage.   [9] **be sped:** succeed.   [10] **heat of choler:** heat of anger.   [11] **In case it be so:** can it be the case.

common multitude. But to speak of true love indeed, there is no jot or part thereof in the society and fellowship of women. Neither do I think that you and such as yourselves, whose affections stand to wives or maidens, do love them no more than a fly loveth milk, or a bee the honeycomb, as caters[12] and cooks who keep fowls in mew,[13] and feed calves and other such beasts fat in dark places,[14] and yet for all that they love them not. But like as nature leads and conducts our appetite moderately, and as much as is sufficient, to bread and other viands,[15] but the excess thereof, which makes the natural appetite to be a vicious passion, is called gourmandize, and pampering of flesh; even so there is naturally in men and women both a desire to enjoy the mutual pleasure of one another, whereas the impetuous lust which cometh with a kind of force and violence, so as it hardly can be held in, is not fitly called love, neither deserveth it that name. For love, if it seize upon a young, kind, and gentle heart, endeth by amity in virtue; whereas of these affections and lusts after women, if they have success and speed never so well, there followeth in the end the fruit of some pleasure, the fruition and enjoying of youth and a beautiful body, and that is all. And thus much testified Aristippus, who when one went about to make him have a distaste and mislike of Lais the courtesan, saying, that she loved him not, made this answer: "I suppose" (quoth he) "that neither good wine, nor delicate fish loves me, but yet" (quoth he) "I take pleasure and delight in drinking the one, and eating the other." For surely the end of desire and appetite is pleasure and the fruition of it. But love, if it have once lost the hope and expectation of amity and kindness, will not continue, nor cherish and make much for beauty sake that which is irksome and odious, be it never so gallant and in the flower and prime of age, unless it bring forth and yield such fruit which is familiar unto it, even a nature disposed to amity and virtue.[16] . . . For surely, more amorous than this man is not he, who not for lucre and profit, but for the fleshly pleasure of Venus, endureth a cursed, shrewd, and froward[17] wife, in whom there is no good nature nor kind affection. . . .

But if we must needs call this passion Love, yet surely it shall be but an effeminate and bastard love, sending us into women's chambers and cabinets.[18] . . . Even so the true and natural love is that of young boys, which sparks not with the ardent heat of concupiscence, as Anacreon[19] sayeth the other of maidens and virgins do. It is not besmeared with sweet ointments,

---

[12] **caters:** caterers.   [13] **mew:** a coop for containing fowls.   [14] **beasts fat in dark places:** refers to the penning and fattening up of birds and animals for eating.   [15] **viands:** foods.   [16] **For surely . . . virtue:** Protogenes argues that friendship and virtue are necessary to love, while lust is not.   [17] **froward:** contrary.   [18] **cabinets:** small private rooms.   [19] **Anacreon:** (582–485 B.C.E.) ancient Greek poet known for his poems in praise of love, wine, and revelry (*EBO*, s.v. Anacreon).

nor tricked up[20] and trimmed; but plain and simple always a man shall see it, without any enticing allurements, in the Philosopher's schools, or about public parks of exercise and wrestling places, where it hunts kindly and with a very quick and piercing eye after none but young striplings and springals,[21] exciting and encouraging earnestly unto virtue, as many as are meet and worthy to have pains taken with them. Whereas the other delicate and effeminate love that keeps[22] home and stirs not out of doors, but keeps continually in women's laps, under canopies, or within curtains in women's beds and soft pallets, seeking always after dainty delights, and pampered up with unmanly pleasures, wherein there is no reciprocal amity, nor heavenly ravishment of the spirit, is worthy to be rejected and chased far away.

[20] **tricked up:** adorned.   [21] **springals:** youths.   [22] **keeps:** stays.

Bibliography

———————————— ❯❮ ————————————

Primary Sources

Agrippa, Cornelius. *Of the Nobility and Excellency of Womankind.* 1529. Trans. David Clapham. London, 1542.

Allen, William. *A True, Sincere, and Modest Defense of English Catholics That Suffer for Their Faith Both at Home and Abroad against a False, Seditious, and Slanderous Libel, Entitled "The Execution of Justice in England."* Ingolstadt, Germany, 1584.

Bacon, Francis. "Of Usury." *The Essays or Counsels, Civil and Moral, of Francis Bacon.* London, 1625.

Becon, Thomas. *The Catechism.* Ed. John Ayre. Parker Society, vol. 3. Cambridge: Cambridge UP, 1844.

Bedell, William. Letter to Adam Newton, Dean of Durham, January 1, 1608. *Two Biographies of William Bedell.* Ed. E. S. Shuckburgh. Cambridge: Cambridge UP, 1902.

Best, George. *A True Discourse of the Late Voyages of Discovery, for the Finding of a Passage to Cathaya. . . .* In *The Principal Navigations, Voyages, Traffics, and Discoveries of the English Nation.* Vol. 3. Ed. Richard Hakluyt. London, 1598–1600.

Bristow, Richard. *Demands to Be Proponed of Catholics to the Heretics.* London, 1596.

Camden, William. *The History of Elizabeth, Queen of England.* Trans. Robert Norton. London, 1630.

Carleton, Dudley. The English Ambassador's *Notes.* Public Record Office, London, State Papers 99, file 8. ff. 340–44 *Venice: A Documentary History.* Ed. David Chambers and Brian Pullan. Oxford: Blackwell, 1992.

Coke, Edward. *The Institutes of the Laws of England.* London, 1642.

——. *The Reports.* 2nd ed. in English. London, 1680.

Coryate, Thomas. *Coryats Crudities: Hastily Gobbled Up in Five Months Travels in France, Savoy, Italy, . . . Switzerland &c.* London, 1611.

Debate on the Usury Bill in the House of Commons. 1571. *The Journals of All the Parliaments during the Reign of Queen Elizabeth, Both of the House of Lords and House of Commons.* Collected by Sir Simonds D'Ewes. London: Paul Bowes, 1682.

A Discovery of the Great Subtlety and Wonderful Wisdom of the Italians, Whereby They Bear Sway over the Most Part of Christendom, and Cunningly Behave Themselves to Fetch the Quintessence out of the People's Purses. London, 1591.

Draxe, Thomas. *The World's Resurrection.* London, 1608.

Elizabeth I. Proclamation 462, "Ordering Peace Kept in London." Hampton Court, August 13, 1559. *Tudor Royal Proclamations.* 3 vols. Ed. Paul L. Hughes and James F. Larkin. New Haven: Yale UP, 1964–69.

——. Proclamation 738, "Establishing Commissions against Seminary Priests and Jesuits." Richmond, October 18, 1591. *Tudor Royal Proclamations.*

——. Proclamation 817, "Banishing All Jesuit and Secular Priests." Richmond, November 5, 1602. *Tudor Royal Proclamations.*

Elyot, Thomas. *The Book Named the Governor.* London, 1531.

Examination of Roderigo Lopez and Others. British Library MS Harley 871, folios 7–64. February 25, 1594.

Fleta. Trans. and ed. H. G. Richardson and G. O. Sayles. Selden Society, Vol. 72. London: Selden Soc., 1952.

Foxe, John. *Acts and Monuments.* 1570. London: Seeleys, 1857.

Holland, Philemon. *The Philosophy Commonly Called the Morals, Written by the Learned Philosopher Plutarch of Chaeronea, Translated out of Greek into English, and Conferred with Latin and French.* London, 1603.

Holinshed, Raphael. *Chronicles of England, Scotland, and Ireland.* 1587. Ed. Henry Ellis. London, 1807.

King James Bible. London, 1611.

Leo, John. *A Geographical History of Africa.* 1526. Trans. John Pory. London, 1600.

The Levant Company's Charter. 1581. In *The Principal Navigations, Voyages, and Discoveries of the English Nation.* Ed. Richard Hakluyt. London, 1589.

Martin, Gregory. *Roma Sancta: The Holy City of Rome, So Called, and So Declared to Be, First for Devotion, Secondly for Charity; in Two Books.* 1581. Ed. George Bruner Parks. Rome: Edizioni, 1969.

Modena, Leon. *The History of Rites, Customs, and Manners of Life, of the Present Jews, throughout the World.* 1637. Trans. Edmund Chilmead. London: J. L., 1650.

Moryson, Fynes. *An Itinerary.* London, 1617.

———. Unpublished chapters of Fynes Moryson's *An Itinerary*. *Shakespeare's Europe*. Ed. Charles Hughes. 2nd ed. New York: Blom, 1967.

Munster, Sebastian. *The Messiah of the Christians and the Jews*. 1539. Trans. Paul Isaiah. London, 1655.

Niccholes, Alexander. *A Discourse of Marriage and Wiving*. London, 1615.

de Nicolay, Nicolas. *The Navigations, Peregrinations, and Voyages Made into Turkey*. 1568. Trans. Thomas Washington. London, 1585.

Parsons, Robert. *A Brief Discourse Containing Certain Reasons Why Catholics Refuse to Go to Church*. London, England, 1580.

Perkins, William. *A Faithful and Plain Exposition upon the Two First Verses of the Second Chapter of Zephaniah Containing a Powerful Exhortation to Repentance*. London, 1606.

da Pisa, Yehiel Nissim. *The Eternal Life. Banking and Finance among Jews in Renaissance Italy*. Trans. and ed. Gilbert S. Rosenthal. New York: Bloch, 1962.

de Pomis, David. *De Medico Hebraeo*. Venice, 1587. *The Jews and Medicine*. Vol. 1 of 3. H. Friedenwald. Baltimore: Johns Hopkins UP, 1944.

Price, Daniel. *The Merchant: A Sermon Preached at Paul's Cross on Sunday the 24th of August Being the Day Before Bartholomew Fair, 1607*. Oxford, 1608.

Sherley, Thomas. "The Profit That May Be Raised to Your Majesty out of the Jews." Historical Manuscripts Commission and Salisbury Manuscripts, vol. 19. 1607. London: MSO, 1965.

Shuckburgh, E. S. ed. *Two Biographies of William Bedell*. Cambridge: Cambridge UP, 1902.

Smith, Thomas. *De Republica Anglorum: The Manner of Government or Policy of the Realm of England*. London, 1583.

Statutes of the Realm. London: G. Eyre and A. Strahan, 1810–22.

Stubbes, Philip. *The Anatomy of Abuses: Containing a Discovery, or Brief Summary of Such Notable Vices and Imperfections as Now Reign in Many Countries of the World; but (Especially) in a Very Famous Island Called Ailgna*. London: Richard Jones, 1583.

Thomas, William. *The History of Italy: A Book Exceeding Profitable to Be Read: Because It Intreateth of the Estate of Many and Divers Commonweals, How They Have Been, and Now Be Governed*. London, 1549.

Usque, Samuel. *Consolation for the Tribulations of Israel*. Ferrara, 1553. Trans. Martin Cohen. Philadelphia: Jewish Pub. Soc. of America, 1965.

Usury Not Exceeding Ten Per Cent Permitted. 13 Elizabeth, c. 8. 1571. *Statutes of the Realm*. Vol. 4. London: G. Eyre and A. Strahan, 1810–22.

Vives, Juan Luis. *A Very Fruitful and Pleasant Book Called the Instruction of a Christian Woman*. 1523. Trans. Richard Hyde. London, 1529.

Wheeler, John. *A Treatise of Commerce*. London, 1601.

Willet, Andrew. *Concerning the Universal and Final Vocation of the Jews That Is to Be Apparent in the Last Days, According to the Extremely Lucid Prophecy of St. Paul*. Trans. Andrew Dinan. Cambridge, 1590.

———. *Tetrastylon Papisticum, That Is the Four Principal Pillars of Papistry*. London, 1593.

Wilson, Robert. *A Right Excellent and Famous Comedy Called the Three Ladies of London.* London, 1584.

Wilson, Thomas. *A Discourse upon Usury by Way of Dialogue and Orations, for the Better Variety, and More Delight of All Those, That Shall Read This Treatise.* London, 1572.

Secondary Sources

Adelman, Howard E. "Leon Modena: The Autobiography and the Man." *The Autobiography of a Seventeenth-Century Venetian Rabbi.* Ed. and trans. Mark R. Cohen. Princeton: Princeton UP, 1988.

Auden, W. H. "Brothers & Others." *The Merchant of Venice: Critical Essays.* Ed. Thomas Wheeler. New York: Garland, 1991.

Baba Metzia. *The Babylonian Talmud.* Ed. I. Epstein. London: Soncino, 1978.

Baker, J. H. *An Introduction to English Legal History.* 3rd edition. London: Butterworths, 1990.

Baker, J. H., and Milsom, S. F. C. *Sources of English Legal History.* London: Butterworths, 1986.

Baron, Salo. *A Social and Religious History of the Jews.* 20 vols. New York: Columbia UP, 1952–1973.

Berger, Harry, Jr. "Marriage and Mercifixion in *The Merchant of Venice:* The Casket Scene Revisited." *Shakespeare Quarterly* 32.2 (Summer 1981): 155–62.

Bietenholz, Peter, ed. *Contemporaries of Erasmus: A Biographical Register of the Renaissance and Reformation.* Toronto: U of Toronto P, 1985.

Boose, Lynda E. "The Comic Contract and Portia's Golden Ring." *Shakespeare Studies* 20 (1987): 241–54.

——. "'The Getting of a Lawful Race': Racial Discourse in Early Modern England and the Unrepresentable Black Woman." *Women, "Race" and Writing in the Early Modern Period.* Ed. Margo Hendricks and Patricia Parker. London: Routledge, 1994.

Botero, Giovanni. *Relations, of the Most Famous Kingdoms and Common-weales thorough the World.* London, 1608.

Braude, Benjamin. "The Sons of Noah and the Construction of Ethic and Geographical Identities in the Medieval and Early Modern Periods." *William and Mary Quarterly,* 3rd ser., 54.1 (1997): 103–42.

Bray, Alan. "Homosexuality and the Signs of Male Friendship in Elizabethan England." *History Workshop* 29 (Spring 1990): 1–19.

Brown, John Russell, ed. *The Merchant of Venice.* London and New York: Methuen, 1985.

Brown, Robert, ed. *The History and Description of Africa.* Leo, Africanus, c. 1492–c. 1550. Trans. John Pory. London: Hakluyt Soc., 1896.

Catholic Encyclopedia On-line. <http://www.newadvent.org/cathen/11611c.htm.>.

Clay, C. G. A. *Economic Expansion and Social Change: England 1500–1700.* 2 vols. Cambridge: Cambridge UP, 1984.

Cohen, Derek. "Shylock and the Idea of the Jew." *Shakespearean Motives*. London: Macmillan, 1988.

Cohen, Martin, trans. *Consolation for the Tribulations of Israel*. Philadelphia: Jewish Pub. Soc. of America, 1965.

Cohen, Walter. "*The Merchant of Venice* and the Possibilities of Historical Criticism." *English Literary History* 49.4 (1982): 765–89.

Connor, Walker. "A Nation Is a Nation, Is a State, Is an Ethnic Group, Is a . . ." *Nationalism*. Ed. John Hutchinson and Anthony D. Smith. Oxford: Oxford UP, 1994.

Cressy, David, and Lori Anne Ferrell. *Religion and Society in Early Modern England*. London: Routledge, 1996.

Danson, Lawrence. *The Harmonies of the Merchant of Venice*. New Haven: Yale UP, 1978.

Dictionary of National Biography (DNB). CD-ROM. Version 1.0. Oxford: Oxford UP, 1995.

Encyclopedia Britannica Online. <http://www.britannica.com/>.

Encyclopedia Judaica. 17 vols. Jerusalem: Keter, 1971.

Fisch, Harold. *The Dual Image: The Figure of the Jew in English and American Literature*. London: World Jewish Cong. (British Section), 1971.

Friedenwald, Harry. *The Jews and Medicine*. 3 vols., Baltimore: Johns Hopkins UP, 1944.

Friedman, Jerome. "Jewish Conversion, the Spanish Pure Blood Laws and Reformation: A Revisionist View of Racial and Religious Antisemitism." *Sixteenth Century Journal* 18.1 (1987): 3–29.

Gaudet, Paul. "Lorenzo's 'Infidel': The Staging of Differences in *The Merchant of Venice*." *The Merchant of Venice: Critical Essays*. Ed. Thomas Wheeler. New York: Garland, 1991.

Greenfield, Liah. *Nationalism: Five Roads to Modernity*. Cambridge: Harvard UP, 1992.

Hall, Kim. "Guess Who's Coming to Dinner? Colonization and Miscegenation in *The Merchant of Venice*." *Renaissance Drama* 23 (1992): 87–111.

Holmer, Joan Ozark. *The Merchant of Venice: Choice, Hazard, and Consequence*. New York: St. Martin's, 1995.

Isaac, Dan. "The Worth of a Jew's Eye: Reflections of the Talmud in *The Merchant of Venice*." *Maarav: A Journal for the Study of the Northwest Semitic Languages and Literatures* 8 (1992): 349–74.

Jeffrey, David Lyle, ed. *A Dictionary of Biblical Tradition in English Literature*. Grand Rapids, Mich.: Eerdmans, 1992.

Jones, Norman. *God and the Moneylenders*. Oxford: Basil Blackwell, 1989.

Kahn, Coppelia. "The Cuckoo's Note: Male Friendship and Cuckoldry in *The Merchant of Venice*." *Shakespeare's "Rough Magic": Renaissance Essays in Honor of C. L. Barber*. Ed. Peter Erickson and Coppelia Kahn. Newark and London: U of Delaware P, 1985.

Katz, David. *Philo-Semitism and the Readmission of the Jews to England*. Oxford: Clarendon, 1982.

——. *Jews in the History of England.* Oxford: Clarendon, 1994.

Korte, Barbara. *English Travel Writing from Pilgrimages to Postcolonial Explorations.* Trans. Catherine Matthias. New York: St. Martin's, 2000.

Leventen, Carol. "Patrimony and Patriarchy in *The Merchant of Venice.*" *The Matter of Difference.* Ed. Valerie Wayne. Ithaca, N.Y.: Cornell UP, 1991.

Lewalski, Barbara. "Biblical Allusion and Allegory in *The Merchant of Venice.*" *Shakespeare Quarterly* 13 (1962): 317–43.

Linder, Amnon. *The Jews in the Legal Sources of the Early Middle Ages.* Detroit: Wayne State UP, 1997.

Marcus, Jacob R. *The Jew in the Medieval World: A Source Book, 315–1791,* Cincinnati: Union of Amer. Hebrew Congregations, 1938.

McPherson, David C. *Shakespeare, Jonson, and the Myth of Venice.* Newark: U of Delaware P, 1990.

Metzger, Mary Janell. "'Now by My Hood, a Gentle and No Jew': Jessica, *The Merchant of Venice,* and the Discourse of Early Modern Identity." *PMLA: Publications of the Modern Language Association of America* 113.1 (1998): 52–63.

Milsom, S. F. C. *Historical Foundations of the Common Law.* 2nd edition. London: Butterworths, 1981.

Mithal, H. S. D., ed. *A Right Excellent and Famous Comedy Called the Three Ladies of London.* Renaissance Imagination, vol. 36. New York and London: Garland, 1988.

Nelson, Benjamin. *The Idea of Usury.* Chicago: U of Chicago P, 1969.

Nerlich, Michael. *Ideology of Adventure: Studies in Modern Consciousness, 1100–1750.* Vol. 1. Minneapolis: U of Minnesota P, 1987.

Newman, Karen. "Portia's Ring: Unruly Women and Structures of Exchange in *The Merchant of Venice.*" *Shakespeare Quarterly* 38.1 (Spring 1987): 19–33.

Nouvelle Biographie Générale. Ed. Dr. Hoefer. 46 vols. Paris: Firmin Didot, 1855–1866.

Oxford Classical Dictionary. Ed. N. G. L. Hammond and H. H. Scullard. Oxford: Clarendon, 1970.

Oxford Companion to English Literature. Ed. Paul Harvey and Dorothy Eagle. 4th ed. Oxford: Oxford UP, 1967.

Oxford English Dictionary On-line. <http://dictionary.oed.com/>.

Oz, Avraham. *The Yoke of Love: Prophetic Riddles in* The Merchant of Venice. Newark and London: U of Delaware P, 1995.

Patterson, Steve. "The Bankruptcy of Homoerotic Amity in Shakespeare's *Merchant of Venice.*" *Shakespeare Quarterly* 50.1 (Spring 1999): 9–32.

Questier, Michael. *Conversion, Politics, and Religion in England, 1580–1625.* Cambridge: Cambridge UP, 1996.

Rackin, Phyllis. "Androgyny, Mimesis, and the Marriage of the Boy Heroine on the English Renaissance Stage." *PMLA: Publications of the Modern Language Association of America* 102.1 (1987): 29–41.

Rosen, Alan. "The Rhetoric of Exclusion: Jew, Moor, and the Boundaries of Discourse in *The Merchant of Venice.*" *Race, Ethnicity, and Power in the*

Renaissance. Ed. Joyce Green MacDonald. Madison, N.J. and London: Fairleigh Dickinson UP, 1997.

Rosenthal, Gilbert S. ed. and trans. *Banking and Finance Among Jews in Renaissance Italy: A Critical Edition of The Eternal Life* (Haye Olam) *by Yehiel Nissim da Pisa.* New York: Bloch, c. 1962.

Roth, Cecil. *A History of the Jews in England.* Oxford: Oxford UP, 1989.

Salingar, Leo. "The Idea of Venice in Shakespeare and Ben Jonson." *Shakespeare's Italy: Functions of Italian Locations in Renaissance Drama.* Ed. Michele Marrapodi, A. J. Hoenselaars, Marcello Cappuzzo, and Santucci F. Falzon. Manchester: Manchester UP, 1993.

Sandys, George. *A Relation of a Journey.* London, 1615.

Schotz, Amiel. "The Law That Never Was: A Note on *The Merchant of Venice.*" *Theatre Research International* 16.3 (1991): 249–52.

Shapiro, James. *Shakespeare and the Jews.* New York: Columbia UP, 1996.

Shell, Marc. "Portia's Portrait: Representation as Exchange." *Common Knowledge* 7.1 (1998): 94–144.

——. "Marranos (Pigs), or from Coexistence to Toleration." *Critical Inquiry* 17 (1991): 306–35.

——. "The Wether and the Ewe: Verbal Usury in *The Merchant of Venice.*" *Kenyon Review* n.s. 1 (1979): 65–92.

Sinfield, Alan. "How to Read *The Merchant of Venice* Without Being Heterosexist." *Alternative Shakespeares, II.* Ed. Terence Hawkes. London: Routledge, 1996.

Smith, Bruce. *Homosexual Desire in Shakespeare's England.* Chicago: U of Chicago P, 1991.

Spinosa, Charles. "The Transformation of Intentionality: Debt and Contract in *The Merchant of Venice.*" *English Literary Renaissance* 24.2 (Spring 1994): 370–409.

Thomas, Keith. *Religion and the Decline of Magic.* London: Penguin, 1988.

Vitkus, Daniel. "Barabas, Shylock and Company: Mediterranean Commerce and Jewish Merchants in the Drama of Early Modern England." Unpublished paper given at the 1999 MLA session "The Representation of Jews on the Medieval and Early Modern Stage: A Roundtable."

Yungblut, Laura Hunt. *Strangers Settled Here Amongst Us.* London: Routledge, 1996.

Acknowledgments
Figure 1. Jacob breeding spotted sheep, from Thomas Cranmer, *Catechismus* (1548). By permission of the Folger Shakespeare Library.
The Merchant of Venice from *The Complete Works of Shakespeare*, 4th ed. Ed. David Bevington. Copyright © 1997 by Addison-Wesley Educational Publishers, Inc. Reprinted by permission of Pearson Education, Inc.
Figure 2. Venetian masque, from Pietro Bertelli, *Diversarum Nationem* (1594). By permission of the Folger Shakespeare Library.
Figure 3. Mercy overcoming revenge, from Richard Day, *A Book of Christian Prayers* (1578). By permission of the Folger Shakespeare Library.
Figure 4. Daniel, with Susannah, judging the elders, from *The Images of the Old Testament* (1549). By permission of the Folger Shakespeare Library.

CHAPTER 1
Figure 5. Venice, from Pietro Bertelli, *Theatrum Urbium Italicarum* (1599). By permission of the Folger Shakespeare Library.
Figure 6. Venetian bride, from Pietro Bertelli, *Diversarum Nationem* (1594). By permission of the Folger Shakespeare Library.
Figure 7. Plaza of St. Mark in Venice, from Pietro Bertelli, *Diversarum Nationem* (1594). By permission of the Folger Shakespeare Library.
Dudley Carleton, The English Ambassador's Notes (1612). Public Record Office, London, State Papers 99, file 8, ff. 340–44. Reprinted by permission of the Public Record Office.
William Bedell, *An English Protestant Looks at Venetian Religious Life*. From letter to Adam Newton, Dean of Durham, January 1, 1608. In E. S. Shuckburgh, *Two Biographies of William Bedell* (New York: Cambridge UP, 1902), pp. 228–29. Reprinted with the permission of Cambridge University Press.
Fynes Moryson, *Unpublished Chapters of Fynes Moryson's Itinerary*, in *Shakespeare's Europe*, 2nd ed., ed. Charles Hughes (New York: Benjamin Blom, Inc., 1967), pp. 290, 408–11, 415, 423, 463, 473, 487. Reprinted by permission of Ayer Company Publishers.
Figures 8 and 9. French attire and English attire, from Andrew Boorde, *The First Book of the Introduction of Knowledge* (1542). The British Library, STC 3383, K1r and A3v. By permission of The British Library.
Figure 10. Jew selling clothes, from Pietro Bertelli, *Diversarum Nationem* (1594). By permission of the Folger Shakespeare Library.
Figure 11. A maiden Jew, from Nicolas de Nicolay, *The Navigations, Peregrinations, and Voyages Made into Turkey* (1585). By permission of the Folger Shakespeare Library.

CHAPTER 2
Figure 12. English Usurer, from John Blaxton, *The English Usurer* (1634). By permission of the Folger Shakespeare Library.
Yehiel Nissim da Pisa, *The Eternal Life*, in *Banking and Finance Among Jews in Renaissance Italy*, trans. and ed. Gilbert S. Rosenthal (New York: Bloch Publishing Company, 1962), pp. 33–36, 40–45, 47–48, 89–91. Reprinted by permission of Gilbert S. Rosenthal.

David de Pomis, *De Medico Hebraeo*, Venice, 1587, in Harry Friedenwald, *Jews and Medicine: Essays* (Baltimore: Johns Hopkins University, 1944), pp. 40–44. © 1944. Reprinted by permission of The Johns Hopkins University Press.

Thomas Sherley, *The Profit That May Be Raised to Your Majesty out of the Jews* (1607), Historical Manuscripts Commission and Salisbury Manuscripts, Volume 19 (London: Her Majesty's Stationery Office, 1965), pp. 473–74. Crown copyright is reproduced with the permission of the Controller of Her Majesty's Stationery Office.

CHAPTER 3

Gregory Martin, *Roma Sancta: The Holy City of Rome, So Called, and So Declared to Be, First for Devotion, Secondly for Charity; in Two Books* (1581), ed. George Bruner Parks (Rome: Edizioni di Storia e Letteratura, 1969), pp. 75–78, 81–82. Reprinted by permission of Edizioni di Storia e Letteratura.

Samuel Usque, *Consolation of the Tribulations of Israel* (1553), trans. Martin A. Cohen (Philadelphia: Jewish Publication Society, 1965, 1977), pp. 182–85, passim. Reprinted by permission of Martin A. Cohen.

Figures 13 and 14. Cranmer's recantation and execution, from John Foxe, *Acts and Monuments* (1583). By permission of the Folger Shakespeare Library.

Fleta, Vol. II, trans. and ed. H. G. Richardson and G. O. Sayles, Selden Society, Vol. 72 (London: Selden Society, 1952), p. 90. Reprinted by permission of the Selden Society.

Figure 15. A physician Jew, from Nicolas de Nicolay, *The Navigations, Peregrinations, and Voyages Made into Turkey* (1585). By permission of the Folger Shakespeare Library.

Figure 16. Lopez conspiring to poison Queen Elizabeth, from George Carleton, *A Thankful Remembrance of God's Mercy* (1624). By permission of the Folger Shakespeare Library.

CHAPTER 4

Figure 17. Two male friends hunting, from *Primer of Salisbury* (1532?). By permission of the Folger Shakespeare Library.

Figure 18. Man and woman wooing, with female attendant, from *Primer of Salisbury* (1532?). By permission of the Folger Shakespeare Library.

Figure 19. Two men riding, with woman, from *Primer of Salisbury* (1532?). By permission of the Folger Shakespeare Library.

Figure 20. Marriage, from *Primer of Salisbury* (1532?). By permission of the Folger Shakespeare Library.

Index

Printed in the United States
By Bookmasters